27. Colloquium der Gesellschaft für Biologische Chemie
29. April - 1. Mai 1976 in Mosbach/Baden

The Immune System

Edited by
F. Melchers and K. Rajewsky

With 104 Figures

Springer-Verlag
Berlin Heidelberg NewYork 1976

Prof. Dr. Fritz Melchers
Basel Institute of Immunology
487 Grenzacherstraße
4058 Basel/Switzerland

Prof. Dr. Klaus Rajewsky
Institut für Genetik
Weyertal 1
5000 Köln/FRG

ISBN 3-540-07976-9 Springer-Verlag Berlin Heidelberg New York
ISBN 0-387-07976-9 Springer-Verlag New York Heidelberg Berlin

Library of Congress Cataloging in Publication Data. Gesellschaft für Biologische Chemie. The immune system. Bibliography: p. Includes index. 1. Immunology-Congresses. I. Melchers, Fritz, 1936—. II. Rajewsky, K., 1936—. III. Title. (DNLM: 1. Lymphatic system-Congresses. 2. Immunogenetics-Congresses. 3. Antibodies-Congresses. 4. Histocompatibility-Congresses. 5. Antigens-Congresses. QW504 M894i 1976). QR180.3.G47 1976. 574.2'9.76-49977.

© by Springer-Verlag Berlin · Heidelberg 1976.

Printed in Germany.

Offsetprinting and bookbinding: Brühlsche Universitätsdruckerei, Gießen.

Preface

The cells of the immune system generate a large variety of binding
sites which differ in their binding specificities and can therefore
react specifically with a large variety of ligands. These binding sites
are part of receptor molecules, enabling the system to react to the
universe of antigens.

The classical antigen receptor is the antibody molecule, and accordingly the first session of this colloquium deals with a classical subject, namely antibody structure. Dramatic recent advances in this field
make it possible to interrelate primary and three-dimensional structure both to each other and to function, i.e. the binding of antigen
and possible reactions occurring in the antibody molecule upon antigen
binding. The latter point is of particular interest since it may be
relevant not only for effector functions of antibodies such as the
binding of complement, but also for the triggering of a lymphocyte
through its antibody receptor for antigen.

Whereas the structural basis of antibody variability is now largely
understood, its genetic basis is still unresolved. This problem is the
subject of the second session, in which we ask the question of how
many structural genes code for antibody molecules and how these genes
are arranged in the genome. Three unlinked clusters of structural genes
appear to collaborate in the coding of antibodies, and we come across
many unorthodox and challenging genetic phenomena such as one polypeptide chain being coded by two or more genes, somatic translocation
of genes, and allelic exclusion. Most provocative still is the question
of whether antibody variability is generated somatically rather than
encoded in the germ line.

How are the genes coding for antibodies expressed at the cellular level?
The third session of the colloquium deals with the immunocompetent
cells, namely the lymphocytes. How do these cells recognize antigen
and how are they triggered by it? How are they differentiated and specialized in terms of receptor specificity and effector function? The
world of lymphocytes is divided in two, the T and B cell populations,
each of which can again be subdivided into functionally distinct sub-
populations. B cells are the precursors of antibody-producing cells.
They carry antibodies as their antigen receptors on the surface. Each
cell is committed to the synthesis of a single pair of heavy and light
chain variable domains, i.e. to one antigen-binding site. T cells are
responsible for reactions of cellular immunity and for the regulation
of lymphocyte functions in both the T and B cell compartment. The antigen receptors of these cells are at present a matter of much debate
in immunology. Most interestingly, structural genes of the antibody
system and also genes in the major histocompatibility complex seem to
control the recognition of antigens by T cells. We are only now beginning to understand how this may be reflected in the molecular structure
of the T cell receptor for antigen. The session on immunocompetent
cells would be incomplete if it did not also deal with experimental
tumors of lymphocytes, which are of highest relevance as model systems
for the analysis of many of the most crucial problems encountered

in the immune system, and if it did not mention the macrophage, whose relation to the immune system is obvious but has also been debated for many decades.

The subject of Session IV is the major histocompatibility complex (MHC). The products of this gene complex, which may harbour hundreds or even thousands of different genes, are of crucial importance for the regulation of T lymphocyte activities. Not only do T cells react most dramatically against products of the MHC (and this is of great practical importance in the field of transplantation), but also, as mentioned above, the MHC appears to control T cell recognition of antigens and may code for parts of the T cell receptor for antigen. Due to recent technological advances, the chemical structure of histocompatibility antigens coded in the MHC is now in the process of being elucidated, and Session IV thus allows us to be present at the stage of research in immunology at which fundamental new insights into the regulation of the immune system appear to be just ahead.

The last session deals with the immune system as a system of interacting cells and molecules. Does one lymphocyte know of the existence of other lymphocytes? It has gradually become clear over the last ten years that virtually all immunological reactions rely on the _interaction_ of lymphocytes. As an extreme view one can imagine that lymphocyte interactions are the driving force of the immune system both during its development in ontogeny and in its reactions to foreign antigens. "Lymphocyte interactions" here implies the specific recognition of one lymphocyte by another. Such interactions could well be mediated by the very same lymphocyte surface receptors which also recognize antigen. The diversity of these receptors resembles the diversity of the antigenic universe, and it is thus not surprising that in the total receptor population of a given immune system each individual receptor finds counterparts which it can specifically "recognize".

Finally, the relevance of immunology for medicine will also be discussed in Session V. It is obvious that this is a subject of enormous practical importance and that the immune system must be of great interest for doctors in almost every field of modern medicine.

Lymphocytes fascinate the immunologist because they produce a large variety of antigen-binding molecules and because they interact with each other to form a differentiated, specialized cell system of the body which deals with the invasion of foreign material and possibly fulfills other important physiological functions. Lymphocytes, however, also obey the general rules of all cells which grow, differentiate and function. They thus attract the attention of all those interested in gene expression, cell differentiation and cell interactions in higher organisms, in particular since they are amongst the best-characterized eucaryotic cells. Many different stages of differentiation of lymphocytes are defined by a wealth of genetic and molecular markers. The most important of these marker molecules, namely immunoglobulin, is well defined at the genetic, structural and functional level, but at the same time fundamental questions concerning this fascinating protein molecule are still open at each of these levels. Most importantly, the cells of the immune system can be easily obtained and maintained as single cells. This is the basis for a quantitative biology of gene expression, cell differentiation and cell interaction which makes the lymphocyte system a powerful model system in general cell biology of higher organisms.

The central role of biochemistry in immunological research appears so obvious from this brief introduction that it needs no further specification. In the frame of the immune system, biochemists have in the past found challenging problems at the level of nucleic acids, proteins and membranes. This Colloquium will, we hope, show that the same is still true at present and will be equally true in the future.

Acknowledgments. Many people have helped us in the organization of this Colloquium: the students, who took care of the projection and recorded the talks and discussions ; the secretaries, who wrote the correspondence and the manuscripts of the book; Professor Zachau and the Gesellschaft für Biologische Chemie, who provided the opportunity to organize a Colloquium on The Immune System; Professor Auhagen and Professor Gibian and their coworkers, who had the largest part of the burden of organizing the Colloquium technically; Springer-Verlag, who helped to publish the proceedings of the Colloquium within 1976. Most of all, the participating speakers, chairmen and discussants.

Mosbach Colloquia are intended to present to the members of the Gesellschaft für Biologische Chemie today's knowledge of a field of modern biology. The Colloquia are, at least in part, planned as teaching endeavors on the highest level. This becomes an almost impossible task in the case of a topic such as The Immune System, which represents a network of genetics, molecular biology, biochemistry, cell biology, physiology, medicine, mathematics and much more.

It could therefore not be our intention to arrange a program which would provide a complete survey of current research in immunology. Rather, we have tried to concentrate on a few fundamental points and on what, as we think, are the most important developments and trends in the field.

We are grateful that almost everyone whom we asked to help us in this task, accepted our invitation, and then presented "his" field of immunology in such a complete and competent way as to make this Colloquium a success.

November, 1976   F. Melchers
                                                        K. Rajewsky

Contents

Contributors

Adam, G.	Fachbereich Biologie, Universität Konstanz, 7750 Konstanz, FRG
Albert, E.D.	Kinderklinik der Universität München, 8000 München, FRG
Amzel, L.M.	Department of Biophysics, Johns Hopkins University School of Medicine, 725 N. Wolfe Street, Baltimore, MD 21205, USA
Berek, C.	Institut für Immunologie und Genetik, Deutsches Krebsforschungszentrum, 6900 Heidelberg, FRG Institut für Genetik, Universität Köln, 5000 Köln, FRG
Black, S.	Institut für Immunologie und Genetik, Deutsches Krebsforschungszentrum, 6900 Heidelberg, FRG Institut für Genetik, Universität Köln, 5000 Köln, FRG
Brownlee, G.G.	MRC Laboratory of Molecular Biology, Hills Road, Cambridge CB2 2QH, England
Chen, B.L.	Department of Biophysics, Johns Hopkins University School of Medicine, 725 N. Wolfe Street, Baltimore, MD 21205, USA
Cheng, C.C.	MRC Laboratory of Molecular Biology, Hills Road, Cambridge CB2 2QH, England
Chiu, Y.Y.	Department of Biophysics, Johns Hopkins University School of Medicine, 725 N. Wolfe Street, Baltimore, MD 21205, USA
Colman, P.M.	Department of Inorganic Chemistry, University of Sydney, N.S.W. 2006, Australia
Cunningham, B.A.	Rockefeller University, New York, NY 1021, USA
Deisenhofer, J.	Max-Planck-Institut für Biochemie, 8033 Martinsried, FRG Physikalisch-Chemisches Institut der Technischen Universität, 8000 München, FRG
Edelman, G.M.	Rockefeller University, New York, NY 1021, USA
Eichmann, K.	Institut für Immunologie und Genetik, Deutsches Krebsforschungszentrum, 6900 Heidelberg, FRG Institut für Genetik, Universität Köln, 5000 Köln, FRG

Eisen, H.N.	Center for Cancer Research, Harvard-M.I.T. Program for Health Science and Technology, Cambridge, MA 02139, USA
Fischer, H.	Max-Planck-Institut für Immunbiologie, Stübeweg 51, 7800 Freiburg, FRG
Günther, E.	Lehrstuhl für Immunbiologie der Universität und Max-Planck-Institut für Immunbiologie, 7800 Freiburg, FRG
Hämmerling, G.	Institut für Immunologie und Genetik, Deutsches Krebsforschungszentrum, 6900 Heidelberg, FRG Institut für Genetik, Universität Köln, 5000 Köln, FRG
Hamlyn, P.H.	MRC Laboratory of Molecular Biology, Hills Road, Cambridge CB2 2QH, England
Henning, R.	Abt. Biochemie II, Universität Ulm, Oberer Eselsberg, 7900 Ulm, FRG
Hess, M.	Max-Planck-Institut für Biochemie, 8033 Martinsried, FRG
Hood, L.	Division of Biology, California Institute of Technology, Pasadena, CA 91125, USA
Hozumi, N.	Basel Institute of Immunology, Grenzacherstrasse 487, 4058 Basel, Switzerland
Huber, R.	Max-Planck-Institut für Biochemie, 8033 Martinsried, FRG Physikalisch-Chemisches Institut der Technischen Universität, 8000 München, FRG
Humphreys, R.E.	Harvard University, The Biological Lab., 16 Divinity Avenue, Cambridge, MA 02138, USA The National Cancer Institute, National Institutes of Health, Bethesda, MD 20014, USA
Jerne, N.K.	Basel Institute of Immunology, Grenzacherstrasse 487, 4058 Basel, Switzerland
Kabat, E.A.	Departments of Microbiology, Human Genetics and Development, and Neurology, Columbia University, The Neurological Institute, Presbyterian Hospital, New York, N.Y. 10032, USA The National Cancer Institute, National Institutes of Health, Bethesda, MD 20014, USA
Kaufman, J.F.	Harvard University, The Biological Lab., 16 Divinity Avenue, Cambridge, MA 02138, USA The National Cancer Institute, Immunology Branch, National Institutes of Health, Bethesda, MD 20014, USA
Mann, D.L.	Harvard University, The Biological Lab., 16 Divinity Avenue, Cambridge, MA 02138, USA The National Cancer Institute, Immunology Branch, National Institutes of Health, Bethesda, MD 20014, USA

Matsushima, M.	Max-Planck-Institut für Biochemie, 8033 Martinsried, FRG Physikalisch-Chemisches Institut der Technischen Universität, 8000 München, FRG
Milner, R.J.	Rockefeller University, New York, N.Y. 1021, USA
Milstein, C.	MRC Laboratory of Molecular Biology, Hills Road, Cambridge CB2 2QH, England
Möller, G.	Division of Immunbiology, Karolinska Institute, Wallenberglaboratory, Lilla Freskati, 104 05 Stockholm 50, Sweden
Palm, W.	Institut für Medizinische Biochemie der Universität Graz, Graz, Austria
Parham, P.	Harvard University, The Biological Lab., 16 Divinity Avenue, Cambridge, MA 02138, USA The National Cancer Institute, Immunology Branch, National Institutes of Health, Bethesda, MD 20014, USA
Du Pasquier, L.	Basel Institute of Immunology, Grenzacherstrasse 487, 4058 Basel, Switzerland
Pecht, I.	Dept. of Chemical Immunology, The Weizmann Institute of Science, Rehovot, Israel
Pernis, B.	Basel Institute of Immunology, Grenzacherstrasse 487, 4058 Basel, Switzerland
Phizackerley, R.P.	Dept. of Biophysics, Johns Hopkins University School of Medicine, Baltimore, MD 21205, USA
Poljak, R.J.	Dept. of Biophysics, Johns Hopkins University School of Medicine, Baltimore, MD 21205, USA
Porter, R.R.	Dept. of Biochemistry, University of Oxford, South Parks Road, Oxford OX1 3QV, England
Potter, M.	Laboratory of Cell Biology, National Cancer Institute, DPHEW, Bethesda, MD 20014, USA
Proudfoot, N.J.	MRC Laboratory of Molecular Biology, Hills Road, Cambridge CB2 2QH, England
Rabbitts, T.H.	MRC Laboratory of Molecular Biology, Hills Road, Cambridge CB2 2QH, England
Rajewsky, K.	Institut für Genetik, Universität Köln, 5000 Köln, FRG
Reske, K.	Rockefeller University, New York, N.Y. 1021, USA
Robb, R.	Harvard University, The Biological Lab., 16 Divinity Avenue, Cambridge, MA 02139, USA The National Cancer Institute, Immunology Branch, National Institutes of Health, Bethesda, MD 20014, USA

Rüde, E. Lehrstuhl für Immunbiologie der Universität und Max-Planck-Institut für Immunbiologie, 7800 Freiburg, FRG

Saul, F. Department of Biophysics, Johns Hopkins University School of Medicine, 725 N. Wolfe Street, Baltimore, MD 21205, USA

Schrader, J.W. Rockefeller University, New York, N.Y. 1021, USA

Schuler, W. Fachbereich Biologie, Universität Konstanz, 7750 Konstanz, FRG

Silver, J. Division of Biology, California Institute of Technology, Pasadena, CA 91125, USA

Springer, T. Harvard University, The Biological Lab., 16 Divinity Avenue, Cambridge, MA 02138, USA
The National Cancer Institute, Immunology Branch, National Institutes of Health, Bethesda, MD 20014, USA

Strominger, J.L. Harvard University, The Biological Lab., 16 Divinity Avenue, Cambridge, MA 02138, USA
The National Cancer Institute, Immunology Branch, National Institutes of Health, Bethesda, MD 20014, USA

Terhorst, C. Harvard University, The Biological Lab., 16 Divinity Avenue, Cambridge, MA 02138, USA
The National Cancer Institute, Immunology Branch, National Institutes of Health, Bethesda, MD 20014, USA

Tonegawa, S. Basel Institute of Immunology, Grenzacherstrasse 487, 4058 Basel, Switzerland

Weiler, E. Fachbereich Biologie, Universität Konstanz, 7750 Konstanz, FRG

Weiler, I.J. Fachbereich Biologie, Universität Konstanz, 7750 Konstanz, FRG

Wigzell, H. Dept. of Immunology, Uppsala University, Biomedical Center, Box 582, 751 23 Uppsala, Sweden

Ysern, X. Dept. of Biophysics, Johns Hopkins University School of Medicine, 725 N. Wolfe Street, Baltimore, MD 21205, USA

Ziffer, J.A. Rockefeller University, New York, NY 1021, USA

Antibody Structure

The Structural Basis of Antibody Specificity

E. A. Kabat

The extraordinary capacity of the antibody-forming mechanism to make combining sites complementary to almost any molecular conformation of matter ranging in length from about 5Å to about 34Å (Kabat, 1976; Goodman, 1975) must ultimately be explained in structural, biosynthetic and genetic terms. While high resolution X-ray crystallography has now elucidated the structures of a number of combining sites including two Fab (Poljak et al., 1973; Segal et al., 1974), one light chain dimer (Schiffer et al., 1973) and an Fv fragment (Epp et al., 1974; see Davies and Padlan, 1976), it will manifestly be impossible to establish by this method the structures of a sufficiently large number of antibody combining sites of various specificities for us to map in detail the fitting of various antigenic determinants with their complementary regions in all antibody combining sites. Recent estimates suggest 10^7 clonotypes (Klinman et al., 1976) of antibody forming cells.

One must thus endeavor to learn the principles on which antibody complementarity is based by developing predictive techniques which if successful might provide a key to elucidating the structure not of a single specific antibody combining site but of a whole group or class of sites. One could then use such new insight for making and testing additional predictions and ultimately discover the basic principles involved.

Among the unique features of antibody molecules which distinguish them from all other proteins is the existence of a variable region. Recognition of this variable region was made possible by the availability (cf. Frisch, 1967; Killander, 1967; Kochwa and Kunkel, 1971) of myeloma globulins and Waldenström macroglobulins from humans and from mice (Potter, 1971) and the finding (Edelman and Gally, 1962) that Bence Jones proteins, useful for over a century in the diagnosis of multiple myeloma, were the light chains of immunoglobulins. It had been known for many years that the urinary Bence Jones proteins differed in their properties from one myeloma patient to another in physicochemical and immunological (Bayne-Jones and Wilson, 1922) properties, and later the myeloma globulins were also shown to differ in physicochemical and immunochemical properties from one myeloma patient to another (see Gutman, 1948; Killander, 1967; Frisch, 1967; Kochwa and Kunkel, 1971; Nisonoff et al., 1975; Kabat, 1976).

When the first sequence data from two Bence Jones proteins became available (Hilschmann and Craig, 1965), it became clear that the molecule could be divided into two regions, a variable region and a constant region. The variable region was so named because it differed in sequence from one Bence Jones protein to another, while the constant region was uniform for each class of Bence Jones protein. Bence Jones proteins had been divided into two classes immunologically (Bayne-Jones and Wilson, 1922; Korngold and Lipari, 1956) and now termed κ and λ. Myeloma globulin and antibodies produced by the usual immunization procedures were shown to have a basic structural similarity, being built up of monomers or oligomers of a basic four chain structure consisting of two identical heavy and two identical light chains. The studies of Porter (1959) and Nisonoff et al. (1959) led

4

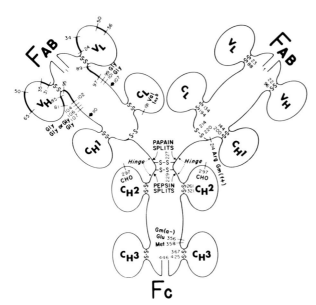

Fig. 1. Schematic view of 4-chain structure of human IgGκ molecule. Numbers on right side: actual residue numbers in protein Eu; Numbers of Fab fragment on left side aligned for maximum homology; light chains numbered as in Wu and Kabat (1970), and Kabat and Wu (1971). Heavy chains of Eu have residue 52A, 3 residues 82A, B, C and lack residues termed 100 A,B,C,D,E and 35 A,B. Thus residue 110 (end of variable region) is 114 in actual sequence. Hypervariable regions: heavier lines. V_L and V_H: light and heavy chain variable region; C_H1, C_H2 and C_H3: domains of constant region of heavy chain; C_L: constant region of light chain. Hinge region in which 2 heavy chains are linked by disulfide bonds is indicated approximately. Attachment of carbohydrate is at residue 297. Arrows at residues 107 and 110 denote transition from variable to constant regions. Sites of action of papain and pepsin and locations of a number of genetic factors are given. (From Kabat, 1973)

to insight into the action of papain and of pepsin in cleaving the molecule into the Fab or Fab' and Fc fragments (Fig. 1), and as the disulfide bonding patterns became known from sequence data, the molecules were seen to have domains each consisting of a disulfide loop of about 65-70 residues and a connecting portion, and each domain was associated with a different function (Fig. 1). The variable region of the light chain and a similar variable region of the heavy chain interacted noncovalently to form a three-dimensional structure with an intact antibody combining site. If light and heavy chains of myeloma globulins or antibodies were separated chromatographically and recombined, an intact combining site would be formed. There also was some degree of specificity of recognition of one heavy chain for its own light chain when given a choice by being presented with a mixture of its own and another light chain (Grey and Mannik, 1965). Another important development was the recognition that some human and mouse myeloma proteins could possess antibody activity, and a host of different specificities have been recognized. To date the crystallographic studies have all been on myeloma proteins and on Bence Jones proteins or their fragments.

While important properties of antibodies are associated with the constant domains, these are in general no different in their genetic control, biosynthesis and evolution from other proteins, so that the discussion of the structural basis of antibody specificity will concentrate on the variable regions of the light (V_L) and heavy (V_H) chains.

5

If one examines a large number of proteins each having a specific receptor site for a given ligand (such as the cytochromes c of different
species) and for which a three dimensional structure has been established by X-ray crystallography, it is very easy to recognize two types
of amino acid residues: those making contact with the ligand and forming the specific binding site, and those which are involved only in
three-dimensional folding i.e. those which play essentially a structural role in permitting the molecule to assume the proper three-dimensional form. For instance, there is among the cytochromes of various
species a preservation of secondary structure which permits the combining site to be in the same place and to function (Takano et al.,
1973). The sequence differences among cytochromes of various species
thus represent what I have called <u>mutational noise</u>; that is, the
permissible mutations, occurring during speciation and evolution, are
only selected provided they permit a functioning combining site to
form; other mutations, if the receptor site must perform some unique
and vital function, would be lethal. It may of course be that some
of the mutations confer some additional advantage to each species and
have thus been selected and preserved, but this is not relevant to
this analysis.

As the number of variable regions of human and mouse Bence Jones proteins and of light and heavy chains of myeloma proteins sequenced began to increase, it became clear that the variable regions could be
divided into subgroups (Hood et al., 1967; Milstein, 1967; Niall and
Edman, 1967), the number varying with the species, and also that certain positions accommodated many more amino acid substitutions, e.g.
showed more variability than others (Milstein, 1967; Kabat, 1968,
1970).

The problem remained of trying to locate and distinguish within the
variable region those residues involved in three-dimensional folding
from those residues making contact with the antigenic determinant.
Reasoning that if one had to generate tens of thousands or even millions of different antibody combining sites, each having a similar
three-dimensional structure, Dr. T.T. Wu and I assumed that one should
find segments of high variability superimposed upon the usual mutational noise associated with species differences. Having compiled a
data bank of the then available sequences of variable regions of human
κ, human λ and mouse κ chains aligned for maximum homology, we defined
a parameter termed variability (Wu and Kabat, 1970) as follows:

$$\text{Variability} = \frac{\text{number of different amino acids at any position}}{\text{frequency of the most common amino acid at that position}}$$

If at any position in the aligned sequences only one amino acid occurs,
e.g. the position is invariant, the numerator and the denominator would
both be one and the variability would be one, the lowest possible value.
If all 20 amino acids were found at a given position the numerator
would be 20, and if they all occurred with equal frequency the denominator would be 1/20 or 0.05 and the variability would be 20/0.05 = 400,
the highest possible value. For instance at position 94 in the original
data, eight and possibly nine amino acids were found, Ile, Leu, Val,
Met, Ala, Arg, Ser, Asp and Asx, a total of 21 sequences had been reported and of these Ser occurred 8 times to give a frequency of 8/21 or
0.38. The variability was thus 8/0.38 or possibly 9/0.38 to give 21.1
or 23.6. When variability so determined was plotted against position,
three segments of hypervariability were seen (Fig. 2) and these were
postulated to be the complementarity-determining segments which form the
antibody combining site. Gaps, considered to be insertion or deletions,
were localized to these hypervariable segments (Wu and Kabat, 1970).

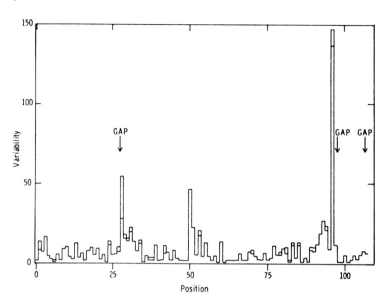

Fig. 2. Variability at different positions for variable region of light chains. GAP indicates positions at which insertions have been found. (From Wu and Kabat, 1970)

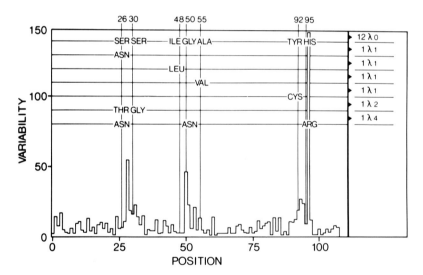

Fig. 3. Location and nature of amino acid substitutions in mouse V_λ region superimposed upon variability plot of Fig. 2. All other residues in sequence were identical. Number of instances of each sequence as well as number of base changes required to obtain it from $\lambda 0$ given. (From Cohn et al., 1974)

Franêk (1969) had approached the problem somewhat differently and found that there were three hypervariable regions based on the occurrence of non-homologous replacements and gaps. The hypervariable re-

7

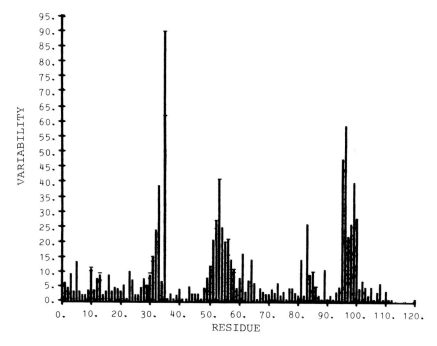

Fig. 4. Variability plot for V_H domain of immunoglobulins. Plot made by PROPHET Computer System by Kabat et al.. (From Kabat, 1976)

gions of the light chain in Figure 2 were 24-34, 50-56 and 89-97, as seen on the left Fab fragment in Figure 1.

When mouse λ chains were discovered and sequenced it became evident (Weigert et al., 1970) that they were extremely restricted in the segments involved in three-dimensional folding, e.g. mutational noise was very low. Indeed 12 mouse λ myeloma light chains had an identical V_L sequence; other mouse λ chains showed only one to three substitutions, and those were in the hypervariable or complementarity-determining segments (Fig. 3; Cohn et al., 1974).

As sequences of heavy chains accumulated, three hypervariable regions were recognized which corresponded to the three light chain hypervariable segments (Fig. 4; Kabat and Wu, 1971; Kehoe and Capra, 1971); there was also higher variability of several residues in the region of 81 to 85 (Capra and Kehoe, 1974, 1975); the three V_H hypervariable regions were displaced several residues away from the disulfide bonds toward the C-terminus (Fig. 1 left Fab fragment) (Kabat and Wu, 1971).

The uniqueness of hypervariable regions of immunoglobulin chains may be seen by comparing the immunoglobulin variability plots with a similar plot made from data on a homologous set of proteins, such as the cytochromes c. Figure 5 shows the variability plot for 67 cytochromes c made by Drs. T.T. Wu, H. Bilofsky and myself from sequences provided by Dr. E. Margoliash of Northwestern University (Kabat et al., 1976b). It is evident that the cytochrome plot (Fig. 5) does not resemble the variability plots for light (Figs. 2 and 3) or for heavy chains (Fig. 4).

8

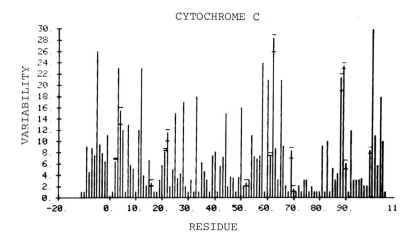

Fig. 5. Variability plot for cytochromes c. (From Kabat et al., 1976b)

The relationship of the hypervariable regions to complementarity-determining residues was supported by numerous studies on affinity labeling. This technique, in which a specific hapten, which possesses
a reactive group, enters its combining site, allowing the reactive
group to form a covalent bond with those amino acids in the site with
which it can react, has been found essentially to label residues in
hypervariable regions of both light and heavy chains (for a review
cf. Givol, 1974), thus suggesting that the walls of the site were
lined with hypervariable residues.

High resolution X-ray crystallographic studies on the immunoglobulin
Fab' fragments, the light chain dimer and the F_V fragment (Poljak et
al., 1973; Segal et al., 1974; Edmundson et al., 1974) confirmed the
prediction (Wu and Kabat, 1970; Kabat and Wu, 1971) that the site was
formed by the hypervariable segments of each chain and that residues
in these segments were indeed complementarity-determining. Moreover
all of the C as well as V domains were formed by two anti-parallel
twisted β sheets, one of four and one of three strands forming a sand-
wich-type structure. The V domains had, in addition, an extra loop
between the two pleated sheets. In the V region of each chain the
hypervariable segments forming the site were at the tip of the domain
(Fig. 6).

Figure 7 shows the orientation in the combining sites of the two Fab'
fragments (Newm and McPC603) for which compounds which bind the site
are known, one binding an hydroxy derivative of vitamin K (Amzel et
al., 1974) and the other binding phosphocholine (Segal et al., 1974).
In both instances the three heavy chain hypervariable segments and
the first and third light chain hypervariable segments form the site
and occupy similar positions. The second hypervariable segment of the
light chain is not involved in these two sites; in the Newm site there
is a deletion of seven residues after the second hypervariable region
(Poljak et al., 1973) and in McPC 603 an insertion of six residues in
the first hypervariable region shields the second hypervariable region from the site (Segal et al., 1974). Thus insertions and deletions
may contribute an additional parameter to the generation of diversity
of antibody combining sites.

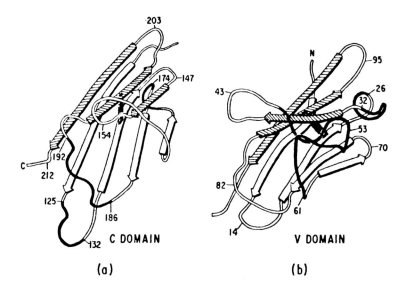

Fig. 6a and b. Comparison of V and C domains of monomer 2 of Bence Jones protein
Mcg. C domain (a) is rotated to approximate orientation of the V domain (b). The
two antiparallel pleated sheets of 3 and 4 strands represented by hatched and white
arrows. Chain segments present in only one domain indicated by solid black lines.
Positions of some residues numbered. Note extra loop including residue 53 in second
complementarity-determining segment of V domain. (From Edmundson et al., 1975)

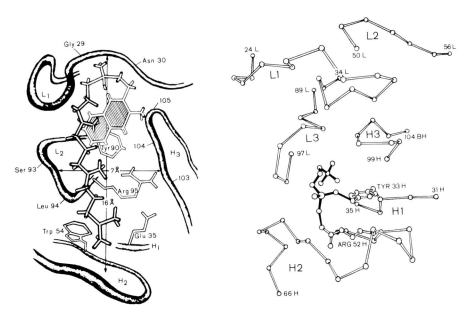

Fig. 7. Binding of Fab' Newm (Amzel et al., 1974) with ligand Vitamin K_1OH (left)
and McPC 603 (Segal et al., 1974) with its hapten, phosphocholine (right). Both
drawings in similar orientations. In Fab' Newm, Vitamin K_1OH lies in a shallow
groove while in McPC 603 phosphocholine lies in a deep cavity (not seen in this
orientation). (From Davies et al., 1975)

In the Bence Jones dimer (Mcg) all three hypervariable regions (Edmundson et al., 1974, 1975) contribute to the binding and are thus complementarity-determining. The Mcg dimer, however, binds a variety of different ligands and thus exhibits much less specificity than do the two Fab fragments, and it is tempting to speculate that a light chain dimer may have been a primitive antibody combining site early in evolution. The binding specificity of the light chain F_V dimer REI is unknown. Since it is clear that all V domains have a similar structure and that specificity resides in complementarity-determining segments, attempts have begun to construct sites of other specificities replacing the complementarity-determining residues of one specificity by those of another (Davies and Padlan, 1976).

While several more immunoglobulin fragments are being studied by X-ray crystallographic methods and will yield additional insight, I should like to outline several approaches which may be of value in predicting the shapes of antibody combining sites and distinguishing in complementarity-determining segments residues which contact the antigenic determinants from those residues which are essentially structural.

The first instance relates to the study of two IgA myeloma proteins with specificity for $\alpha 1 \rightarrow 6$-linked dextrans (Cisar et al., 1974). One of these, W3129, could be shown by oligosaccharide inhibition assays to have a site complementary to isomaltopentaose, while the other, QUPC 52, had a site complementary to isomaltohexaose. As would be expected, each gave a linear Scatchard plot in equilibrium dialysis (Cisar et al., 1975), but we were astonished to find that the site complementary to the hexasaccharide had a lower binding constant, 8.4×10^3, while the site complementary to a pentasaccharide gave an association constant of 1×10^5. This was quite different from findings with antibodies to dextran in which the higher binding correlated with site size (Harisdangkul and Kabat, 1972; Cisar et al., 1975).

Accordingly we attempted both by inhibition assays of precipitation and by competition experiments in equilibrium dialysis to determine the contribution of each sugar residue to the binding in W3129 and QUPC 52. The results are shown in Table 1. While binding of methyl αDglucoside and isomaltose contributed 56 and 60 % of the total binding energy in W3129, in the QUPC52 site they were very weakly bound and contributed 5 % or less of the total binding energy. These results clearly indicate that the binding sites of these two proteins differed greatly. The W3129 site in which methyl αDglucoside contributed most of the binding energy had to be so shaped that it would hold on to a least the non-reducing one or two residues in three dimensions, e.g. part of the site is a cavity into which at least these terminal residues fit. On the other hand the QUPC52 site would have to be a groove open at both ends into which the linear chain of dextran could fit, perhaps resembling the lysozyme site which attacks internal residues in the linear polymer which makes up the bacterial cell wall (Phillips, 1966).

Independent evidence for the differences in shape of these two types of site was established using a linear dextran synthesized chemically by Ruckel and Schuerch (1967) which consisted of linear chains of about 200 $\alpha 1 \rightarrow 6$-linked glucoses. This linear dextran did not precipitate myeloma antidextran W3129 since, having only one terminal non-reducing end, it was essentially monovalent. Indeed it inhibited precipitation of W3129 by dextran on a molar basis equally to IM5 as if only the five glucoses at the terminal nonreducing end were filling the site. However, this linear dextran precipitated QUPC52 as would be expected if the site were a groove, and reaction could occur with any six glucoses along the linear chain; thus toward QUPC52, the linear dextran is multivalent.

Table 1. Differences between Myeloma Antidextrans W3129 and QUPC52. (From Cisar
et al., 1975)

	W3129	QUPC52
Oligosaccharide best fitting site	IM5[a]	IM6[a]
Association content K^a M^{-1}	1×10^5	8×10^3

Relative contribution to total binding of best oligosaccharide (%)

	W3129	QUPC52
methyl αDglucoside	56,56[b]	5 < 0
IM2	61,50	5
IM3	91,94	72,64
IM4	91,94	88,80
IM5	100,100	96,92
IM6	95,100	100,100

[a] IM2, IM3 etc. denote numbers of sugar residues in these oligosaccharides of the
isomaltose series (α1 → 6 linked).

[b] First value obtained by competition in equilibrium dialysis, second by inhibition
of precipitation.

When antibodies to dextran were examined with the linear dextran, they
were found to be mixtures of antibody molecules with both kinds of
sites, as were rabbit antibodies to isomaltotrionic and isomalto-
hexaonic acids coupled to BSA (Cisar et al., 1975). Thus a new dimension
of antibody heterogeneity was revealed toward single antigenic determinants and these two myeloma antidextrans perhaps represent the maximum differences in site structure. Obviously if one determines the sequences of these two myeloma proteins, succeeds in crystallizing them
and studies them by high resolution X-ray crystallography, one could
learn a great deal about contacting residues in these two kinds of
sites, which might well lead to an understanding of all types of carbo-
hydrate-specific sites.

A second approach results from the data bank of variable region sequences being accumulated in PROPHET computer system (Raub, 1974) by
Drs. Wu, Bilofsky and myself. At present we have compiled complete and
partial sequence data on 165 heavy and 373 light chains which will be
published shortly (Kabat et al., 1976b). With this data bank, we were
able to calculate the probability on a random basis that any two residues in complementarity-determining segments would be associated with
one another and then select by computer instances in which the association was greater than expected by chance.

Table 2 gives the frequency data used at positions 32, 33, 34 and 35
of the heavy chain (Kabat et al., 1976a). Thus for example, the probability that a Phe at position 32 would be associated with a Tyr at
position 33 on a random basis would be 9/49 x 12/47 = 0.047 and with
47 sequences about 2.2 such pairs should be found. For Phe 32 and Glu
35 random probability would be 9/49 x 7/41 = 0.03 and only one such pair
would be expected. When the computer picked out these actual pairs,
seven, six and six proteins were found in which Phe 32-Tyr, 33, Phe
32-Glu 35 and Tyr 33-Glu 35 were associated. All seven were phospho-

Table 2. Frequency of amino acid residues at positions 32, 33, 34 and 35. (From Kabat et al., 1976a)

	Position			
	32	33	34	35
Trp	O	3	O	O
Ile	O	O	2	1
Tyr	20	12	O	4
Phe	9	1	1	1
Pro	O	O	O	O
Leu	O	1	2	O
Val	1	2	4	1
Lys	O	O	O	3
Met	O	O	36	O
Cys	O	O	O	2
Ala	2	7	O	3
Arg	O	1	1	2
Thr	3	5	O	1
Ser	7	3	1	7
Gly	2	8	O	1
His	O	O	O	2
Asp	3	2	O	O
Asn	1	O	O	1
Glu	1	O	O	7
Gln	O	O	O	O
Glx	O	1	O	1
Asx	O	1	O	4
	49	47	47	41

Data on completely and partially sequenced proteins at these positions are included.

choline-binding mouse myeloma proteins, but one CBPC 2 had residue 35 determined only as Glx.

Since one of these (McPC 603) had been studied by high resolution X-ray crystallography (Segal et al., 1974), it had been established that Tyr 33 and Glu 35 were contacting residues (Fig. 7) for the phosphocholine; the role of Phe 32 is not known but it could be conformational.

All seven phosphocholine-binding myeloma proteins had a methionine at position 34, but this residue was not selected as occurring more frequently than chance since it occurred in 29 other heavy chains. Met 34 is thus a structural rather than a contacting element in the complementarity-determining segment, and its role essentially is to position

Tyr 33 and Glu 35 to make contact with the phosphocholine; the X-ray structure shows this to be correct (Fig. 7). It has thus been possible in at least this one instance to distinguish contacting from structural residues within a complementarity-determining segment.

It was especially significant that when the heavy chain of a human phosphocholine-binding myeloma (Riesen et al., 1975) protein was sequenced (Riesen et al., 1976), it had Phe-Tyr-Met atpositions 32, 33, and 34; position 35 was Asp instead of Glu. It had a lower binding constant than four of the five phosphocholine-binding myeloma proteins which have been studied (Chesebro et al., 1973).

A search of our variable region data bank and of the Dayhoff Atlas of Protein Sequence and Structure Vol. 5 and Supplement I (Dayhoff, 1972, 1973) showed that the sequence Phe-Tyr-Met-Glu occurred only in phosphocholine-binding myeloma proteins and only at these positions in the first complementarity-determining segment. The closest sequences found were Phe-Tyr-Met-Ser at positions 32-35, Phe-Tyr-Thr-Glu at positions 58-61 and Ala-Tyr-Met-Glu at positions 78-81 of the heavy chains of proteins Tro (Kratzin et al., 1975), MU (Shinoda, 1973), and EU (Cunningham et al., 1970) and Phe-Tyr-Met-Ala at positions 68-71 of cytochrome c3, Phe-Tyr(Asp Glu) at positions 69-72 in hemerythrin and Phe-Thr-Met-Glu at position 148-151 of glutamic dehydrogenase (Dayhoff, 1972). Although Phe-Tyr-Met-Glu would be only 1 of 160,000 possible tetrapeptides, its exclusive association with phosphocholine binding suggests that it could play a unique role. If this is so and if characteristic sequences in complementarity-determining segments are uniquely associated with certain antibody combining sites, more such associations should result as further sequences are accumulated on antibodies of a given specificity. One would certainly like to know whether protein Tro binds phosphocholine. It may also become possible to predict which residues contribute to idiotypic specificity or play a conformational role.

It should be emphasized that the selection of these residues in the phosphocholine-binding myeloma proteins was possible not because of their sequences but because the data bank permitted the computation of random probabilities of association of the various residues. It is perhaps fortunate that a substantial number of variable region sequences were determined on human and mouse myeloma proteins not known to possess antibody activitiy, since these data may have approximated more closely the random set of sequences needed.

Thus it has not been possible from present sequence data on homogeneous rabbit antibodies to the type-specific pneumococcal (Jaton, 1975; Margolies et al., 1975) and group-specific streptococcal C polysaccharides (Chen et al., 1975) to draw inferences as to contacting residues in complementarity-determining segments. The number of different antibody sequences even to these relatively simple antigens appears to be quite large, the sequence differences associated with specificity quite subtle, and one does not have a random sample of other rabbit sequences.

Another important clue to the mechanism of generation of antibody diversity was the finding (Wu et al., 1975) that a λ II (Vil) and a λ V (Mcg) Bence Jones protein despite 21 amino acid differences in their variable regions had identical first complementarity-determining segments of 14 residues. If a different structural gene is responsible for each subgroup, the findings favor insertion of information for the hypervariable or complementarity-determining segments. Episomal insertion of the complementarity-determining segments (Kabat, 1970; Wu and Kabat, 1970) was proposed to account for generation of diversity and to provide evolutionary stability for the antibody-forming mechanism. The insertional hypothesis favors incorporation during embryogenesis of the complementarity-determining sequences so that each B clonotype would have the information for one or a few specificities. The existence of subgroups into which a given set of complementarity-determining sequences would be inserted to produce a given specificity would provide a mechanism for protecting the system. Thus loss of the structural gene for a given subgroup would not limit the capacity of the system to form antibody. The recent hybridization findings (Farace et al., 1976) that there are only one or a few genes for a mouse κ myeloma light chain (MOPC 41) were also considered as supporting an episomal concept, as would the similar findings on mouse λ light chain (Leder et al., 1975).

More recent suggestions (Capra and Kindt, 1975) have favored a more limited episomal mechanism involving some selection of sequences for certain complementarity-determining segments in relation to subgroups for the production of certain specificities, rather than the more generalized version (Wu and Kabat, 1970), because of the association of cross-idiotypic specificities of anti-IgG immunoglobulins with the V_KIIIb subgroup (Kunkel et al., 1974). There are also differences in numbers of residues in complementarity-determining segments which appear to correlate with subgroup. However cross-reacting idiotypes of cold agglutinins with I and i specificity were asssociated with IgM cold agglutinins of different V_K subgroups and one IgM with a Vλ chain (Lecomte and Feizi, 1975).

While the question cannot be resolved at present and while the episomal theory is not favored by many workers (Cohn et al., 1974), it is important to point out that the sequence data are extremely one-sided. Thus the λ II and λ V proteins with identical first complementarity-determining sequences were found among 95 first complementarity-determining sequences including human, mouse, rabbit, rat, pig, and shark light chains. This obviously suggests that one should look for more such instances. The accumulated data on heavy chain V region sequences (Kabat et al., 1976b) reveal that there are very serious problems arising in terms of the nature of the immunoglobulin chains being sequenced. This is most strikingly seen in the one-sided distribution of heavy chain sequences. A summary is given in Table 3. It is evident that the favoring of V_HIII chains for sequencing and the extreme paucity of sequences of complementarity-determining segments are giving a very one-sided distribution to the data. Searches for identities in sequences of complementarity-determining segments of different heavy chain subgroups are essential for testing insertional theories. Obviously the chance of finding such insertions is minimal if almost all sequences are from the V_HIII subgroup.

A recent and most unexpected finding by Kobzik et al. (1976) may prove of great importance for an understanding of genetic mechanisms. A human IgM immunoglobulin with anti-i specificity was found to show the same idiotypic specificity when the intact immunoglobulin, the separated heavy chain and the separated light chain were assayed, all three giving almost complete inhibition. Moreover affinity gels made with intact IgM, with μ heavy chain and with κ light chain absorbed all of the antiidiotypic antibody. Since the sequences of the structural portions of the variable regions of the heavy and light chains are quite different and since idiotypic specificity is generally associated with the complementarity-determining segments, the findings suggest that the heavy and light chains may prove to have identities in some residues on one or more complementarity-determining segments. Since the V_H and V_L genes are unlinked and may not even be on the same chromosomes, the findings might strongly support an episomal

Table 3. Numbers of heavy chain variable regions and of complementarity-determining segments for which sequences are available. (Compiled from Kabat et al., 1976b)

	No. of proteins on which some sequence data are reported	No. of complete or almost complete V region sequences	No. of additional complementarity-determining regions sequenced		
			1	2	3
Human					
$V_H I$	22	1	2[a]	1	0
$V_H II$	8	5	2[a]	0	0
$V_H III$	58	11	16[b]	1	1
$V_H IV$	5	0	0	0	0
Mouse					
$V_H I$	4	1	2	1	0
$V_H II$	3	1	0	0	0
$V_H III$	26	6	7[b]	0	0
$V_H IV$	1	0	0	0	0
$V_H V$	11	0	6[b]	0	0
Rabbit	10	2	3	3	2
Total	148	26	38	6	3

[a] partial

[b] many partial

mechanism. Indeed if the same sequence can be inserted into the V_H and V_L regions, generation of diversity on an episomal basis will be substantially simplified. Unfortunately one does not know how many residues are involved in an idiotypic determinant, so that it is still possible that only a few amino acids in the heavy and light chain complementarity-determining residues of each V region are the same. The findings of Hopper et al. (1976) that a human IgMλ and an IgGκ from the same individual have the same idiotypic specificities even though the IgM heavy chains is $V_H III$ and the IgG heavy chain is $V_H I$, if their complementarity-determining segments prove similar, could also support the generalized episomal concept. Obviously the sequence data are awaited with great interest.

Acknowledgments. Work of the laboratories at Columbia is sponsored by a grant from the National Science Foundation (BMS-72-02219 A03). Work with the Prophet Computer System is sponsored by the National Cancer Institute, National Institutes of Allergy and Infectious Diseases, National Institute of Arthritis, Metabolic and Digestive Diseases, National Institute of General Medical Sciences, Division of Research Resources Contract NO1-RR4-2147 National Institutes of Health.

References

Amzel, L.M., Poljak, R.J., Saul, F., Varga, J.M., Richards, F.F.:
Proc. Nat. Acad. Sci. 71, 1427-1430 (1974)
Bayne-Jones, S., Wilson, D.W.: Bull. Johns Hopkins Hospital 33, 119-
125 (1922)
Capra, J.D., Kehoe, J.M.: Proc. Nat. Acad. Sci. 71, 845-848 (1974)
Capra, J.D., Kehoe, J.M.: In: Advances in Immunology. Dixon, F.J.,
Kunkel, H.G. (eds.). New York: Academic Press 1975, Vol. XX, pp.
1-40
Capra, J.D., Kindt, T.J.: Immunogenetics 1, 417-427 (1975)
Chen, K.C.S., Kindt, T.J., Krause, R.M.: J. Biol. Chem. 250, 3289-
3296 (1975)
Chesebro, B., Hadler, N., Metzger, H.: In: Third International Con-
vocation on Immunology. Pressman, D., Tomasi, T.B., Jr., Grossberg,
A.L., Rose, N.R. (eds.). Basel: S. Karger 1973, pp. 205-217
Cisar, J.O., Kabat, E.A., Liao, J., Potter, M.: J. Exp. Med. 134,
159-179 (1974)
Cisar, J., Kabat, E.A., Dorner, M.M., Liao, J.: J. Exp. Med. 142,
435-459 (1975)
Cohn, M., Blomberg, B., Geckeler, W., Raschke, W., Riblet, R., Weigert,
M.: In: The Immune System, Genes, Receptors, Signals. Sercarz, E.E.,
Williamson, A.R., Fox, C.F. (eds.). New York: Academic Press 1974,
pp. 89-117
Cunningham, B., Rutishauser, U., Gall, W.E., Gottlieb, P.D., Waxdal,
M.J., Edelman, G.M.: Biochemistry 9, 3161-3170 (1970)
Davies, D.R., Padlan, E.A., Segal, D.: In: Contemporary Topics in
Molecular Immunology. Inman, F.P., Mandy, W.J. (eds.). New York:
Plenum Press 1975, Vol. IV, pp. 127-155
Davies, D.R., Padlan, E.A.: Roy Soc. Med. Symp., Rockefeller University,
Oct. 20-22, 1975. New York: Raven Press, in press (1976)
Dayhoff, M.O.: Atlas of Protein Sequence and Structure. Washington D.C.:
Nat. Biomed. Res. Found. (1972) Vol. V and (1973) Supplement 1.
Edelman, G.M., Gally, J.A.: J. Exp. Med. 116, 207-227 (1962)
Edmundson, A.B., Ely, K.R., Abola, E.E., Schiffer, M., Panagiotopoulos,
N.: Biochemistry 14, 3953-3961 (1975)
Edmundson, A.B., Ely, K.R., Girling, R.L., Abola, E.E., Schiffer, M.,
Westholm, F.A., Fausch, M.D., Deutsch, H.F.: Biochemistry 13,
3816-3827 (1974)
Epp, O., Colman, P., Fehlhammer, H., Bode, W., Schiffer, M., Huber,
R.: Europ. J. Biochem. 45, 513-524 (1974)
Farace, M.-G., Aellen, M.-F., Briand, P.-A., Faust, C.H., Vassalli,
P., Mach, B.: Proc. Nat. Acad. Sci. 73, 727-731 (1976)
Franêk, F.: In: Developmental Aspects of Antibody Formation and Struc-
ture. Sterzl, J., Riha, I. (eds.). Prague: Academia 1969, Vol. I,
pp. 311-313
Frisch, L. (ed.): Immunoglobulins. Cold Spring Harbor Symp. Quant.
Biol. 32, 1-619 (1967)
Givol, D.: London: Academic Press. Essays in Biochemistry 10, 73-103
(1974)
Goodman, J.W.: In: The Antigens III. Sela, M. (ed.). New York: Academic
Press 1975, pp. 127-187
Grey, H.M., Mannik, M.: J. Exp. Med. 122, 619-632 (1965)
Gutman, A.B.: Advan. Protein Chem. 4, 155-250 (1948)
Harisdangkul, V., Kabat, E.A.: J. Immunol. 108, 1232-1243 (1972)
Hilschmann, N., Craig, L.C.: Proc. Nat. Acad. Sci. 53, 1403-1409 (1965)
Hood, L., Gray, W.R., Sanders, B.G., Dreyer, W.J.: Cold Spring Harbor
Symp. Quant. Biol. 32, 133-146 (1967)
Hopper, J.E., Noyes, C., Heinrikson, R., Kessel, J.W.: J. Immunol.
116, 743-746 (1976)
Jaton, J.-C.: Biochem. J. 147, 235-247 (1975)

Kabat, E.A.: Proc. Nat. Acad. Sci. 59, 613-619 (1968)

Kabat, E.A.: Ann. N.Y. Acad. Sci. 169, 43-54 (1970)

Kabat, E.A.: In: 3rd Intern. Convocation on Immunol. Pressman, D.,
    Tomasi, T.B., Jr., Grossberg, A.L., Rose, N.R. (eds.). Basel:
    S.Karger 1973, pp.4-30

Kabat, E.A.: Structural Concepts in Immunology and Immunochemistry,
    2nd ed. New York: Holt Rinehart and Winston 1976

Kabat, E.A., Wu, T.T.: In: Immunoglobulins. Kochwa, S., Kunkel, H.G.
    (eds.). New York: Ann. N.Y. Acad. Sci. 190, 1971, pp. 382-393

Kabat, E.A., Wu, T.T., Bilofsky, H.: Proc. Nat. Acad. Sci. 73, 617-
    619 (1976a)

Kabat, E.A., Wu, T.T., Bilofsky, H.: Variable regions of immunoglobulin
    chains. Tabulations and analyses of amino acid sequences. Bolt
    Beranek and Newman, Inc. Cambridge, Mass. (1976b)

Kehoe, M., Capra, J.D.: Proc. Nat. Acad. Sci. 68, 2019-2021 (1971)

Killander, J. (ed.): γ-Globulins. 3rd Nobel Symp. New York: Inter-
    science 1967, pp. 17-643

Klinman, N.R., Press, J.L., Sigal, N.H., Gearhart, P.J.: In: Generation
    of Diversity - A New Look. Cunningham, A. (ed.). London: Academic
    Press 1976, pp. 127-149

Kobzik, L., Brown, M.C., Cooper, A.G.: Proc. Nat. Acad. Sci. 73, 1702-
    1706 (1976)

Kochwa, S., Kunkel, H.G. (eds.): Immunoglobulins. New York: Ann. N.Y.
    Acad. Sci. 1971

Korngold, L., Lipari, R.: Cancer 9, 262-272 (1956)

Kratzin, H., Atlevogt, P., Ruban, E., Kortt, A., Staroscik, K.,
    Hilschmann, N.: Z. Physiol. Chem. 356, 1337-1342 (1975)

Kunkel, H.G., Winchester, R.J., Joslin, F.G., Capra, J.D.: J. Exp.
    Med. 139, 128-136 (1974)

Lecomte, J., Feizi, T.: Clin. Exp. Immunol. 20, 287-302 (1975)

Leder, P., Honjo, T., Swan, D., Packman, S., Nau, M., Norman, B.:
    In: 7th Ann. Miami Winter Symp. Jan. 13-17. New York: Academic
    Press 1975, pp.173-188

Margolies, M.N., Cannon III, L.E., Strosberg, A.D., Haber, E.: Proc.
    Nat. Acad. Sci. 72, 2180-2184 (1975)

Milstein, C.: Nature (Lond.) 216, 330-332 (1967)

Niall, H.D., Edman, P.: Nature (Lond.) 216, 262-263 (1967)

Nisonoff, A., Wissler, F.C., Woernley, D.L.: Biochem. Biophys. Res.
    Commun. 1, 318-322 (1959)

Nisonoff, A., Hopper, J.E., Spring, S.: The Antibody Molecule. New:
    York: Academic Press 1975

Phillips, D.C.: Scientific American 215, 78-90 (1966)

Poljak, R.J., Amzel, L.M., Avey, H.P., Chen, B.L., Phizackerley, R.P.,
    Saul, F.: Proc. Nat. Acad. Sci. 70, 3305-3310 (1973)

Porter, R.R.: Biochem. J. 73, 119-127 (1959)

Potter, M.: In: Immunoglobulins. Kochwa, S., Kunkel, H.G. (eds.).
    New York: Ann. N.Y. Acad. Sci. 1971, 190, pp. 306-321

Raub, W.F.: Federation Proc. 33, 2390-2392 (1974)

Riesen, W.F., Braun, D.G., Jaton, J.-C.: Proc. Nat. Acad. Sci. 73,
    2096-2100 (1976)

Riesen, W.F., Rudikoff, S., Oriol, R., Potter, M.: Biochemistry 14,
    1052-1057 (1975)

Ruckel, E.R., Schuerch, C.: Biopolymers 5, 515-523 (1967)

Schiffer, M., Girling, R.L., Ely, K.R., Edmundson, A.B.: Biochemistry
    12, 4620-4631 (1973)

Segal, D.M., Padlan, E.A., Cohen, G.H., Rudikoff, S., Potter, M.,
    Davies, D.R.: Proc. Nat. Acad. Sci. 71, 4298-4302 (1974)

Shinoda, T.: Biochem. Biophys. Res. Commun. 52, 1246-1251 (1973)

Takano, T., Kallai, O.B., Swanson, R., Dickerson, R.E.: J. Biol. Chem.
    248, 5234-5255 (1973)

Weigert, M.G., Cesari, I.M., Yonkovich, S.J., Cohn, M.: Nature (Lond.)
    228, 1045-1047 (1970)

Wu, T.T., Kabat, E.A.: J. Exp. Med. 132, 211-250 (1970)
Wu, T.T., Kabat, E.A., Bilofsky, H.: Proc. Nat. Acad. Sci. 72, 5107-
    5110 (1975)

Structure and Specificity of Immunoglobulins

R. J. Poljak, L. M. Amzel, B. L. Chen, Y. Y. Chiu, R. P. Phizackerley, F. Saul, and X. Ysern

In this presentation we will briefly review our results on the study of the amino acid sequence and the three-dimensional structure of the Fab fragment from a human myeloma protein, IgG1 New (λ, Gm 1+, 3-, 4-, 5-). This structure has been determined by X-ray diffraction techniques to a nominal resolution of 2 Å and partially refined using "model building" and "real space" procedures (Diamond, 1966, 1971). The complete amino acid sequence of the L and Fd chains has been elucidated and correlated with the Fourier map of the electron density to provide a three-dimensional high resolution model.

The amino acid sequences of the L and Fd chains are given in Figure 1. The V_L sequence of IgG New has a unique feature consisting of a deletion of seven amino acid residues right after the second "hypervariable" region. By analogy with other alterations in the structure of light (L) and heavy (H) chains from human myelomas and from murine tumor cell lines this deletion can be interpreted as probably due to chromosomal alterations taking place during the establishment or progression of the neoplastic process. An insertion of residues designated 27a, b, and c in the first hypervariable region of V_L New occurs in many other human λ chains; a deletion in the third hypervariable region of V_L New (two residues around position 95) is also frequently observed in human λ chains. The amino acid sequence of V_H New, recently completed, closely resembles that of many other human V_H sequences, in particular those that are classified under subgroup II.

The most important conclusions that were obtained from a study of the three-dimensional model of Fab New will be summarized below.

The Fab structure consists of four globular "homology subunits", each containing the amino acid sequence of a homology region, i.e., V_H, V_L, C_H1 and C_L. In the homology subunits over 50 % of the amino acid residues are folded in two irregular, roughly parallel β-sheets surrounding a tightly packed core of hydrophobic side chains, which include the invariant intrachain disulfide bond. The homology subunits share the same pattern of polypeptide chain folding or "immunoglobulin fold". However, an additional segment of polypeptide chain, not present in C_H1 and C_L, is found in the V_H and V_L subunits (see Fig. 2). The presence of a common pattern of polypeptides chain folding is in agreement with the proposal of gene duplication events which gave rise to immunoglobulin genes (Hill et al., 1966). Although the alignment of amino acid sequences between V_H and V_L and between the various C subunits is straightforward by sequence comparison, a putative homology between V and C subunits is more difficult to recognize. This is due in part to the fact that the hydrophobic amino acid side chains which provide the interchain contacts and those that face out to the solvent are at different positions along the primary structure in the V compared to the C subunits, and in part, to the additional length of polypeptide chain present in V_H and V_L. The difficulties in the alignment of the V with the C sequences can be solved by matching their similar three-dimensional structures (Poljak et al., 1974). Using this structural alignment and a minimum mutation distance method Moore and

V_L `- - - - Z S V L T Q P P S V S G A P - G Q R V T I S C T G S S N I G A G N H V K W Y Q Q L P G T A P K - L L I F H N N A R * * * *`
(1 ... 10 ... 20 ... 27 a b c 30 ... 40 ... 50)

V_H `- - - - Z V Q L E Q S G P G L V R P - S Q T L S L T C T V S G S T F S N D - Y Y T W V R Q P P G R G L E W I G Y V F Y H G T S D T D`
(1 ... 10 ... 20 ... 30 ... 40 ... 50 ... 60)

C_L `Q P K A A P S V T L F P P S S E E L Q A N K A T L V C L I S D F Y P G A V - T V A W K - - A D S S`
(110 ... 120 ... 130 ... 140 ... 150)

C_H1 `A S T K G P S V F P L A P S S K S T S G G T A A L G C L V K D Y F P E P V - T V S W N - - S G`
(120 ... 130 ... 140 ... 150 ... 160)

V_L `* * * - - F S V S K S G - - - - - - - S S A T L A I T G L Q A E D E A D Y Y C Q S Y D R S L R * * - V F G G G T K L T V L R`
(61 ... 70 ... 80 ... 90 ... 100 ... 109)

V_H `T P L R S R V T M L V N T - S - - - - - K N Q F S L R L S S V T A A D T A V Y Y C A R N L I A G - C I D V W G Q G S L V T V S S`
(70 ... 80 ... 90 ... 100 ... 110 ... 117)

C_L `- P V K A - - G V E T T T P S K Q S N N K Y A A S S Y L S L T P E Q W K S H K S Y S C Q V T H - - E G S T - V E K T - V A P T E C S`
(160 ... 170 ... 180 ... 190 ... 200 ... 210 ... 214)

C_H1 `- A L T S - - G V H T F P A V L Q S S G L Y S L S S V V T V P S S S L G T - Q T Y I C N V N H K P S N T K - V D K R - V E P K S C`
(170 ... 180 ... 190 ... 200 ... 210 ... 220)

Fig. 1. Amino acid sequences of homology regions of Fab New aligned by comparison of their three-dimensional structures. Hyphens (-) indicate gaps introduced to maximize alignment of three-dimensional structures of V_L, V_H, C_L, and C_H1. Asterisks (*) indicate amino acid deletions in sequence of V_L (New). V_L and C_L sequences from Chen and Poljak (1974), V_H from Nakashima et al. (1976), C_H1 from Edelman et al. (1969). For definition of amino acid symbols, see Dayhoff (1972)

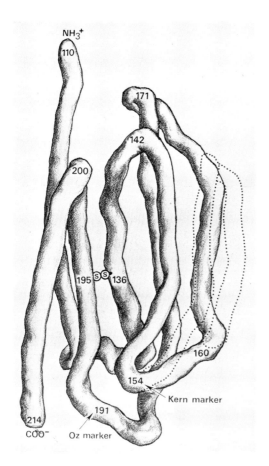

Fig. 2. Diagram of immunoglobulin fold. Folding of polypeptide chain in C_L and C_H1 subunits indicated by solid lines; broken lines indicate additional loop of polypeptide chain characteristic of V_L and V_H. Numbers designate L (λ) chain residues

Goodman (1976) have shown true homology between V and C sequences. Since the amino acid sequence of β_2-microglobulin, a ubiquitous membrane component of mammalian cells (associated with histocompatibility antigens), is highly homologous to that of C_H1 and C_L (Peterson et al., 1972) it is reasonable to expect that it shares the basic polypeptide chain folding of the immunoglobulin homology subunits. This basic folding has recently been observed by Richardson et al. (1975) in the Cu, Zn superoxide dismutase enzyme from bovine erythrocytes. Thus we are presented with the possibility that the immunoglobulin genes may have derived from other widespread, earlier genes which specified functions not directly related to self, non-self recognition.

Different patterns of inter H-L chain disulfide bonds that have been determined by amino aicd sequence analysis of immunoglobulins (IgA, IgG, IgM from different species) can be readily explained by the spatial proximity of the homologous residues in the human Fab New structure. This observation, taken together with the existence of a basic immunoglobulin fold common to the V and C subunits, suggests that the structure observed for Fab New is shared by the Fab regions of all immunoglobulins. This postulate has received support by the finding that a human V_K (Epp et al., 1974) fragment and a mouse myeloma Fab fragment (IgA, K; Segal et al., 1974) share the immunoglobulin fold and the same overall structure as determined for Fab New.

22

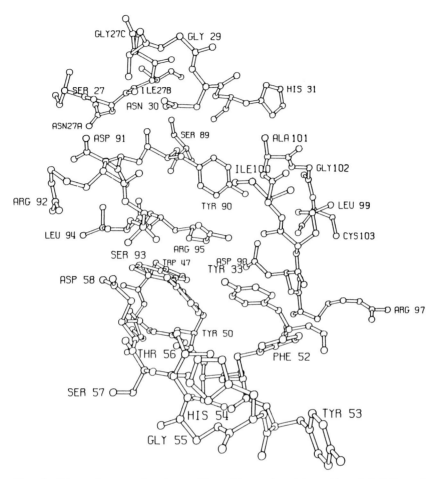

Fig. 3. View of some of amino acid residues at active site of IgG New. Residue numbers for V_L (27 to 31 and 89 to 95) and V_H (33, 47, 50 to 58 and 97 to 103) correspond to those of Figure 1

Hypervariable Regions and the Combining Site

The segments of polypeptide chain which correspond to the hypervariable regions of the L chain and to the first, second and fourth hypervariable regions of the H chain occur at one end of the molecule, fully exposed to solvent and in relatively close spatial proximity. These hypervariable segments describe a cavity or pocket which in the case of Fab New is about 15 x 6 Å with a depth of about 6 Å (see Fig. 3). This pocket has the general appearance of a ligand combining site generally resembling those determined for several enzymes by X-ray diffraction studies. Thus, the expectations created by the results of the amino acid sequence studies, in which the hypervariable regions were envisioned as determining the unique antigen binding specificity of every immunoglobulin, were given a firm structural basis by the crystallographic study of Fab New. Both the H and the L chains contribute to the combining site. In the H chain, positions 85-90 which constitute a third hypervariable region (Kehoe and Capra, 1971) are not close to the other hypervariable regions of the H or L chain and do not directly contribute to the combining site. However, as in the case of the other V_H and V_L hypervariable segments, they occur at an out-side bend of the polypeptide chain where there are less spatial con-straints than in other regions of the structure. To a more limited extent, murine V_K sequences also display variation in the homologous region, positions 74-85 (see Fig. 3, Cohn et al., 1974).

As stated above, the hypervariable regions contribute to the confor-mation of combining sites which are unique for each immunoglobulin molecule. In H and L chains, hypervariability refers to both the num-ber and the type of amino acid side chains which make the antigen com-bining site. For example, the combining site of IgG New consists of a shallow groove or pocket, about 15 x 6 Å with a depth of about 6 Å; the combining site of McPC 603, a mouse myeloma IgA with anti-phos-phoryl choline activity is deeper, about 12 Å (Segal et al., 1974). This increased depth is partly due to the presence of a longer, pro-truding first hypervariable region in the McPC 603 L chain. Since a pattern of insertions and/or deletions is typical of the hypervariable regions of H and L chains significant differences will occur at the antigen combining sites of immunoglobulins. A large number of recog-nition specificities can be incorporated in this structural design while maintaining a constant three-dimensional immunoglobulin fold. Furthermore, unrestricted pairing of H and L chains could give rise to a large number of specificities with a smaller number of H and L chains, thus allowing further paucity in structural design. In several in vitro recombination studies (see Discussion in Manjula et al., 1976) it has been shown that heterologous L and H chains recombine to re-constitute a native immunoglobulin molecule. The possibility that pairing of H and L chains is restricted to mutually compatible "sub-groups" of H and L chains has been submitted to experimental test (Stevenson and Mole, 1974) and to analysis in terms of amino acid se-quence of immunoglobulin classes and subgroups (Capra and Kehoe, 1975). Study of the three-dimensional model of a human Fab' indicates that the close interactions between V_L and V_H involve the side chains of residues at positions 35, 37, 42, 43, 86, and 99 (V_L) and at positions 37, 39, 43, 45, 47, 94, and 107 (V_H; sequence numbers are as given in Fig. 1). These positions are occupied by constant residues or by fairly conservative replacements in human V_L and V_H sequences as well as in V_L and V_H sequences that have been studied from other animal species (see Dayhoff (1972), for a compilation of these sequences). It is striking that human κ and λ chains are very similar in their amino acid sequences at these positions: almost all human L chains have Tyr (35), Gln (37), Ala (42), Pro (43), Tyr (86) and Phe (99). L chain positions 88 and 90 occur at the active site and also appear to provide for V_L - V_H interactions, but the importance of these in-teractions is more difficult to evaluate because they could be affected to some degree by the conformation of the last hypervariable region of V_H; even so there are preferred residues at these positions: Gln (88) and Trp/Tyr (90). Although preferred associations could occur by mech-anisms under genetic control, the interactions between V_H and V_L, which should provide the stereochemical basis for the association of H, L chain pairs, appear to be independent of H and L chain "subgroups" or even L chain class, so that any L chain has the potential to pair with a given H chain.

The concept of subgroups of κ and λ chain sequences can, however, be correlated with the functional specificity of immunoglobulins. Since such subgroups are defined by (or can be correlated with) the presence of insertions and/or deletions in the hypervariable regions (in par-

ticular the first hypervariable region) they define differen conformations of the active site. For example, human κ chains of subgroup
III ($V_{\kappa III}$) containing "insertions" of up to six amino acid residues
in their first hypervariable region will delineate a "deep" active
site such as that observed in the murine myeloma protein McPC 603
(Segal et al., 1974). By contrast a κ chain of subgroup I or a human
λ chain will delineate a more shallow active site as seen in Fab New
(Poljak et al., 1973). It is interesting to observe that murine λ
chains appear to lack the pattern of deletions and/or insertions at
the hypervariable regions characteristic of murine λ chains and human
κ and λ chains. Murine λ chains would thus make a smaller contribution
to the diversity of antibody specificity so that the number of genes
encoding λ chains, or their expression in mice, would be limited relative to κ chains as reflected in the $\lambda/_{\kappa}$ serum ratio (Weigert et al.,
1970).

On the basis of the structural model presented here, amino acid sequence changes in the hypervariable regions, affecting V_L, V_H, or both,
can easily be recognized by a molecule of complementary structure or
anti-idiotypic antibody. Not all the idiotypic antigenic determinants
will necessarily be active in binding an antigen; however it is possible to visualize that binding of an antigen or a suitable hapten at
the combining site will block recognition of the immunodeterminant
groups of the combining site region by an anti-idiotypic antibody
(Brient and Nisonoff, 1970; Carson and Weigert, 1973).

The characterization of the combining site of Fab New was further advanced by a study of the crystalline complex formed with the ligand
Vit. K_1OH (3-(3'-hydroxy 3',7',11',15', tetramethyl hexadecyl) 2-
methyl 1,4 naphtoquinone) (Amzel et al., 1974). This compound was found
to bind to Fab' New with an affinity constant of 1.7×10^5 liters/mol.
An X-ray diffraction study to a resolution of 3.5 Å showed that Vit.
K_1OH binds to the active site of Fab' New. The quinone moiety of Vit.
K_1OH makes close contact with the phenolic ring of L chain Tyr 90,
with the backbone and side chain of H chain residue 100 and with the
backbone of L chain residues 29 and 30. The phytyl tail of Vit. K_1OH
is bent at positions 3' and 4', making close contacts with L chain
residues Gly 29 and Asn 30. From that point it proceeds in an extended
form making close contacts with L chain residues 93 and 94 and H chain
residue 100. Towards its end it interacts with the side chain of H
chain Trp 47 and with H chain residues 50 and 58. The extensive interaction of Vit. K_1OH with the active site of Fab' New gives a general
indication of the type of contacts that could be involved in the formation of antigen-antibody complexes.

No major conformational changes in the Fab' fragment were observed
after the binding of Vit. K_1OH to crystalline Fab' New. However this
result does not exclude the possibility of conformational changes
following the formation of antigen-antibody complexes in solution.
Analysis of the structure of Fab' New suggests flexibility in the molecule compatible with possible conformational changes (Poljak et al.,
1974). Fab' New consists of four homology subunits in a tetrahedral
arrangement covalently linked in pairs by linear stretches of polypeptide chain (switch regions). In these switch regions the H chain
is bent to a larger extent than the L chain, so that V_H and C_H1 are
closer to each other than V_L and C_L. Furthermore, two light chains of
identical sequence in a crystalline L chain dimer (Schiffer et al.,
1973) were found to adopt different conformations in the switch regions,
so that one of them appears similar to the L chain of Fab' New and
the other to the Fd' fragment. These observations suggest that a conformational change could take place by a hinge-like movement at one
or both switch regions resulting in a change of the relative orienta-

tion of the V and C domains. In support of this postulate, recent work by Colman et al. (1976) indicates that the V_H - C_H and the V_L - C_L subunits can occur in different relative spatial orientations than those observed in Fab New, McPC 603 and the λ chain dimer Mcg.

Crystalline ligand-Fab New complexes will be the object of continued study by X-ray diffraction using improved phases obtained from the refined three-dimensional model of Fab New.

Acknowledgments. This research was supported by Grants AI 08202 from the National Institutes of Health, E-638 from the American Cancer Society, and N.I.H. Research Career Development Award AI 70091 to R.J.P.

References

Amzel, L.M., Poljak, R.J., Saul, F., Varga, J.M., Richards, F.F.: Proc. Nat. Acad. Sci. 71, 1427-1430 (1974)

Brient, B.W., Nisonoff, A.: J. Exp. Med. 132, 951-962 (1970)

Capra, J.D., Kehoe, J.M.: J. Immunol. 114, 678-681 (1975)

Carson, D., Weigert, M.: Proc. Nat. Acad. Sci. 70, 235-239 (1973)

Chen, B.L., Poljak, R.J.: Biochemistry 13, 1295-1302 (1974)

Cohn, M., Blomberg, B., Geckeler, W., Raschke, W., Riblet, R., Weigert, M.: In: The Immune System. Genes, Receptors, Signals. Sercarz, E.E., Williamson, A.R., Fox, C.F. (eds.). New York: Academic Press 1974, pp. 89-117

Colman, P.M., Deisenhoffer, J., Huber, R., Palm, W.: J. Mol. Biol. 100, 257-278 (1976)

Dayhoff, M.O. (ed.): Atlas of Amino Acid Sequence and Structure, Vol. V, Nat. Biomed. Res. Found. Silver Spring, Md. (1972)

Diamond, R.: Acta Cryst. 21, 253-266 (1966)

Diamond, R.: Acta Cryst. A27, 436-452 (1971)

Edelman, G.M., Cunningham, B.A., Gall, W.E., Gottlieb, P.D., Ruttis-hauser, U., Waxdal, M.J.: Proc. Nat. Acad. Sci. 63, 78-85 (1969)

Epp, O., Colman, P., Fehlhammer, H., Bode, W., Schiffer, M., Huber, R., Palm, W.: Europ. J. Biochem. 45, 513-524 (1974)

Hill, R.L., Delaney, R., Fellows, R.E., Jr., Lebowitz, H.E.: Proc. Nat. Acad. Sci. 56, 1762-1769 (1966)

Kehoe, M.J., Capra, D.J.: Proc. Nat. Acad. Sci. 68, 2019-2021 (1971)

Manjula, B.N., Goodman, M.: J. Mol. Evol., in press 1976

Nakashima, Y., Konigsberg, W., Chen, B.L., Poljak, R.J.: The amino acid sequence of the V_H homology region of IgG New. Manuscript in preparation (1976)

Peterson, P.A., Cunningham, B.A., Berggard, I., Edelman, G.M.: Proc. Nat. Acad. Sci. 69, 1697-1701 (1972)

Poljak, R.J., Amzel, L.M., Avey, H.P., Chen, B.L., Phizackerley, R.P., Saul, F.: Proc. Nat. Acad. Sci. 70, 3305-3310 (1973)

Poljak, R.J., Amzel, L.M., Chen, B.L., Phizackerley, R.P., Saul, F.: Proc. Nat. Acad. Sci. 71, 3440-3444 (1974)

Richardson, J.S., Thomas, K.A., Rubin, B.H., Richardson, D.C.: Proc. Nat. Acad. Sci. 72, 1349-1353 (1975)

Schiffer, M., Girling, R.L., Ely, K.R., Edmundson, A.B.: Biochemistry 12, 4620-4631 (1973)

Segal, D.M., Padlan, E.A., Cohen, G.H., Rudikoff, S., Potter, M., Davies, D.R.: Proc. Nat. Acad. Sci. 71, 4298-4305 (1974)

Stevenson, G.T., Mole, L.E.: Biochem. J. 139, 369-374 (1974)

Weigert, M.G., Cesari, I.M., Yonkovich, S.J., Cohn, M.: Nature (Lond.) 228, 1045-1047 (1970)

X-Ray Diffraction Analysis of Immunoglobulin Structure

R. Huber, J. Deisenhofer, P. M. Colman, M. Matsushima, and W. Palm

Abstract

The crystal structures of a human IgG antibody molecule Kol and a human Fc fragment have been determined at 4 Å and 3.5 Å resolution respectively by isomorphous replacement.

The electron-density maps were interpreted in terms of immunoglobulin domains based on the Rei and McPC 603 models (Kol) and by model-building (Fc).

The Kol Fab parts have a quarternary structure different from the fragments. The Fc part C-terminal to the hinge is disordered in the Kol crystals. There is no longitudinal V-C contact in Kol in contrast to Fab fragments. It is suggested that the Kol molecule is flexible in solution, while fragments are rigid.

In the Fc fragment both C_H3 and C_H2 show the immunoglobulin fold. The C_H3 dimer aggregates as C_H1 - C_L while C_H2 are widely separated from each other. The carbohydrate bound to Fc is in fixed position. From these structures it is suggested that antibody molecules are inherently flexible but become rigid upon interaction with antigen, which triggers the formation of all longitudinal contacts.

Some crystallographic and structural details of the X-ray diffraction studies of the complete IgG molecule Kol and of an Fc fragment have been described (1, 2). Kol is a human myeloma immunoglobulin (3, 3a, 3b) IgG1 that is 'normal' by several criteria: molecular weight of intact material and heavy and light chain components, and splitting into Fab and Fc fragments (1). Further experiments by SDS gel electrophoresis (4) under reducing and nonreducing conditions of the Kol protein and the Kol Fc fragment demonstrated the presence of disulfide bonds between heavy chains and between heavy and light chains.

The Fc fragment was obtained by plasmin digestion from pooled human serum. Partial amino acid sequence analysis indicated the presence of a normal hinge region (2). The material is inhomogeneous (5). The main source of heterogeneity might be due to allotypic differences and to variations in the carbohydrate moiety.

The crystallographic analysis of the Kol protein was pursued to 5 Å resolution by isomorphous replacement (1) and later extended to 4 Å resolution. The interpretation of the electron density map in terms of domain structure succeeded using the domain Patterson map interpretation procedure (1). The electron-density map was analyzed in terms of the known structure of the V and C parts of the Fab fragment McPC 603 (6) and the Rei fragment (8). It was evident that there are differences in domain tertiary structure between the Kol protein and McPC 603. We have not yet tried to analyze these in detail, but it

L   R

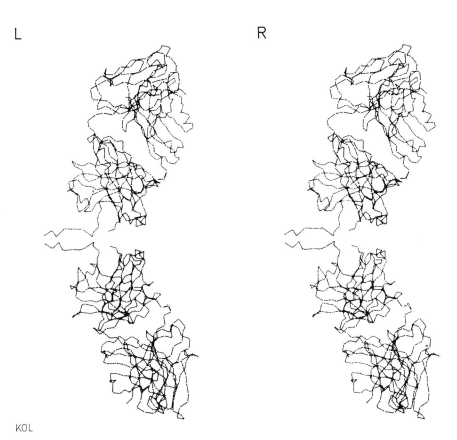

KOL

Fig. 1. Stereogram of Kol (Fab)$_2$ part. C$^\alpha$ atom positions derived by fitting McPC 603 (6) V and C dimers to the Kol electron-density map. Coordinates of C-terminal residues of heavy and light chain and hinge peptide obtained by interpretation of 4 Å Fourier map

is clear that they do not affect the later discussion on longitudinal domain interfaces. A tentative interpretation in the 4 Å map of the C$^\alpha$ chain folding is given for the heavy chain from residue 213 to the Cys-Pro-Pro-Cys-Pro segment and for the light chain from residue 209 to the heavy-light chain disulfide.

The model discussed in the following is based on the fitting of the McPC 603 model to the Kol electron-density and including the few C$^\alpha$ positions obtained directly from the Fourier map.

The Fc fragment crystal structure was analyzed at 4 Å resolution (2) and later at 3.5 Å resolution, using isomorphous phases. The electron-density map was interpreted using the domain Patterson map interpretation procedure with respect to the C$_H$3 dimer. The C$_H$2 domains were analyzed directly in the electron density map (2). A complete atomic model of one chain was constructed based on the 3.5 Å resolution map, using a Richard's comparator. The atomic coordinates were subjected to a model-building procedure (11). The second chain was created by applying the known local symmetry. We have not yet analyzed the deviations from the local symmetry, but they appear to be small.

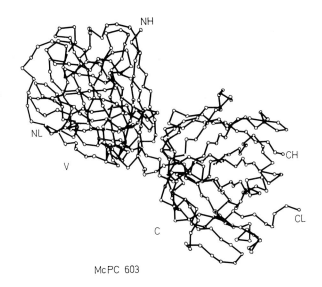

McPC 603

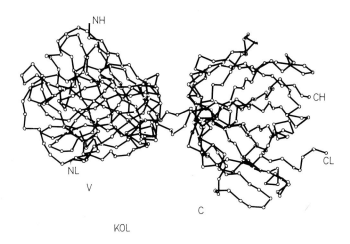

KOL

Fig. 2. Comparison of McPC 603 Fab fragment and Kol Fab part. Fab angle changes by 50°. There is a V_H–C_H1 contact in fragment but not in intact molecule

The Kol Molecule

The C^α carbon atom positions of the (Fab)$_2$ part of the Kol molecule are shown in the stereogram (Fig. 1). The Fab arms subtend an angle of 125°. The tips of the Fab arms are about 146 Å apart. Apart from the slight differences in domain tertiary structure, the V_L–V_H and C_L–C_H1 dimers appear to be indistinguishable between Kol and McPC 603. In particular, the geometry of aggregation within V and C dimers is very similar. There is a difference, however, in the relative orientation of V and C dimers, which can be described by a rotation about an axis through the switch peptides at residues 118-119 (heavy chain) and 109-110 (light chain) (Fig. 2). This, of course, requires a conformational change in the switch peptide segment, which might be accomplished with minimal perturbation as the segment is in an extended conformation. Figure 2 compares the McPC 603 Fab fragment and the Kol Fab part as seen down the switch peptide rotation axis. V_L and V_H as well as C_L and C_H1 are related by local diads. These diads subtend an angle (Fab angle) of 120° in the McPC 603 fragment and 170° in the Kol protein, demonstrating the major change in quarternary structure: 'bending the elbow'.

It is already evident from Figure 2 that in the McPC 603 molecule some residues of V_H and C_H1 come close to each other. There are 6 C^α-C^α distances smaller than 6.5 Å (formed by segments 8-10 in V_H and 152 and 207-208 in C_H1) and 41 below 10 Å. There are, of course, several contacts between residues of the switch peptide and C_H1 not included. We believe that these are irrelevant for the later discussion. Equivalent residues of V_L C_L of the light chain adopting the heavy chain conformation are in contact in the Bence Jones protein dimer structure (12 and pers. comm. A. Edmundson). In contrast, due to the asymmetry of the molecule, V_L and C_L are much further separated in the McPC 603 molecule. There are 2 C^α distances below 6.5 Å and 28 below 10 Å.

Through the widening of the Fab angle from 120° to 170° in Kol there is no contact between C^α atoms of V_H and C_H1 smaller than 7 Å; 5 distances are below 10 Å. Also, V_L and C_L have no contact (6 distances below 10 Å). In comparison, the lateral V_H-V_L and C_L-C_H1 contacts in the McPC 603 Fab fragment involve 9 and 7 C^α distances respectively closer than 6.5 Å and about 120 distances below 10 Å for both contacts. The C_H3-C_H3 contact in Fc involves 14 C^α distances smaller than 6.5 Å and about 150 below 10 Å. According to this admittedly very crude criterion, V-V, C-C and V_H-C_H1 contacts might not be too different in strength in the Fab fragment structures while there is no appreciable longitudinal interaction in the Kol Fab parts. In rabbit light chains a disulfide bond links residues 80 and 172 (13). A sufficiently close approach in the Kol model occurs between C^α atoms of residues 81 and 171 of 9 Å. This distance is 7 Å in the McPC 603 model. The disulfide linkage therefore does not oppose the Fab angle ('elbow') bending considerably. The two Fab arms in the Kol protein have no contact with each other, except the covalent linkage of the hinge. There is no approach of C^α of the two C_L domains closer than 12.5 Å.

No interpretable electron-density is found in the Kol Fourier map which could be assigned to the Fc part (1). The Fourier map contains significant density in the region where the Fc stem must be located, with peak heights about 2.5 times the root-mean-square-error in electron-density. This is to be compared with the density in the Fab parts, with peak heights about four times the error level. We have concluded that the Fc stem is disordered in the Kol crystals, C-terminal to the hinge peptide. Indeed, the (Cys-Pro-Pro-Cys)$_2$ peptide of the hinge on the molecular and crystallographic diad is only 50 Å away from the crystallographic 3_2 axis. The Fc fragment molecule is too long and too thick to be accomodated on the diad axis (1, 2). It would interfere with crystallographically related Fc parts. On the other hand, the very close packing in the Kol crystal structure around the hinge peptide (see Fig. 6 in (1)) does not allow a close approach of the C_H2 domains of Fc to (Fab)$_2$, but requires a rather extended hinge segment.

We constructed a 'minimum disorder' Kol molecule in the following way: the Fc fragment model was placed with its local diad on the crystallographic diad of the Kol lattice; starting position was the rigid model described below. Fc was rotated azimuthally so that there was minimal overlap with (Fab)$_2$ parts from crystallographically related molecules (about 90° from the starting position). It had then to be shifted by 15 Å along the diad away from the Fab parts to avoid overlap. Such a molecule would penetrate the 3_2 axis and interfere with crystallographically related Fc parts. Fc was therefore bent about an axis

L   R

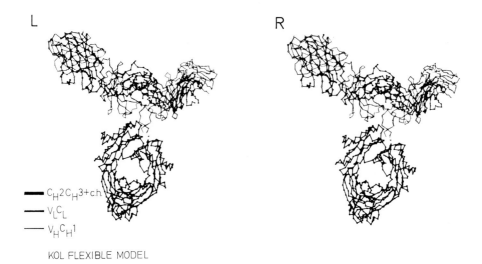

━━━ C_H2C_H3+c.h.
━━ V_LC_L
── V_HC_H1

KOL FLEXIBLE MODEL

Fig. 3. Stereogram of 'minimum disorder' model of Kol molecule to be regarded as snap shot of flexible molecule. No contacts between $(Fab)_2$ and Fc. Structure of Fc part derived from Fc fragment, although Fc thought to be flexible between C_H2 and C_H3 in Kol molecule. Hinge peptide drawn as observed in Kol $(Fab)_2$ and Fc fragment. To make covalent linkage, hinge bent back between C_H2 domains has to be pulled out

close to the top of the molecule and perpendicular to the a b plane of the Kol lattice minimally to avoid this steric interference. It is clear that these three operations are not independent. It appears that a correct, but untractable three-dimensional treatment would not change our model in a substantial way. Figure 3 represents the model.

It is clear that there are no contacts between $(Fab)_2$ and Fc, except the covalent linkage through the hinge peptide, which must have a rather extendend conformation. We will describe below that in the Fc fragment the N-terminus appears to be bent back in between the C_H2 domains. In the 'minimum disorder' model the wide separation of $(Fab)_2$ and Fc requires this piece of chain to be pulled out. These observations of absent longitudinal interactions between V_H and C_H1 and between C_H1 and C_H2 as well as missing lateral interactions between both C_L parts in the Kol molecule in comparison with the existing V_H-C_H1 interactions in free Fab fragments will play an important role in later discussions.

The Fc Fragment Structure

Figure 4 is a stereo-plot of the C^α carbon atom positions of the Fc fragment. The molecule has the shape of a mickey mouse with the compact C_H3 dimer forming the head and the C_H2 domains forming the ears. The C_H3 dimer pairing is closely similar to the C_H1-C_L pairs found in Fab fragments.

C_H2 and C_H3 show the common immunoglobulin fold. They are connected by a loosely folded segment from Ser (337) to Gln (342) which is susceptible to proteolytic attack (14, 15). The longitudinal C_H2-C_H3 con-

L   R

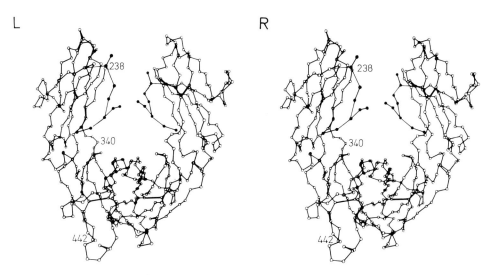

Fig. 4. Stereogram of C$^\alpha$ carbon positions (— o —) and centers of carbohydrate hexose units (— • —) of the Fc fragment. The interpretation is uncertain N-terminal to residue 238. This segment has been omitted

tact is characterized by 3 C$^\alpha$ distances less than 6.5 Å (and 43 below 10 Å), not considering contacts made by the connecting peptide. It comprises the segments 247-253 and 310-314 (C$_H$2) and 376-379 and 428-433 (C$_H$3) respectively. These segments are homologous to those mediating the V$_H$-C$_H$1 contact in Fab fragments described above.

Comparative drawings of C$_H$3 and C$_H$2 are shown and discussed in a later paragraph. The electron-density at the C-terminus fades away after Ser (442) indicating mobility from this residue on. The N-terminus appears to be folded back from Pro (238) on and in intimate contact with the carbohydrate moiety discussed below. This mostly unresolved contact makes an interpretation of the peptide chain conformation around the hinge region difficult at present. A possible chain tracing has the segment from Gly (236) to Leu (234) in contact with one branch of the carbohydrate chain (16). The (Cys-Pro-Pro-Cys)$_2$ segment appears to be disordered. This is not unexpected, as in the Kol crystals the (Cys-Pro-Pro-Cys)$_2$ segment ist fixed while the Fc part is disordered. A detailed description of the molecule including side chains has to await further refinement of the crystal structure.

The electron density assigned to the bound carbohydrate could accommodate a branched carbohydrate chain of the following arrangement

$$
\begin{array}{c}
\text{H} \\
| \\
\text{Asn} - \text{H} - \text{H} - \text{(H)} - \text{H} - \text{H} \begin{array}{c} \nearrow \text{H} - \text{H} \\ \searrow \text{H} - \text{H} \end{array}
\end{array}
$$

(H is a hexose unit; the residues are: NAG: N-acetylglucosamine, MAN: mannose; Gal: galactose; FUC: fucose (16).

Such a chain would be longer than indicated by carbohydrate sequences of myeloma proteins (16). We have to expect inhomogeneity of the carbohydrate in our material, which might make the detailed analysis difficult. The crystallographic analysis of the carbohydrate structure requires further refinement and model building.

The carbohydrate moiety covers a large part of the C face of the C_H2 domain. Our preliminary present model indicates that the carbohydrate covers apolar residues on the C face; these include: Phe (241), Phe (243), Val (262), Val (264). There might be a close contact with Leu (234, 235) of the N-terminal segment.

The site of attachment of the carbohydrate moiety at Asn (297) is part of a reverse turn (type RTI (17)). The four residues in the turn are Tyr-Asn-Ser-Thr. In general carbohydrate attachment preferentially occurs at amino acid sequences X-Asn-X-Thr(Ser) (see review (18)). A statistical analysis of protein conformation showed that there is a strong preference of Asn and Thr (Ser) to be part of reverse turns (17). Pancreatic trypsin inhibitor (PTI) and ribonuclease are molecules of known three-dimensional structure (19, 20, 21). There exist closely related variants of these molecules (colostrum inhibitor (22) and ribonuclease B (23) which carry carbohydrate. In all cases the carbohydrate attachment site is an Asn residue that is part of a turn. Sugar transferases appear to have sequence and structure specificity.

Domain Tertiary-Structure Comparison
=====================================

A set of comparative drawings of the individual domains in the Fab fragments McPC 603 and New and the Bence Jones proteins Mcg and Rei has been published (6). A detailed comparison with the chain folding in the (Fab)$_2$ part of the Kol molecule has to await higher resolution data of the Kol crystals. The preliminary comparison based on the 5 Å resolution map of Kol revealed some differences which were summarized (1). It seems premature to discuss these in detail.

Figure 5 compares the individual domains of Rei V_L, McPC 603 C_H1, Fc C_H2, and Fc C_H3. The most prominent difference between V and C_H1 domains is the presence of the loop at 40 (Rei enumeration) in V domains which is absent in C_H1. In addition there is a short loop in the V domain around residue 52 which is absent in C_H1. The C_H1 domain has a very long loop around residue 180 (McPC 603 enumeration) which is part of the concave surface involved in C-C aggregation. This loop is short in V domains and the whole surface in planar. It is clear that Fc C_H3 has the characteristics of a C domain: the long loop involved in the C-C interface is present, with the segment 399-402 forming the corner. V-characteristic loops are absent. C_H2, however, has some of the features of a V domain: a loop homologous to the loop at 40 is present in C_H2 with a hair-pin bend involving residues Val-Asp-Gly-Val (278-282). This loop is longer by 2 steps (4 residues) in V domains. There is a vestigial loop from Asn (286) to Lys (290) in C_H2 homologous to the loop at 52 in V domains. The loop homologous to the long 180 loop in C_H1 and C_H3 domains is two steps shorter in C_H2 but also 2 steps longer than in V domains. Asn (297) at the hairpin bend ending this loop is the site of attachment of the carbohydrate.

This raises the interesting question of the evolutionary origin of the C_H2 segment. While amino acid sequence homologies have been interpreted as indicating a close relation of C_H2 to other C segments and only a distant relation to V segments, the structural comparison favours an intermediate structure. This matter should be reconsidered.

There is a remote possibility that the type of aggregation influences the chain folding, the C_H2 conformation being characteristic of a

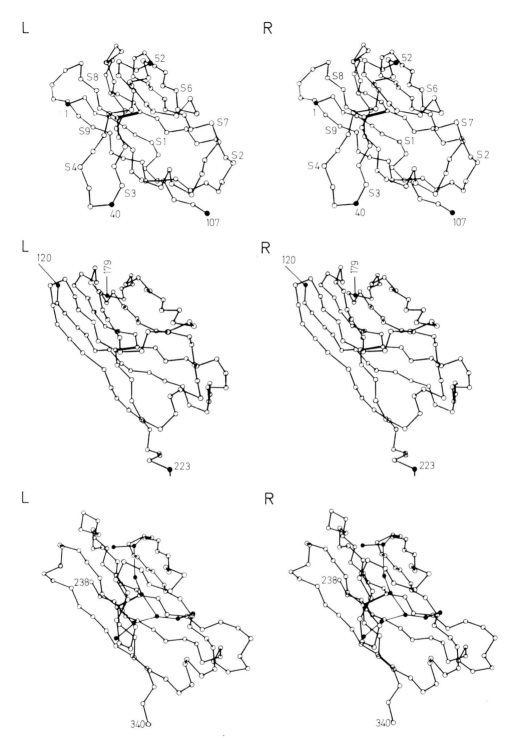

Fig. 5 continued and legend see page 34

L 339   R 339

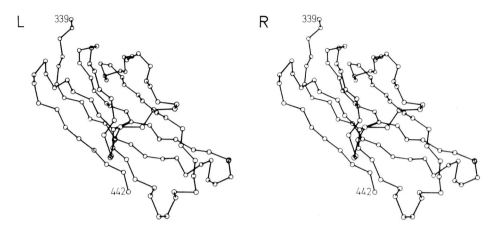

442   442

Fig. 5. Stereograms of individual domains of Rei V_L, κ-type, McPC 603 C_H1, Fc C_H2, Fc C_H3

free domain, and the other C segments being deformed by their C-C contacts.

Domain-Domain Interactions

Lateral Interactions

The lateral V-V (V_H-V_L, V_L-V_L) and C-C (C_L-C_H1, C_L-C_L) interactions observed in Fab fragments (24, 25) and Bence Jones proteins (26, 9) have been described in great detail and will not be considered here. The relative arrangement of the two C_H3 domains observed in the Fc fragment is closely similar to the C_L-C_H1 aggregation observed in McPC 603 (2). Residues lying close to the local diad in the Fc fragment are: Leu (351), Pro (353), Thr (366), Leu (368), Tyr (407), Thr (394), Pro (395). The C_H2 domains have no contact with each other apart from the covalent linkage through the hinge. It is an intriguing question why C_H2 forms neither the familiar C-C nor the V-V contact. It has been discussed that the tertiary structure of the C_H2 domain does not resemble a C but also lacks some features of a V domain. This might prevent dimerization. In the presence of the spatially fixed carbohydrate moiety, formation of the C-C contact is impossible. Furthermore, some of the hydrophobic residues in C_H3 which lie close to the local diad, and are probably important for the contact, are replaced by polar groups in C_H2. A most conspicuous change intolerable in a C-C contact is Tyr 407 (C_H3) for Arg 301 (C_H2). Formation of a V-V contact of C_H2 would be more probable in view of the more pronounced similarities with a V domain. However, here also several of the contact residues (9), conserved in V segments, are exchanged for polar residues. Among these are Gln (89) and Tyr (87) (Rei), which are replaced by Lys (322) and Lys (320) (C_H2).

Longitudinal Interactions

In the Kol protein and the 'minimum disorder' model derived from it, no longitudinal contacts are observed between V_H and C_H1, V_L and C_L,

C_H1 and C_H2. There are also no contacts between the Fab arms. In view of this lack of stabilization along the chain, it is difficult to assume that the Kol molecule is rigid in solution. We presume rather that there exists flexibility between V and C dimers (elbow V-C-flexibility), between the Fab arms ((Fab)$_2$-flexibility) and between Fc and (Fab)$_2$ (Fc-(Fab)$_2$-flexibility). Flexibility between Fc and (Fab)$_2$ could explain the weak, uninterpretable electron-density for Fc in the Kol crystals.

The relative positions of V and C dimers and Fab arms are frozen in the Kol crystal lattice dictated by crystal packing requirements. Even the C_H2 segments might sit flexibly on the C_H3 dimers (C_H2-C_H3-flexibility) in the Fc part in the Kol protein, in contrast to their fixed position in the fragment. We have, of course, no evidence for this as the Fc part in the Kol crystals has not yet been analyzed. Rigidity without longitudinal contact between the domains could only be brought about if the switch peptides and the hinge peptide adopted rigid conformations. These segments are in an extended conformation, largely exposed to solvent. Inherent rigidity of these peptide chains is highly improbable.

The model which we would like to propose for the Kol protein and for other unliganded (without antigen) IgG molecules is a highly flexible one.

The degree of segmental flexibility appears more limited for V-C (rotation about the 'switch axis') (1) than for (Fab)$_2$ or Fc-(Fab)$_2$, which are connected by far longer extended polypeptide chains.

In obvious contrast the Fab fragments (24, 25) and the closely similar Bence Jones protein dimer (26) appear to be rigid molecules. In the three structures, V and C dimers subtend a similar Fab (elbow) angle of about 120°. Crystal packing is different in the three crystal structures and would be unlikely to have the same effect on the Fab angle if it were variable. Indeed there exists a longitudinal contact between V_H and C_H1 in these fragments. In size this contact surface is smaller than the lateral C-C or V-V contacts, but it has been found in other systems that small contact surfaces may mediate strong binding (i.e. PTI-inhibitor-trypsin (27)).

In the Fc fragment C_H2 and C_H3 form a contact as described. The linking peptide is not in a completely extended conformation but is irregularly bent between residues Lys (338) and Gly (341). The presence of two (identical) contacts in a crystallographically nonidentical environment suggests a rigid Fc fragment, although it would be desirable to confirm this by structure analyses of different crystal forms. Loosening of the C_H2-C_H3 contact and a stretching out of the linking peptide appear possible. As the C_H2 parts are not packed together, the flexibility might be greater than in Fab parts.

It is remarkable and deserves further consideration that homologous segments mediate the close V-C contacts in the Fab fragment and the close C_H2-C_H3 contacts in the Fc fragment: V_H: residues 8-10, C_H1: residues 152 and 207-208 (McPC 603) C_H2: residues 247-253 and 310-314, C_H3: residues 376-379 and 428-433.

These contacts have been inferred from C^α distance calculations. The actual bonding interactions are, of course, made by side chains and/ or polar atoms of the back-bone. These segments contain residues highly conserved in immunoglobulin sub-classes and classes. Of particular interest is the conservation of the three hydrophobic residues in C_H2 (Leu-Met-Ile (251-253)) in all IgG sub-classes, but also in IgM C_H3 (Ile-Phe-Leu (352-354)) (28) IgA C_H2 (Leu-Leu-Leu (29) and IgE C_H3 (Leu-Phe Ile (339-341)) (30). We note in passing that this observation adds to the evidence that in IgM and IgE, which contain an additional constant domain, C_H3 resembles C_H2 in IgG. In C_H1 and C_H3 ('top' contact area) the contact residues are from the bends C-terminal to the two intradomain Cys residues. In V_H and C_H2 ('bottom' contact area) these are predominantly residues of the bend N-terminal to the first intradomain Cys residue. It is of further interest that the contact segments described for C_H1 and C_H3 when projected into a V segment coincide with the first and third hypervariable segments forming the antigen-binding surface in V dimers. This suggests the presence of a general binding area in immunoglobulin domains. The residues forming this area must of course be variable in V domains to meet the requirements of complementarity to various antigens, but conserved in C domains for interdomain interaction.

This concept of homologous interdomain longitudinal contacts will be extended in a speculative way to the C_H1-C_H2 contact between (Fab)$_2$ and Fc in later discussion. In an allosteric, cooperative immunoglobulin model which we will propose, the signal transfer through the domains from the general binding area (top contact) to the opposite end of the domains (bottom contact) occurs along homologous segments in all domains.

We have presented crystallographic indications that the Kol molecule (an unliganded IgG molecule) in solution is flexible in the Fab parts (V-C-bending) and has no interaction between the Fab arms ((Fab)$_2$-bending) and between Fab and the Fc stem ((Fab)$_2$-Fc bending). We also speculate that Fc itself is flexible. If there is a preferred conformation in solution (perhaps resembling the fragment structures), its stabilization energy must be low as it is perturbed by crystal packing forces in the Kol lattice.

The rigidity of Fab and Fc fragments and the segmental flexibility in the complete molecule Kol might be happenstance and future studies might provide a better statistical sample of the whole population of molecules.

The following model-building studies and discussion is based on the hypothesis that a) rigidity and segmental flexibility are indeed inherent properties of fragments and complete, unliganded molecules, respectively, and b) the rigid Fab fragment structures are characteristic for the liganded molecules. This is primarily based on the observation that Fab fragments show no gross structural change upon hapten binding (24, 25). We are aware of the fact that different explanations for this observation are possible.

We constructed, therefore, a hypothetical liganded antibody molecule by combining the fragment molecules. Starting with the Kol (Fab)$_2$, we set the Fab elbow angle to 120°, which is characteristic for Fab fragments.

The hinge peptide which is well defined in the Kol crystals and less well ordered, but apparently bent back in Fc fragment crystals, was superimposed with the local diads coinciding. Fc was then rotated azimuthally in order to avoid overlap between C_H1 and C_H2 of the same molecule. Only a very narrow range of the azimuthal angle turned out to be allowed. The resulting model is shown in Figure 6. C_H1 and C_H2 are close at segments 132-133, 194-200 (C_H1) and 329-330, 265-268 (C_H2).

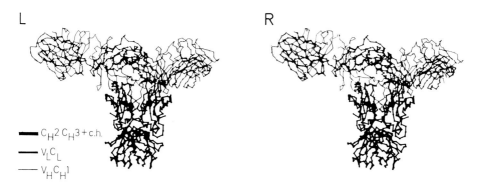

L   R

— $C_H2\ C_H3$ + c.h.
— $V_L C_L$
— $V_H C_H1$

KOL RIGID MODEL

Fig. 6. Stereogram of hypothetical liganded IgG molecule. Molecule T-shaped with Fc forming stem. Fc and (Fab)$_2$ in close contact. Hinge peptide folded back between the C_H2 parts

These include the general binding segments discussed above and comprise conserved residues in immunoglobulin sub-classes and classes. The hinge peptide is bent back in between the C_H2 domains and forms numerous internal contacts and with both domains and the carbohydrate.

The important aspect of this model is that all longitudinal interdomain contacts are closed, resulting in a rigid structure. (Fab)$_2$ bending is prohibited by the $C_H1\ C_H2$ interaction.

We suggest that antigen binding causes a stiffening of the flexible antibody molecule by formation of the longitudinal interdomain contacts. The largest structural change involves the retraction of the hinge peptide inbetween the C_H2 globules to allow the $C_H1\ C_H2$ approach. This model allows deletion of a large proportion of the hinge segment as observed in two crystalline antibodies (32, 33), without change in quarternary structure, in contrast to the flexible model with the more extended hinge segment. The hinge deletion forces C_H1 and C_H2 to form the contact. This could induce an overall quarternary structural change to the rigid conformation. The Fab bending produces a model more like a T, compared with the Y of the Kol molecule (Fig. 6; 59).

In the usual terminology of allosteric cooperative interactions the flexible model would represent the tense state (T) which changes to the relaxed state (R) upon antigen binding. Antigen binding triggers the formation of the longitudinal interdomain contacts. The T state quarternary structure is maintained only in the intact molecule. Fragment structures adopt a conformation characteristic for the R state of the intact molecule.

It is remarkable that electron micrographs of hapten-crosslinked IgG molecules show a short Fc stem and no break between (Fab)$_2$ and Fc in accord with the rigid model proposed (31).

Segmental flexibility between Fab and Fc parts of antibody molecules is experimentally established by spectroscopic and hydrodynamic observations (34, 35). Electron micrographs also show variable angles between the Fab arms of antibody molecules (31, 36). Here, however, variability might be caused by the drastic, denaturing conditions under which these experiments are performed. The beautiful electron micrographs of an Fc fragment (37) clearly show variability in the attachment of the C_H2 domains to C_H3, while the model obtained by crystallography favours a rigid structure (2), as discussed. Hydrogen-exchange experiments as well as experiments with limited proteolysis of IgM and their Fab fragments may be interpreted in terms of a less flexible fragment structure (38, 39).

There are several observations of structural changes in antibody molecules upon antigen binding. Circular polarization of fluorescence changes in complete antibodies upon antigen binding, but not in Fab fragments (40, 41). An antigen-induced volume contraction has been found by small angle X-ray scattering in complete antibodies, but not in Fab fragments (42, 43). Antibodies appear to lose flexibility upon antigen binding (44, 45) and to gain conformational stability against denaturation (46). The structural difference in the Fab part between complete antibody and Fab fragment (elbow bending) might manifest itself in a change of the characteristics of ligand binding. Observations of differences in rate constants of fragments and complete antibodies are controversial (47, 48). There are apparently no significant hapten binding affinity differences between Fab fragments and complete antibodies, indicating that there is no cooperation between the two hapten binding sites in an antibody molecule (49). This does not contradict the ligand-induced structural change proposed, as the energy required for the conformational change might be provided by an increased ligand affinity with both effects canceling out. Furthermore, there are observations on IgM molecules of differences in thermodynamic parameters for haptens and antigens (50).

Immunoglobulins have a number of effector functions of which the binding of complement has been extensively studied. It is well documented that the C1 binding site is located on the C_H2 domains (51, 52, 53, 54). C_H2, Fc and IgG have intrinsic complement binding properties which are, however, greatly enhanced by specific antigen binding or heat-aggregation (55, 56). It has been suggested that this might be due to proximity of several complement-fixation sites and/or to structural changes (for review see 57, 58). The rigid, liganded antibody molecule that we propose could enhance C1 binding for several reasons:

a) In the unliganded, flexible antibody molecule the C1 binding site might be transiently occluded by the Fab arms and possibly also by C_H3. The long extended hinge peptide in the unliganded Kol molecule allows the Fab arms far-reaching wagging motions to cover part of C_H2.

b) Protein-protein complex formation requires freezing out of rotational and translational degrees of freedom. A component with internal flexibility is therefore less favorable for complex formation than a rigid molecule.

c) The formation of longitudinal contacts might induce a structural change in C_H2 favoring C1 binding.

Acknowledgments. Extensive use has been made of the PROTEIN program system written by Dr. Wolfgang Steigemann in this laboratory. Others we wish to thank include Drs. G. Schwick and H. Haupt, Behring Werke, Marburg, for providing the Fc material and Frau K. Epp for X-ray technical assistance.
The financial assistence of the Deutsche Forschungsgemeinschaft and Sonderforschungs-bereich 51 is gratefully acknowledged.

Abbrevations Used: IgG: Immunoglobulin with γ-heavy chain; Fab: antigen-binding fragment consisting of light chain and half of the heavy chain (V_L, V_H, C_L, C_H1);

Fc: C-terminal half of the heavy chain with the inter-heavy chain disulfide bond intact; V_L: variable half of light chain; C_L: constant half of light chain; V_H: variable part of heavy chain; C_H1, C_H2, C_H3: the three constant homology regions of the heavy chain; hinge peptide: the segment connecting C_H1 and C_H2 and containing the inter-heavy chain disulfide linkage; switch peptides: the segments connecting V and C parts comprising residues at 110 (light chain) and 119 (heavy chain).
Amino acid sequence numbers in the Kol Fab part refer to the McPC 603 molecule (6) and to the Rei protein V_L dimer (7, 8, 9). In the Fc fragment the amino acid sequence numbers are based on the Eu amino acid sequence (10).

References

1. Colman, P.M., Deisenhofer, J., Huber, R., Palm, W.: J. Mol. Biol. 100, 257-282 (1976)
2. Deisenhofer, J., Colman, P.M., Huber, R., Haupt, H., Schwick, G.: Hoppe-Seyler's Z. Physiol. Chem. 357, 435-445 (1976)
3. Palm, W.: Hoppe-Seyler's Z. Physiol. Chem. 355, 877-880 (1974)
3a. Palm, W.: Ibid. 357, 795-798
3b. Palm, W.: Ibid. 357, 799-802
4. Weber, K., Osborn, M.: J. Biol. Chem. 244, 4406-4412 (1969)
5. Haupt, H., Heide, K.: Klin. Wochenschr. 47, 270-272 (1969)
6. Davies, D.R., Padlan, E.A., Segal, D.M.: Annu. Rev. Biochem. 44, 639-667 (1975)
7. Palm, W., Hilschmann, N.: Hoppe-Seyler's Z. Physiol. Chem. 356, 167-191 (1975)
8. Epp, O., Colman, P.M., Fehlhammer, H., Bode, W., Schiffer, M., Huber, R., Palm, W.: Europ. J. Biochem. 45, 513-524 (1974)
9. Epp, O., Lattman, E.E., Schiffer, M., Huber, R., Palm, W.: Biochemistry 14, 4943-4952 (1975)
10. Rutishauser, U., Cunningham, B.A., Bennet, C., Konigsberg, W.H., Edelman, G.M.: Biochemistry 9, 3171-3181 (1970)
11. Diamond, R.: Acta Cryst. Sect. A 21, 253-266 (1966)
12. Schiffer, M., Girling, R.L., Ely, K.R., Edmundson, A.B.: Biochemistry 12, 4620-4631 (1973)
13. Strosberg, A.D., Fraser, K.J., Margolies, M.N., Haber, E.: Biochemistry 11, 4978-4985 (1972)
14. Connell, G.E., Porter, R.R.: Biochem. J. 124 (1971)
15. Kehoe, J.M., Bourgois, A., Capra, J.D., Fougereau, M.: Biochemistry 13, 2499-2504 (1974)
16. Kornfeld, R., Keller, J., Baenziger, J., Kornfeld, S.: J. Biol. Chem. 246, 3259-3268 (1971)
17. Crawford, J.L., Lipscomb, W.N., Schellmann, C.G.: Proc. Natl. Acad. Sci. USA 70, 538-542 (1970)
18. Marshall, R.D.: Ann. Rev. Biochem. 41, 673-702 (1972)
19. Huber, R., Kukla, D., Rühlmann, A., Epp, O., Formanek, H.: Naturwissenschaften 57, 389-392 (1970)
20. Kartha, G., Bello, J., Harker, D.: Nature (Lond.) 213, 862-865 (1967)
21. Wyckoff, H.W., Hardmann, K.D., Allewell, N.M., Inagami, T., Tsernoglou, D., Johnson, L.N., Richards, F.M.: J. Biol. Chem. 242, 3749-3753 (1967)
22. Tschesche, H., Klauser, R., Cechová, D., Yanáková, V.: Hoppe-Seyler's Z. Physiol. Chem. 356, 1759-1764 (1974)
23. Jackson, R.L., Hirs, C.H.W.: J. Biol. Chem. 245, 637-653 (1970)
24. Amzel, L.M., Poljak, R.J., Saul, F., Varga, J.M., Richards, F.E.: Proc. Natl. Acad. Sci. USA 71, 1427-1430 (1974)
25. Segal, D.M., Padlan, E.A., Cohen, G.H., Rudikoff, S., Potter, M., Davies, D.R.: Proc. Natl. Acad. Sci. USA 71, 4298-4302 (1974)

26. Schiffer, M., Girling, R.L., Ely, K.R., Edmundson, A.B.: Biochemistry 12, 4620-4631 (1973)
27. Huber, R., Kukla, D., Bode, W., Schwager, P., Bartels, K., Deisenhofer, J., Steigemann, W.: J. Mol. Biol. 89, 73-101 (1974)
28. Watanabe, S., Barnikol, H.U., Horn, J., Bertram, J., Hilschmann, N.: Hoppe-Seyler's Z. Physiol. Chem. 354, 1505-1509 (1973)
29. Kratzin, H., Altevogt, P., Ruban, E., Kortt, A., Staroscik, K., Hilschmann, N.: Hoppe-Seyler's Z. Physiol. Chem. 356, 1337-1342 (1975)
30. Bennich, H., Bahr-Lindström, H.: Progress in Immunology II. Bent, L., Holborow, J. (eds.). Amsterdam, Oxford: North-Holland Publ. 1974, Vol. I, pp. 49-58
31. Valentine, R.C., Green, N.M.: J. Mol. Biol. 27, 615-617 (1965)
32. Lopes, A.D., Steiner, L.A.: Federation Proc. 32, 1003 (1973)
33. Fett, J.W., Deutsch, H.W., Smithies, O.: Immunochemistry 10, 115-118 (1973)
34. Yguerabide, J., Epstein, H.F., Stryer, L.: J. Mol. Biol. 51, 573-590 (1970)
35. Noelken, M.E., Nelson, C.A., Buckley, C.E., Tanford, C.: J. Biol. Chem. 240, 218-224 (1965)
36. Feinstein, A., Rowe, A.J.: Nature (Lond.) 205, 147-149 (1965)
37. Pinteric, L., Painter, R.H., Connell, G.E.: Immunochemistry 8, 1041-1045 (1971)
38. Ashman, R.F., Kaplan, A.P., Metzger, H.: Immunochemistry 8, 627-641 (1971)
39. Ashman, R.F., Metzger, H.: Immunochemistry 8, 643-656 (1971)
40. Schlessinger, J., Steinberg, I.Z., Givol, D., Hochman, J., Pecht, I.: Proc. Natl. Acad. Sci. USA 72, 2775-2779 (1975)
41. Jaton, J.C., Huser, H., Braun, D.G., Givol, W., Pecht, I., Schlessinger, J.: Biochemistry 14, 5312-5315 (1975)
42. Pilz, I., Kratzky, O., Licht, A., Sela, M.: Biochemistry 12, 4998-5005 (1973)
43. Pilz, I., Kratky, O., Licht, A., Sela, M.: Biochemistry 14, 1326-1333 (1975)
44. Tumerman, L.A., Nezlin, R.S., Zagyanski, Y.A.: FEBS Lett. 19, 290-292 (1972)
45. Warner, C., Shumaker, V.: Biochemistry 9, 451-459 (1970)
46. Cathou, R.E., Warner, T.C.: Biochemistry 9, 3149-3155 (1970)
47. Kelly, K.A., Schon, A.H., Froese, A.: Immunochemistry 8, 613-625 (1971)
48. Levison, S.A., Hicks, A.N., Portmann, A.J., Dandliker, W.B.: Biochemistry 14, 3778-3786 (1975)
49. Nisonoff, A., Wissler, F.C., Woernley, D.L.: Arch. Biochem. Biophys. 88, 241-245 (1960)
50. Brown, J.C., Koshland, M.E.: Proc. Natl. Acad. Sci. USA 72, 5111-5115 (1975)
51. Connell, G.E., Porter, R.R.: Biochem. J. 124 (1971)
52. Colomb, M., Porter, R.R.: Biochem. J. 145, 177-183 (1975)
53. Kehoe, J.M., Bourgois, A., Capra, J.D., Fougereau, M.: Biochemistry 13, 2499-2504 (1974)
54. Yasmen, D., Ellerson, J.R., Dorrington, K.J., Painter, R.H.: J. Immunol. 116, 518-526 (1976)
55. Goers, J.W., Shumaker, V.N., Glovsky, M.M., Rebek, J., Müller-Eberhard, H.J.: J. Biol. Chem. 250, 4918-4925 (1975)
56. Sledge, C.R., Bing, D.H.: J. Biol. Chem. 248, 2818-2823 (1973)
57. Cathou, R.E., Dorrington, K.J.: In: Subunits in Biological Systems. Timasheff, S.N., Fasman, G.D. (eds.). New York, Basel: Marcel Dekker 1975, Vol. VII, Part C, pp. 91-224
58. Nisonoff, A., Hoper, J.E., Spring, S.B.: In: The Antibody Molecule. Dixon, F.J., Jr., Kunkel, H.G. (eds.). New York, San Francisco, London: Academic Press 1975
59. Sarma, V.R., Silverton, E.W., Davies, D.R., Terry, W.D.: J. Biol. Chem. 246, 3753-3759 (1971)

Recognition and Allostery in the Mechanism of Antibody Action

I. Pecht

Abstract

The binding equilibria of antibody and hapten are usually characterized by a single step mechanism. This simple mechanism enabled the detailed analysis of molecular recognition in kinetic terms. Still this mechanism does not always prevail and a number of homogeneous immunoglobulins of known specificity, which exhibit conformational transitions induced by hapten binding, have by now been observed. One of these immunoglobulins (MOPC 460) is found to exist in two conformational states, both when free or hapten bound. The transition between the different states is induced by the hapten binding.

Following the circular polarization of the intrinsic fluorescence of the antibodies, the induction of conformational changes upon antigen binding has been shown to be transmitted to the Fc domains. The examination of these spectroscopic changes has now been extended and correlated with the induction of complement binding. For antibodies of the IgG class a correlation was found only when divalent or polyvalent antigens were bound, whereas for IgM class antibodies of the same specificity, binding of the monovalent antigenic determinant was sufficient for both activation of complement and induction of the changes in circular polarization of luminescence (CPL).

These observations support the notion that the interaction between the antigen binding site and the effector sites of antibodies have an allosteric nature. The significance of the requirement of antigen divalency for IgG may be an inherent part of this nature so as to accommodate an appropriate angular relation between the different domains of the immunoglobulin molecule.

Introduction

Antibody molecules are the main molecular reagents of the immune system. They combine in their structure the capacity for both antigen recognition and the communication with other, nonantigen-specific components of the body's defence system (1). The specific recognition of antigens is known to be carried out by the variable, N-terminal region of the immunoglobulin molecule. Here, the combination of the hypervariable loops constructed on the framework of nonhypervariable sequences forms the antigen combining site (2, 3). This unique architecture enables the production of the wide diversity of specificities using an economic set of building blocks. The communication of the combining site with distal domains of the immunoglobulin is essential for the performance of its effector functions, i.e. triggering the variety of activities which are independent of their antigen specificity yet are induced by the foregoing antigen recognition-binding step (1). These functions of the immunoglobulins vary with their class and accordingly the site

at which interaction with the respective partners take place is located at different domains. These were localized in the constant region domains, more specifically at the Fc part of the molecule (1, 4).
Thus, a significant body of evidence has accumulated for the role of
the CH_2 domains in the binding of complement. The CH_3 domain was found
to be involved in the interactions with cells, the triggering of his-
tamine release from mast cells and probably binding to macrophages (1,4).

In our earlier work we tried to resolve and define the kinetic parameters for the process of antigen binding by its antibody. This yielded
a detailed picture of the dynamics of recognition and the different
forces involved in it. Recently we were also able to obtain kinetic
evidence for conformational transitions involved in the antibody-hapten
binding equilibria. Using a spectroscopic approach, independent evidence for antigen binding-induced conformational changes in the antibody emerged and was correlated with the complement binding capacity.
The results support the notion that allosteric effects along with
changes in the angular relation between the Fab and Fc are operative
in activating antibodies of the IgG class. Antibodies of the IgM class,
being of a pentameric structure, seem to function without the requirement for multipoint attachment to the antigen, and probably only an
allosteric effect induced upon binding a monovalent antigen is sufficient to induce the effector activity of complement binding.

Kinetic Analysis of the Antibody-Antigen Recognition

The process of molecular recognition of a ligand by a protein molecule,
whether a substrate by the enzyme or a hormone by the receptor, is a
dynamic series of multicenter interactions. The appropriate spatial
arrangement of subsites in the binding site, each presenting some type
of binding force, produces the ultimately observed affinity and specificity. It has to be stressed, however, that the individual interactions
ought to be relatively labile so as to allow the rapid scanning of the
different configurations and thus lead to fast recognition of the
proper ones (5). The early kinetic studies of the interactions between
antibodies (Ab) and their respective haptens (H) led to the conclusion
that recognition is characterized by a single step

binding equilibrium: $Ab + H \underset{k_{21}}{\overset{k_{12}}{\rightleftharpoons}} AbH$ (6, 7). This simple mechanism

seemed to leave no kinetic support for further steps in the AbH complex
involved in triggering the effector functions of the antibody. It should
however be emphasized at this point that in those early investigations,
heterogeneous antibody populations were studied. Furthermore, it was
only the hapten binding which was examined. Later studies illustrated
the importance of examining homogeneous sites with antigenic determinants of larger sizes and of varying chemical nature. Still it is the
mechanistic simplicity of the hapten binding to a large number of
antibodies that opened up the way for the detailed kinetic studies of
their recognition process. The opportunity for systematic investigation
of antibody combining sites and meaningful correlation with a structural
model became feasible only with the advent of homogeneous immunoglobulins of known specificities (8). One of the first of such immunoglobulins was protein 315, an IgA excreted by the MOPC 315 tumor which has
been shown by Eisen et al. to bind nitroaromatic derivatives (9). The
chemical relaxation study of the binding equilibrium of this protein
with 2,4-dinitrophenyl derivatives again proved the single step binding

Table 1. Specific rates of interaction and equilibrium constants (K) of different 2,4-dinitroaniline derivatives and MOPC 315

		$k_{12} \times 10^{-8}$ $(M^{-1} sec^{-1})$	k_{21} (sec^{-1})	k_{12}/k_{21} $10^{-6} (M^{-1})$	$K \times 10^{-6}$ (M^{-1})
CH_3	$= R$ [a]	3.5 ± 0.2	370 ± 30	0.94	
$(CH_2)_3CH_3$	"	2.5 ± 0.2	79 ± 10	3.20	3.4
$(CH_2)_5CH_3$	"	1.2 ± 0.1	123 ± 18	1.0	1.1
CH_2COO^-	"	1.7 ± 0.2	1340 ± 100	0.13	0.2
$CH_2CH_2COO^-$	"	1.8 ± 0.3	285 ± 50	0.64	
$CH_2CH_2NH_3^+$	"	0.7 ± 0.1	735 ± 65	0.089	0.13
$CH(CH_3)_2$	"	3.7 ± 0.2	108 ± 9	3.4	
$C(CH_3)_3$	"	3.6 ± 0.3	58 ± 6	6.1	
$CH_2phenyl$	"	1.8 ± 0.2	47 ± 9	3.8	
$(CH_2)_3COO^-$	"	3.8 ± 0.2	45 ± 9	8.4	
$(CH_2)_3NH_3^+$	"	0.8 ± 0.1	850 ± 100	0.098	
$(CH_2)_5COO^-$	"	1.2 ± 0.1	60 ± 10	2.0	
$(CH_2)_6NH_3^+$	"	0.9 ± 0.1	170 ± 10	7.5	
cyclohexyl	"	2.6 ± 0.2	116 ± 10	2.2	
$(Lys)_n$-lysyl- ε-N-DNP					
$n = 0$		1.4 ± 1.4	53 ± 5	2.6	
1		0.87 ± 0.08	35 ± 4	2.5	
2		0.81 ± 0.06	27 ± 4	3.0	
6		0.37 ± 0.06	29 ± 4	1.3	
8		0.33 ± 0.03	32 ± 5	1.0	

[a] R - substituent of 2,4-dinitroaniline.

mechanism to be operative (10). The specific rates of binding were of the order of $1-3 \times 10^8$/M/sec and were later shown to vary only to a limited extent, whereas the specific rates of dissociation exhibited pronounced variation as structural changes were introduced into the hapten (11, 12). This behavior is illustrated by a sample of data presented in Table 1. The specific rates of binding and dissociation of a series of 2,4-dinitroaniline derivatives to the combining site of protein 315 show instructive trends as the side chains substituted at the aniline nitrogen are varied in number, size, hydrophobicity, electrostatic charge and sign as well as distance from the nitrophenyl ring.

As indicated above, the variation in the rate of binding is quite limited, especially when no charge is carried on the hapten's side chain. It is due to the fact that these rather fast rates are close to the limits of diffusion control, and that minor barriers have to

Table 2. Specific rates of binding and dissociation of isomaltose oligosaccharides to fractionated antidextran and of oligoalanines to fractionated anti-polyalanine antibodies

Hapten	antibody fraction	$k_{12}M^{-1}sec^{-1}$	$k_{21}sec^{-1}$	$k_{12}/k_{21}M^{-1}$
IM_4-NPF1	IM_3 eluate	$(4.4 \pm 0.3) \cdot 10^5$	3.9 ± 0.6	1.1×10^5
IM_6-NPF1	Im_3 eluate	$(6.1 \pm 0.3) \cdot 10^5$	1.7 ± 0.4	3.6×10^5
$D(Ala)$-R	$D(Ala)_3$ eluate	$(5.7 \pm 0.6) \cdot 10^6$	2.0	2.9×10^6
$D(Ala)$-R	$D(Ala)_4$ eluate	$(3.0 \pm 0.3) \cdot 10^6$	1.5	2.0×10^6
$D(Ala)_2$-R	$D(Ala)_3$ eluate	$(4.3 \pm 0.4) \cdot 10^6$	1.8	2.4×10^6
$D(Ala)_2$-R	$D(Ala)_4$ eluate	$(1.7 \pm 0.2) \cdot 10^6$	3.8	4.5×10^5

IM = isomaltose; NPF1 = 1 (m-Nitrophenyl) flavazole.
    (See Harisdangkul, V., Kabat, E.A.: J. Immunol. 108, 1232 (1972)).
R = -HN-$(CH_2)_2$-NH-dimethylamino napththalene sulphonyl.
    (Licht, A., Pecht, I.: to be published).

be overcome by the encounter complex between hapten and site prior to the formation of the final complex, that no pronounced differences are observed in the rates. These small barriers may become more pronounced when the side chain is an extended hydrophobic chain, or even more so when it carries a net electric charge. Thus in the case of protein 315, which was found to have a relatively high negative surface charge, the presence of positive charges on the hapten causes a decrease in the specific rate of its binding (cf. Table 1). When antibodies with different specificities are compared by the kinetic approach, some show significantly slower rates of binding down to the range of 10^5 to 10^6/M/sec (13). These rates are characteristic for the binding of larger haptens, usually oligomers of saccharides or amino acids, and contrast with those observed for smaller and rigid haptens such as aromatic derivatives (Table 2). The markedly different rates reflect the increasing activation barriers which such oligomeric ligands (lacking preferred conformation) have to cross in order to achieve the proper multicenter attachment to their respective combining site.

Returning to Table 1 and comparing the specific rates of dissociation from the site on protein 315, it becomes evident that they span a wider range and are rather sensitive to changes in the structural features of the hapten. This becomes even more apparent if we consider that the changes are confined to substituents on the amino group only. These specific rates of dissociation are the reciprocal values of the residence times of the hapten in the combining site and therefore constitute a most sensitive and direct measure of the complementarity and strength of interaction between them. We have therefore used this approach systematically for the "kinetic mapping" of the combining site and gained insight into the structural features of the site of protein 315 (11, 12). It is worth noting that these features determined kinetically show good agreement with a three-dimensional model constructed recently for this site (14) using coordinates of protein 603, which are obtained from crystallographic data and the known amino acid sequence of protein 315.

The kinetic examination of the interaction between antibodies specific
for oligomers of saccharides or amino acids (Table 2) raises another
noteworthy point, namely that the significantly slower rate of the
binding of these ligands to their combining sites is accompanied by
an approximately parallel decrease in the rate of dissociation. Therefore the equilibrium constants (obtained by the kinetic ($k_{12}/k_{21} = K$)
or static measurements) are similar to those observed for antibodies of
different specificities, which exhibit faster rates of binding, approaching the diffusion controlled limit, yet also have faster rates
of dissociation. Thus increase in affinity of the latter type of sites
can only be attained by a further decrease in the rates of dissociation
of the respective haptens.

Resolution of Further Elementary Steps in Hapten-Antibody Equilibrium

While the relative mechanistic simplicity exhibited by a large number
of antibody-hapten equilibria made feasible the detailed analysis of
the interactions with the combining sites by the kinetic approach, a
rather significant question remained open: the molecular events occurring upon ligand binding to the site, which are expected to cause the
initiation of the antibody effector functions, seemed to have no expression detectable by kinetic methods. It has been argued with a certain degree of justification that, in contrast to the analogous combining sites of enzymes where chemical changes in the substrate are
expected to be catalyzed (and therefore the transitions $E + S \rightleftharpoons ES$
$\rightleftharpoons ES' \rightleftharpoons EP \rightleftharpoons E + P$ are obvious), no isomerization of the antibody site-hapten complexes is required. This and other points have been
discussed in an extensive and critical review (1) of this subject
which analyzed the problem in terms of the experimental evidence for
three models: (a) the allosteric scheme, where antigen binding at its
site causes transformation at the distal effector sites; (b) the distortive model, which at least in principle is similar to the allosteric
one, yet requires interaction with a di- or polyvalent antigen in order
to change the relative spatial location of the Fab and Fc fragments;
(c) the associative model, which also requires interactions of the
antibody (Ab) with di- or multivalent antigen (Ag), but puts the emphasis on the formation of Ab-Ag aggregates where multipoint attachment
of the third partner, e.g. the C1 component of the complement system
may take place (15).

As already stated, the hitherto observed single step antibody-hapten
binding mechanism was used as evidence against the possibility of
allosteric interactions in immunoglobulins (1). It was only recently
that kinetic evidence was found for structural transitions in immunoglobulins induced upon hapten binding. The significance of these observations in terms of triggering the above mentioned effector functions of antibodies have still to be examined critically.

One of the main motivations for studying the kinetics of interaction
between homogeneous antibodies and their ligands was the expectation
of detecting elementary steps occurring after the binding, which may
be unresolved in relaxation spectra of the normal, heterogeneous populations (10). We have examined by now more than eight different
homogeneous immunoglobulins of known specificities to haptens of different chemical nature, and indeed for at least two of these evidence
for further kinetic steps is observed. This is the case with XRPC 24
and MOPC 460, both murine IgA molecules. The first has specificity
for oligogalactose (16) and the second for ortho di- or -trinitro

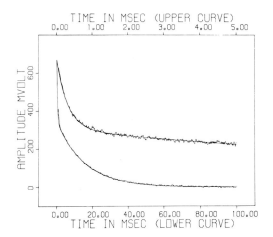

TIME IN MSEC (UPPER CURVE)

AMPLITUDE MVOLT

TIME IN MSEC (LOWER CURVE)

Fig. 1. Computer drawn relaxation trace observed with protein 460 and ε-N-(2,4-dintirophenyl) lysine: Noisy trace represents observed data, smooth trace calculated using fitted parameters. To show the two relaxation times, two different sweep rates were used as indicated on upper and lower scales. [Protein 460] = 1.65 x 10^{-6}M sites, [DNP-Lys] = 1.93 x 10^{-5}M

aromatic derivatives (17). So far we have analyzed in detail the behavior of the equilibrium between the intact monomeric MOPC 460 and ε-N-(2,4-dinitrophenyl)-lysine in terms of the concentration dependence of its relaxation times and amplitudes (18). This enabled us to formulate the detailed mechanism of interactions between this immunoglobuline and its hapten. We also examined the kinetic behavior of the Fab fragment of protein 460 with DNP-lysine as well as of the intact monomer with two further haptens: 2,4-dinitronapht-1-ol and 7-nitro benz-2-oxa 1,3-diazol (19) (all containing the structural element of an ortho dinitro aromatic ring). The latter hapten enabled us to follow the chemical relaxation of its binding equilibrium to protein 460 via three different spectral probes: (1) fluorescence quenching of the protein; (b) fluorescence quenching of the intrinsic emission of the hapten (which is a rare case of a fluorescent nitroaromatic derivative (19)); and (c) via the new absorption bands formed by the complexes (18). All three probes gave identical results, an observation which was very helpful in the analysis of the two relaxation times exhibited by this system.

The faster of the two distinct relaxation times observed is in the 0.2 - 1.0 msec range (τ_F) and the slower (τ_S) is in the 8 - 20 msec range (Fig. 1). Whereas τ_F^{-1} increased linearly with rising hapten concentration, τ_S^{-1} decreased to a limiting value (Fig. 2). As the τ_F is kinetically and thermodynamically uncoupled from τ_S we analyzed it separately and have shown that it is a result of the bimolecular step of hapten- immunoglobulin association and monomolecular dissociation. That the slow relaxation time stems in the protein was shown by the facts that it can be followed via protein fluorescence also in the absence of hapten and that it is found to vary with all three different haptens examined. Now the two simplest schemes which may account for these kinetic observations are: one which involves protein-hapten binding followed by an isomerization step of the complex (I) and one which consists of the hapten binding equilibria and of the isomerization of the free immunoglobulin (II).

$$(I) \quad \begin{array}{c} H + T_0 \rightleftharpoons T_1 \\ \Updownarrow \\ R_1 \end{array} \qquad\qquad (II) \quad \begin{array}{c} T_0 \\ \Updownarrow \\ H + R_0 \rightleftharpoons R_1 \end{array}$$

(where T_0 and R_0 are two conformers of the free protein, T_1 and R_1 are conformers of the complexed protein, and H is the hapten). The first

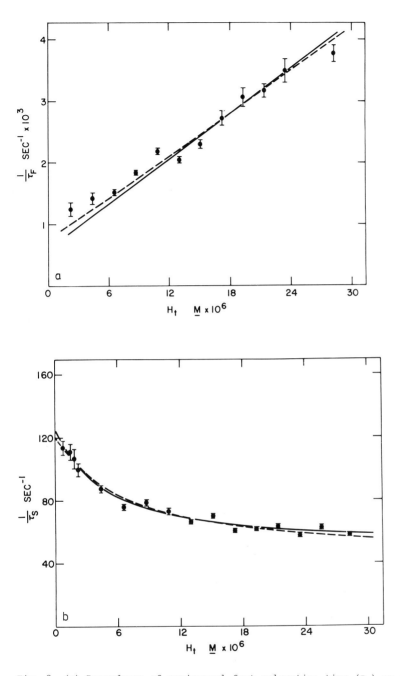

Fig. 2. (a) Dependence of reciprocal fast relaxation time (τ_F) on total DNP-Lys concentrations. [Protein 460] = 1.7 x 10⁻ M sites. Broken and full lines represent fit parameters for schemes II and III respectively (see text). Bars represent approximate standard deviations in group of jumps at given concentration. (b) Dependence of reciprocal slow relaxation time (τ_S) on total DNP-Lys concentrations. Other details are identical to 2a

mechanism is ruled out since the observed decrease in τ_s^{-1} is incompatible with the expected increase (18). The second mechanism seems to fit qualitatively the relaxtion time data, yet when a detailed analysis of both the relaxation times and amplitudes was carried out, we had to discard this two step mechanism also. The next simplest scheme that had to be considered combines the elements of the two simpler mechanisms. It involves four steps: two interconvertible conformational states of the free and hapten bound immunoglobulin, each form binding the hapten with a different affinity. In this closed square mechanism (III), the hapten binding causes the transition in the equilibrium between the two bound states toward the one of better binding.

$$
\text{(III)} \qquad
\begin{array}{ccc}
H + R_0 & \rightleftharpoons & R_1 \\
\Updownarrow & & \Updownarrow \\
H + T_0 & \rightleftharpoons & T_1
\end{array}
$$

The behavior of all four relaxation paramaters (two relaxation times and two amplitudes) were found to fit that predicted by scheme (III) and it is therefore considered to be the simplest mechanism for describing the protein 460-hapten equilibrium. We achieved its full, self consistent, thermodynamic and kinetic characterization without any simplifying assumptions, and also incorporated in this scheme static data from equilibrium binding and microcalorimetric measurements (18). This mechanism of hapten-immunoglobulin interactions is in line with the allosteric model for the initiation of effector functions, a model which assumes heterotropic interactions between the antigen binding site and effector sites at distal domains mediated by structural transitions in the immunoglobulin. The significance of this conformational transition in protein 460 is now being investigated in terms of one effector function, namely its complement binding capacity by the alternative pathway.

There are at least two important questions arising from the kinetic behavior of this system. The fact that such conformational equilibria are observed only with a small number of immunoglobulin hapten systems may be the result of the limitations of the hapten-site interactions. This point is related to other experiments described later in this paper. The difference in free energy of hapten binding between the two conformers of protein 460 amounts to 1.4 Kcal/mol. This relatively small difference may be another reflection of the fact that it is only the hapten that binds to the site, and it would be of interest to examine carrier-hapten derivatives for their energetics of binding to the two states of this protein.

Spectroscopic Evidence for Antigen-Induced Conformational Changes in Antibodies

The wide range of physical and physicochemical approaches applied to the effort of resolving the triggering signal conveyed from the combining sites at the Fab to the effector sites on the Fc yielded only limited evidence which was considered insufficient for adoption of an allosteric type of action mechanism (1). Recently, however, the application of the newly developed method for measurement of the circularly polarized component in the light emitted by fluorescence molecules (20, 21) to the study of antibodies, has produced evidence for conformational changes in both Fab and Fc upon antigen binding (22,

23). We have now extended these spectroscopic studies and also correlated them with measurements of triggering of one effector function
of antibody-antigen complexes, i.e. with complement fixation (24).

The circular polarization of luminescence (CPL) is the emission analog
of circular dichorism; while the latter reflects the chirality of the
chromophore or its environment as expressed by preferential absorption
of left or right circularly polarized light, the former is an expression of the asymmetry of the fluorophore, i.e. the chromophore in its
excited state (20). Thus in the case of proteins in general and immunoglobulins in particular, it detects even small changes in the
asymmetry of the emitting (mainly tryptophane) residues. Thus the
multitude of tryptophans distributed throughout the elongated IgG molecules serve as "built-in" probes for structural changes in their
neighborhood. The examination of the CPL of rabbit IgG of several
specificities, prior to and after binding their respective antigens,
has shown that marked changes were induced. This was observed when excess multivalent antigen (RNase or poly A-L) reacted with its respective antibody or when a monovalent antigenic determinant ("loop" peptide of lysozyme) binds a both equivalence or higher ratios to its
antibody (22). A most significant aspect of these findings is that
though spectroscopic changes are observed in both intact and Fab fragments of the antibody, they are expressed in different parts of the
emission band. Thus the changes in the intact IgG cannot be accounted
for, even qualitatively, by those occurring only in the Fab fragments.
This clearly indicates that further changes take place in the intact
antibody upon antigen binding, i.e. they are due to the presence of
the Fc domains and most probably proceed in them. Indeed even mild
reduction and alkylation of the antibody under conditions which are
known to cause the opening of the hinge-interheavy disulfide bridges
abolishes the extra spectral changes assigned to the Fc (22).

By now these CPL studies have been extended to homogeneous rabbit
antibodies raised against two types of antigens: type III Pneumococcal
and Micrococcus lisodeicticus polysaccharides and the complexes formed
with a series of the oligosaccharides of different size derived from
their respective antigens (23, 25). Whereas the short oligosaccharides
induce CPL changes limited to the Fab domains only, when the hapten
reaches a size larger than 16 units, spectral changes which may be
assigned to the Fc are observed. The significance of this requirement
of minimal size which is also observed with other antibodies will be
discussed below.

The spectroscopic evidence for structural changes induced by antigen
binding and being communicated to the Fc domain raised the immediate
question of their functional relevance. This is now being examined
on selected antibody-antigen systems where the spectroscopic changes
and complement activation can be correlated (24). The loop peptide
of lysozyme (residues 64-80 of hen egg-white lysozyme) is a particularly convenient and effective antigenic determinant (26) for carrying
out this correlation: (a) having no aromatic residues in its structure
it neither absorbs nor emits in the spectral range of our interest;
(b) it is a well defined monovalent determinant from which, however,
dimers can readily be prepared with appropriate spacer groups; also
polyvalent derivatives of it can be synthesized (26, 27); (c) it constitutes a fragment of a natural antigen (lysozyme) which is significantly larger, yet is still monovalent in terms of its loop determinant (28); (d) specific antibodies with restricted structural and
functional heterogeneity and relatively high affinity (3 x 10^6/M (28))
are induced to the loop in rabbits.

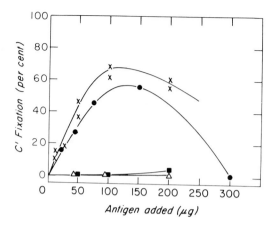

Fig. 3. Complement fixation by anti-loop antibodies as effected by different loop derivatives: Loop ■——■; bis-loop ●——●; lysozyme △——△; multiple loops attached to A-L x——x. Antiserum dilution was 1/200

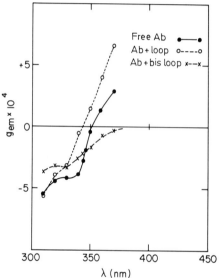

Fig. 4. Circular polarization spectra of fluorescence of anti-loop antibodies (6.5 mg/ml) ●——● and of their complexes with loop (1.5 equivalents) o----o and with (same equivalents) bis-loop x---x

As described earlier, the loop was found to cause CPL changes in both Fab and Fc domains upon binding to its specific antibodies (22). The capacity of the (monovalent) loop to induce complement fixing activity upon binding to its antibodies was found however to be negative. In contrast, the complexes formed between these antibodies and a dimeric derivative prepared by coupling two loops via their carboxyl termini with 1,9 diamino nonane (bis-loop) were almost as effective in causing complement fixation as the polyvalent antigen, loop A-L (24) (Fig. 3). Also in the CPL spectra of complexes formed between antiloop antibodies and the loop or bis-loop distinct differences are observed (Fig. 4). In agreement with previous measurements, the binding of the loop causes a marked shift in the emission anisotropy factor

$(g_{em} = \frac{\Delta f}{2f}$, where Δf is the circularly polarized component and f the

total emission intensity) toward positive values in the longer wavelength half of the emission band (340 - 370 nm). This is contrasted by the effect of binding the bis-loop to these antibodies, which causes an inversion of the CPL in the 340 – 370 nm range: values of g_{em} more negative than those of the free antibody are obtained in that range for this complex (Fig. 4). The reversibility of the equilibrium is illustrated by the fact that adding an excess of the monovalent loop to the latter complex causes the reversal of the CPL spectrum to that of the complex with the loop.

The nature of the complexes formed between the anti-loop antibodies and the bis-loop was systematically studied in the analytical ultra-centrifuge, by both sedimentation velocity and differential sedimenta-tion modes. Throughout the range of protein and bis-loop concentrations examined (0.3-0.6 mg/ml by the photoelectric scanner and 5-7 mg/ml with the schlieren optics at bis-loop concentration ranging from half equivalence to a six-fold excess over antibody sites), no significant concentration of oligomeric species was observed. Furthermore, in the differential sedimentation experiment where antibody complexes, with either the loop or the bis-loop, were examined at a 1/1.4 ratio, no difference between the two samples within 1 % could be resolved. This behavior is consistent with a preferential formation of monomeric com-plexes with the bis-loop. Yet in this case where a spacer of more then 35 Å separates the two loops, it is most probably an intramolecular ring closed complex which has a pronounced thermodynamic advantage over the opened or dimeric complexes (29, 30).

Although we find that the binding of a monovalent, large or preferably conformationally stable, antigenic determinant induces changes in the CPL spectrum of the antibody which occur also in the Fc domains, for antibodies of IgG class this is not sufficient for the induction of the effector function of this antibody-antigen complex. It is only the divalent derivative (bis-loop) which is capable of triggering this function. We may therefore conclude that both the binding of the antigenic determinant and a change in the angular relation between the Fab and the Fc are required for activating the effector functions of antibodies of the IgG class.

More recent experiments where anti-loop antibodies of the IgM class were examined lend further support to the notions described above. Thus with the macroglobulins, complement activation was found to occur even with the monovalent antigenic reagents, lysozyme or even the loop peptide alone. This is illustrated in Figure 5 where the complement fixing capacity of the anti-loop IgM induced by binding different derivatives carrying the loop peptide is shown. The marked feature is the efficacy of lysozyme, which is similar to that of the poly-valent antigen loop A-L. Thus with this class of immunoglobulins a monovalent antigenic determinant (again of significant size and con-formation) is sufficient for inducing the complement binding capacity. This observation is in line with the recent report of Brown and Kosh-land who used a monohapten (lactose)-carrier conjugate to show acti-vation of complement binding by the anti-hapten antibodies of the M class (31). The loop anti-loop system deserves special notice because: (a) the monovalent determinant itself is also effective in complement activation upon binding to the antibodies; (b) a well-defined, mono-valent macromolecular derivative of the loop is made available by nature in the form of lysozyme.

The CPL spectra of the IgM anti-loop antibodies are shown in Figure 6. The binding of the loop causes pronounced changes throughout the whole emission band which serve as independent evidence for the structural changes induced by the binding of the antigen to the antibody. A more specific assignment of these spectral changes to domains of the anti-body molecule is presently being examined.

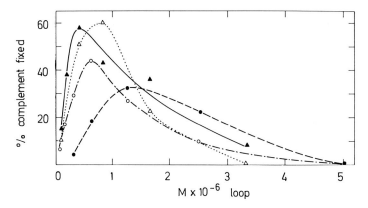

Fig. 5. Complement fixation by IgM anti-loop antibodies as induced by binding:
Loop o---o; Lysozyme ▲———▲; bis-loop o-·-·-o; multi-loop A-L Δ·····Δ

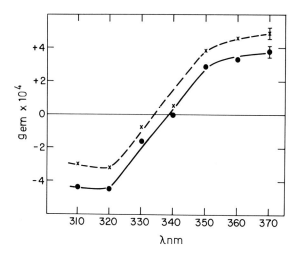

Fig. 6. Circular polarized
fluorescence spectra of IgM
anti-loop antibodies in the
absence ●———● and presence
x---x of an equivalent concentration of loop

The finding that a hapten carrier conjugate effectively induces complement activation upon binding to IgM anti-hapten antibodies has already been interpreted as evidence for the induction of conformational
changes (31). The present results offer an independent corroboration
of this idea and the CPL data supply a correlation with an intrinsic
probe of the protein.

In conclusion, the data obtained with the anti-loop antibodies of both
G and M class support the concept that the binding of an antigenic
determinant of defined conformation elicits a structural transition
in the antibody. This transition is necessary but not sufficient for
inducing the effector functions of the IgG class, yet is sufficient for
that of the IgM class. For activation of complexes formed with antibodies of the first class, divalency of the antigenic determinant is
required, probably for achieving an appropriate angular relation between the Fab and Fc so as to expose regions which are accessible in
antibodies of the M class by virtue of the pentameric ring structure
or their extra domain.

The separate notions of allosteric or distortive mechanism of antigen triggering of antibody effector function (1) seem to merge and to form a combined single model.

This view receives some support from recent structural studies by Huber et al. of the X-ray diffraction of the whole immunoglobulin and its Fc fragment (32, 33). The relative position of the Fc domains in the intact molecule is found to occupy different orientations and also an extra flexibility is found which coincides with the switch regions between the V and C domains of the Fab arms. The different possible angles taken up at the two types of hinge regions are proposed to carry the allosteric message from one domain to the other (cf. Huber et al., this volume, p. 26).

References

1. Metzger, H.: Advan. Immunol. 18, 169 (1974)
2. Davies, D.R., Padlan, E.R., Segal, D.M.: Contemp. Topics Mol. Immunol. 4, 127 (1975)
3. Poljak, R.: Advan. Immunol. 21, 1 (1975)
4a. Edelman, G.M., Gall, W.E.: Ann. Rev. Biochem. 38, 415 (1969)
4b. Kehoe, M.J., Bourgois, A., Capra, J.D., Fougereau, M.: Biochemistry 13, 2499 (1974)
4c. Yasmeen, D., Ellerson, J.R., Dorrington, K.J., Painter, R.H.: J. Immunol. 116, 518 (1976)
5. Eigen, M.: In: Fast Reactions and Primary Processes in Chemical Kinetics. Claessen, S. (ed.). Stockholm: Almqvist and Wiskell 1968, p. 333
6. Froese, A., Sehon, A.H., Eigen, M.: Can. J. Chem. 40, 1786 (1962)
7. Day, L.A., Sturtevant, J.M., Singer, S.J.: Ann. N.Y. Acad. Sci. 103, 619 (1963)
8. Eisen, H.N., Simms, E.S., Potter, M.: Biochemistry 7, 4126 (1968)
9. Eisen, H.N., Michalides, M.C., Underdown, B.J., Schulenberg, E.P., Simms, E.S.: Federation Proc. Fed. Am. Soc. Exp. Biol. 29, 78 (1970)
10. Pecht, I., Givol, D., Sela, M.: J. Mol. Biol. 68, 241 (1972)
11. Pecht, I., Friedman, S., Haselkorn, D.: FEBS Letters 24, 331 (1972)
12. Haselkorn, D., Friedman, S., Givol, D., Pecht, I.: Biochemistry 13, 2210 (1974)
13. Pecht, I.: In: The Immune System. Sercarz, E., Fox, F. (eds.). New York: Academic Press 1974, p. 15
14. Padlan, E.A., Davies, D.R., Pecht, I., Givol, D., Wright, C.: Symp. Quant. Biol., Vol. XLI (in press)
15. Reid, K.B.M., Porter, R.R.: Contemp. Topics Mol. Immunol. 4, 1 (1975)
16. Jolly, U.E., Rudikoff, S., Potter, M., Glaudemans, C.P.J.: Biochemistry 12, 3039 (1973)
17. Jaffe, B.M., Simms, E.S., Eisen, H.N.: Biochemistry 10, 1693 (1971)
18. Lancet, D., Pecht, I.: Proc. Nat. Acad. Sci. 73 Oct. (1976)
19. Lancet, D., Pecht, I.: Israel J. Med. Sci. 11, 1393 (1975)
20. Steinberg, I.Z.: In: Concepts in Biochemical Fluorescence. Chen, R., Edelhoch, H. (eds.). New York: Marcel Dekker 1975
21. Steinberg, I.Z., Schlessinger, J., Gafni, A.: In: Peptides, Polypeptides and Proteins. Blout, E.R., Bovey, F.A., Goodman, M., Lotan, N. (eds.). New York: John Wiley and Sons 1974, pp. 351-369
22. Schlessinger, J., Steinberg, I.Z., Givol, D., Hochman, J., Pecht, I.: Proc. Nat. Acad. Sci. 72, 2775 (1975)

23. Jaton, J.C., Huser, H., Braun, D., Givol, D., Pecht, I., Schlessinger, J.: Biochemistry 14, 5308 (1975)
24. Pecht, I., Ehrenberg, B., Calef, E., Arnon, R.: submitted for publication. Biochem. Biophys. Res. Comm.
25. Schreiber, A.B., Pecht, I., Gafni, A., Schindler, M., Strossberg, A.D.: Arch. Int. Biochim. 84, 86 (1976)
26. Arnon, R., Sela, M.: Proc. Nat. Acad. Sci. 62, 163 (1969)
27. Maron, E., Schiozawa, C., Arnon, R., Sela, M.: Biochemistry 10, 763 (1971)
28. Pecht, I., Maron, E., Arnon, R., Sela, M.: Europ. J. Biochem. 19, 368 (1971)
29. Crothers, D.M., Metzger, H.: Immunochemistry 5, 171 (1972)
30. Goers, J.W., Schumaker, V.N., Glovsky, M.M., Rebek, J., Müller-Eberhard, H.J.: J. Biol. Chem. 250, 4918 (1975)
31. Brown, J.C., Koshland, M.E.: Proc. Nat. Acad. Sci. 72, 5111 (1975)
32. Colman, P.M., Deisenhofer, J., Huber, R., Palm, W.: J. Mol. Biol. 100, 257-282 (1976)
33. Deisenhofer, J., Colman, P.M., Huber, R., Haupt, H., Schwick, G.: Hoppe-Seyler's Z. Physiol. Chem. 357, 435-445 (1976)

Antibody Structural Genes

Structure and Genetics of Antibodies

R. R. Porter

The structure of immunoglobulin raises several obvious questions
about the structural genes coding for their peptide chains.

1. Are there two genes coding for each peptide chain, one for the
   variable region and one for the constant region?

2. How are the 20-30 genes coding for the constant regions of the
   different classes, subclasses, types and subtypes associated on
   the chromosomes?

3. Is there a small number of structural genes in the germ line coding
   for the variable regions, which is increased during development, or
   are there separate genes coding for each of the large number of different variable region sequences?

The first major difficulty in answering these question lies in obtain-
ing conclusive evidence that the allelic property being followed is a
genetic marker of the structural genes being investigated. In almost
every case the antigenic specificity of the immunoglobulin is followed.
This specificity certainly derives from their structure, but the complexity has required that the specificity be correlated with changes
in amino acid sequences in different sections of the peptide chains
of the immunoglobulin.

The second major difficulty arises because although many polymorphic
forms of the immunoglobulin have been identified, only one genetic
marker has been associated with the variable region - that of the rabbit heavy chains. Relevant information is however being gained on the
idiotypic variants of the variable regions, but there is little evidence available on the structural basis of these idiotypic specificities
(see e.g. Hopper, 1975; Hopper et al., 1976).

The answers to the first two questions appear to be straight-forward
and I will summarise them briefly; the third remains unresolved and I
will discuss this more fully.

Two Genes - One Polypeptide Chain

Todd showed in 1963 that the 'a' locus allotypic specificities of rabbit immunoglobulin were present on both γ and μ chains and it was
shown subsequently that they were associated also with the α chain
and ε chain, i.e. the 'a' locus allotypes are found on four peptide
chains with obviously different amino acid sequences. Subsequent work
showed that the 'a' allotypic specificities are dependent on structures
in the variable region. Crossing over between these specificites and
genetic markers of the heavy chain constant regions has been found.

The existence of the variable region and constant region in itself
suggested that the heavy chains must be coded by two separate genetic

58

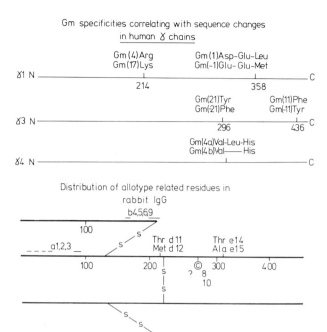

Fig. 1. Gm specificities
correlating with sequence
change in human γ chains

Fig. 2. Distribution of allotype related residues in rabbit IgG. The only single residue changes identified are
those correlating with d 11
and 12, e 14 and 15 in Fc
region

mechanisms. This has been supported by the occurrence in two or three
cases of a patient with a double myeloma where variable regions identical in sequence were found joined to different classes of heavy chain
of immunoglobulin in the serum of the same patient.

The only explanation of these results appears to be that genes coding
for the constant regions are joined, probably before transcription,
to distinct genes coding for the variable regions.

Mapping of Structural Genes

Here allocation of the antigenic specificity to the different sections
of peptide chain became essential, and for the constant region genes
this seems to be straightforward though not perhaps formally conclusive.
Much evidence has accumulated for the human and rabbit immunoblobulins
and I will give only a few examples.

The Gm specificities are found on the different subclasses of the γ
chains and in many cases they are on the Fc portion and hence must be
dependent on structural differences within the constant region. Correlation of specificity and sequence are apparent for Gm (1)(-1)(4)
(4a)(4b)(11)(-11)(17)(21)(-21) in which a change in sequence of only
one or two residues in the complete sequence is found (Fig. 1). Similarly, in rabbit γ chains, where no subclasses have been detected,
the d11 and d12 and e14 and e15 specificities correspond to single
residue changes (Fig. 2). d11 is a rare example where a fairly small
peptide has been obtained which contains the sequence variant and can
be shown to carry the allotypic specificity. The peptide is obtained
from a peptic digest of rabbit IgG and contains short sequences coming
from both heavy and light chains (Fig. 3) and hence presumably retains
some of the steric structure of these sections in the whole molecule.
Surprisingly the equivalent peptide carrying the d12 specificity however does not retain antigenic activity (Burnette and Mandy, 1975).

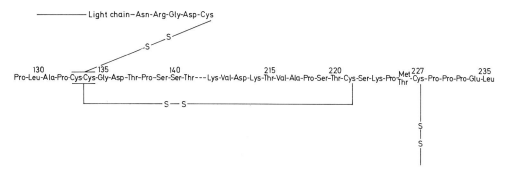

<u>Fig. 3.</u> Petide carrying the d 11 allotypic specificity of rabbit IgG. This peptide
isolated from d 11 specific IgG (threonine in positions 226 retains allotypic specificity). Equivalent peptide d 12 (methionine at position 226) does not retain specificity

The importance of steric structure in determining specificity is shown
by the Km (1),(1,2) and (3) specificity present on human kappa chains.
These depend on changes in residue at two positions well separated in
sequence, Km (1) being $Val^{153} Leu^{191}$, Km (1,2) $Ala^{153} Leu^{191}$, and Km 3
$Ala^{153} Val^{191}$ (Milstein et al., 1974). The crystallographic structure of
the Fab fragment shows that these two positions are adjacent but there
is little prospect of isolating a peptide retaining them in the arrangement found in the native molecule.

If it is accepted that the correlation of sequence and specificity
proves a direct relation between specificity and primary structure,
then the genetic markers of the constant region may be accepted as
markers of structural genes. There may, however, be exceptions to this
such as the 'b' locus specificity of the rabbit κ chain where differences in 20 or more positions are being found. Such a change could
not arise from a single mutation and it remains to be established
that the structures determining the four specificities, b4, 5, 6 and
9 are true alleles (i.e. coded for by alternative forms of the same
nucleotide sequence) as earlier data suggested (see Strosberg et al.,
1974).

With these reservations, breeding experiments and population studies
have shown that the genes coding for the heavy chain constant regions
of each class and subclass are closely linked. Genes coding for the
kappa chains are not linked to those of the lambda chain, and neither
are linked to the genes of the heavy chains. The three linkage groups
are probably on separate chromosomes.

More detailed mapping of the human heavy chain cistrons has come from
population studies which revealed a Lepore type γ chain which was a
$γ^3$ - $γ^1$ combination (Natvig et al., 1971). This and the finding of
other rare examples of crossovers has lead to a postulated order of
γ chain cistrons $γ^4$, $γ^2$, $γ^3$, $γ^1$ (Yount et al., 1970).

The Number of Structural Genes Coding for Variable Region Sequences

Other approaches to this key question in understanding the genetic
origin of variable region sequences will be discussed in this volume,
but in the present context the argument has centred on the problem

60

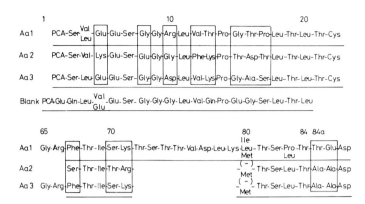

Fig. 4. Allotype related residues in variable regions of rabbit heavy chains. Boxed
residues are allotype related. Sequences underlined have been shown to be present
in variable regions of both γ and α chains

as to whether the 'a' locus specificities of rabbit heavy chains are
markers of the structural genes coding for this region. Two other (x
and y) loci have been found for rabbit V_H genes but the 'a' locus
specificities are on the majority of immunoglobulin molecules (60-90 %)
in most sera. If the 'a' specificities are structural gene markers it
is likely that there is only a small number of genes coding for V_H
in the germ line, as multiplication to millions of copies makes it
likely that the markers would be scrambled during crossovers through
innumerable generations. Yet 'a' specificities are found in all rabbits - wild as well as in laboratory colonies - in different parts
of the world.

The 'a' locus antigenic specificities are complex with subspecificities
detectable for each (Oudin, 1960; Kindt et al., 1973; Horng et al.,
1976). Similarly sequence studies of the V_H region have shown a comparable complexity of allotype-related residues. This is true whether
using pooled normal IgG where only the relatively constant residues
may be identified unequivocably, or using homogeneous anti-polysac-
charide IgG where the distinction between idiotypic and allotypic variation is difficult to make.

The 'a' specificities are in the variable region as they are present
in the V_H fragment isolated by papain digestion of the Fd fragment
(Mole et al., 1975), and sequence correlation with these specificities
must be in the V_H sequence, as sequences in the constant region are
identical for the different specificities (Pratt and Mole, 1975).

Combination of results from several laboratories suggests the correlation between allotypic specificity and sequence shown in Figure 4.
While the differences are consistent, the residues shown as allotype
related sometimes have alternatives in these positions, minor changes
could be missed and the antigenic sites may include adjacent more
variable positions. All this could give rise to the multiple serological specificities. Further, any antiserum to 'a' locus allotypes may
contain antibodies to only part of these complex antigenic determinants
and display overlapping specificities. In many positions the residues
of two specificities are the same and, depending on the size of the
antigenic site recognised, some cross-reactivity is possible. That the

PRONASE DIGESTION					
PCA peptides of Aa1,Aa3 γchain and (Fab')₂ α,and SP					
(Yields as moles peptide/mole polypeptide chain)					
	γ chain		α chain		
	Aa1	Aa3	Aa1	Aa3	SP
PCA-Ser-Val-Glu	0·56	–	0·61	–	–
PCA-Ser-Leu-Glu	0·14	0·50	0·15	0·43	–
PCA-Glu-Gln	–	0·26	–	0·27	–
PCA-Ser-Ser	–	–	–	–	0·65
Total	0·70	0·76	0·76	0·70	0·65

Fig. 5. Pyrollidine carboxylic acid N terminal peptides from γ chain and (Fab')₂ α chain of Aa1 and Aa3 allotypic specificities. Note that results from γ chains and α chains are the same. PCA-Ser-Ser reported previously as present in chains is found only in secretory piece (SP)

13 positions indicated do reflect allotypic specificity is supported by the recent finding that the a1 and a3 allotype specific γ chains and α chains from rabbit colostral secretory IgA have the same sequence in these positions (Johnson and Mole, in preparation). The only difference between serum IgG and colostral sIgA found was in the much higher content of blank (x and y locus) α chains. This was over 60 % in the a3 IgA preparations. The finding of an earlier report of a unique N terminal sequence in the α chains (Wilkinson, 1969) has been shown to be due to contamination of α chain by secretory piece (Fig. 5) (Johnston and Mole, 1975).

Though the correlation of sequence in these 13 positions with 'a' specificities is strong, the number excludes their origin by single point mutation. Doubt as to whether a1, 2 and 3 are allelic has now been raised by the finding in a single rabbit immunised with Micrococcus Lysodiecticus of all three alleles of the 'a' locus and of three 'b' locus specificities in certain breeds (Strosberg et al., 1974). The specificities were shown to be carried by different molecules. Low levels of a 3rd specificity in heterozygous rabbit have also been reported (Kindt, 1975).

With the validity of the 'a' locus alleles in doubt as well as the difficulty of relating specificity to sequence in such a complex situation, an alternative approach was desirable. Mole (1975) has now identified chemically a double residue change in positions 84 and 85. This is shown in Figure 6. Thr-Glu in a1 sequences becomes Ala-Ala in a2, a3 and possibly in blank (x and y locus) molecules. Though correlating with 'a' locus specificities it is unlikely to contribute to them as it is identical in a2 and a3 and possibly blank molecule. A peptide Ts4 could be obtained from a CNBr and tryptic digest of the γ chain which contained a half cystine residue at position 92. This was reacted with ¹⁴C-iodoacetate and hence after chromatographic separation the Ts4 peptide could be identified by autoradiography giving a much simpler picture than if all the peptides were made visible. Mole worked out a thin layer system which distinghuised clearly between the two alternative forms of the Ts4 peptide (Fig. 7) and showed that in numerous sera all antibody or pooled IgG preparations contained one or both forms of the peptide. When breeding experiments were carried out, F1 hybrids showed the presence of both peptides, and backcrosses gave mixtures of homozygotes and heterozygotes consistent with Mendelian inheritance of simple allelic forms. Thus evidence is given of a genetic marker dependent directly on primary structure. This Ts4 marker is closely linked to the apparent locus containing the genes determining the 'a' specificities, but may not be identical to it as a2 and a3 specificities and blank molecules have the same sequence.

```
                  Ile   80                  85                    90
Aa1(K)   Lys- Leu - Thr-Ser-Pro-Thr-Thr-Glu-Asp-Thr-Ala-Thr-Tyr-Phe-Cys-Ala-Arg

                                Leu
Aa1(M)      -Met - Thr-Ser- Pro -Thr-Thr-Glu ─────────────────────────

Aa2         -Met - Thr -Ser-Leu-Thr-Ala-Ala ──────────────────────────

Aa3         -Met - Thr -Ser-Leu-Thr-Ala-Ala ──────────────────────────
```

Fig. 6. Sequence of Aa1, Aa2 and Aa3 of region 78-94 in rabbit γ chain. Two forms of this peptide are found in Aa1 as differing at the N terminal depending on presence of methionine at position 79 or lysine at position 78. Sequence of Aa1 differs from Aa2 and Aa3, with Thr-Glu in place of Ala-Ala at positions 84 and 85

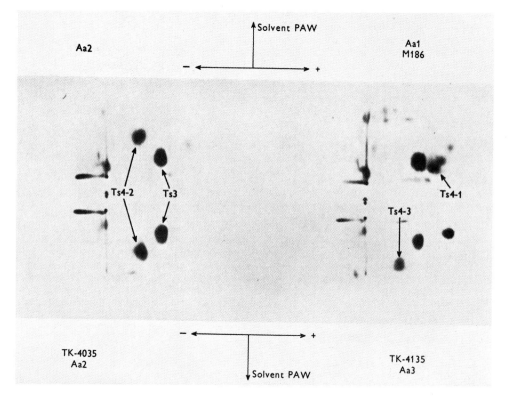

Fig. 7. Tryptic peptide maps by autoradiography of Ts4 peptide (position 79 or 80 to 95) from Aa1, Aa2 and Aa3 specific rabbit γ chains. Ts4-2 and Ts4-3 peptides are identical, while Ts4-1 peptide is distinct. Ts3 peptide present in each map comes from constant region and is a valuable marker

As the situation in the variable region is so complex it is unwise to be dogmatic, but the Ts4 alleles behave as a structural gene marker. If only the double position change is considered then there is no difference between the Ts4 marker and say Gm(1) which is associated with a double change of residue (Asn-Glu-Leu) and other specificities GM(-1) are the same (Glu-Glu-Met) in these positions whether on IgG1 or other subclasses. There seems no reason to accept Gm(1) as a structural gene marker but not Ts4, except that many other sequence changes occur on molecules carrying the Ts4 marker. The critical test of allelism in which the 'a' locus allotypes now appear to fail (i.e. the presence of 3 alleles in one individual animal) cannot be applied to Ts4 where only 2 alternatives are known. In any case Bodmer (1973) has suggested that there may be examples in highly polymorphic systems where genes for apparent alleles are always present and the true alleles are those of a controller gene which determines which structural genes are repressed or depressed.

In the case of the Ts4 marker it is simpler to assume at present that there are two allelic structural genes stable over a long period. Hence it is likely that there is only a small number of genes carrying the Ts4 marker in the germ line and that the many other sequence changes found in the variable regions are likely to have arisen by some type of somatic change.

Answers to the three questions posed at this position in time appear to be:

1. There is good evidence that there are independent genes coding for variable and constant regions of immunoglobulin chains. Episomal insertion into the variable region gene to give the hypervariable regions has been suggested (Wu and Kabat, 1970; Capra and Kindt, 1975), but no direct structural evidence so far bears on this possibility.

2. The genes coding for the constant regions within each of the heavy chains and the kappa and lambda chains are closely linked but each group is well separated from the other.

3. With reservation, the survival of the 'a' locus specificities, and much more directly, the Ts4 locus through many generations is good evidence that there are likely to be a small number of genes coding for the variable region in the germ line and that these are increased greatly in number in all or, more probably, in different cells during development by a somatic mechanism.

A full discussion of these topics will be found in four recent reviews: Kunkel and Kindt (1975); Kindt (1975); Nisonoff et al. (1975); Knight and Hanly (1975).

References

Bodmer, W.F.: Transpl. Proc. 5, 1471 (1973)
Burnette, S.K., Mandy, W.T.: Immunochemistry 12, 861 (1975)
Capra, J.D., Kindt, T.J.: Immunogenetics 1, 417 (1975)
Hopper, J.E.: J. Immunol. 115, 1101 (1975)
Hopper, J.E., Noyes, C., Heinrikson, R., Kessel, J.W.: J. Immunol. 116, 743 (1976)
Horng, W.J., Knight, K.L., Dray, S.: J. Immunol. 116, 117 (1976)
Johnstone, A.P., Mole, L.E.: Nature (Lond.) 151, 337 (1975)
Kindt, T.J.: Advanc. Immunol. 21, 35 (1975)
Kindt, T.J., Seide, R.K., Tack, B.F., Todd, C.W.: J. Exp. Med. 138, 33 (1973)
Knight, K.L., Hanley, W.C.: In: Current Topics in Molecular Immunology. Inman, F.P., Mandy, W.J. (eds.). New York: Plenum Press 1975

Kunkel, H.G., Kindt, T.J.: In: Immunogenetics and Immunodeficiency. Benacerraf, B. (ed.). Lancaster: MTP 1975

Milstein, C.P., Steinberg, A.G., McLaughlin, C.L.: Nature (Lond.) 248, 160 (1974).

Mole, L.E.: Biochem. J. 151, 351 (1975)

Mole, L.E., Geier, M.D., Koshland, M.E.: J. Immunol. 114, 1442 (1975)

Natvig, J.B., Michaelson, T.E., Kunkel, H.G.: J. Exp. Med. 133, 1004 (1971)

Nisonoff, A., Hopper, J.E., Spring, S.B.: The Antibody Molecule. New York: Academic Press 1975

Oudin, J.: J. Exp. Med. 112, 107 (1960)

Pratt, D.M., Mole, L.E.: Biochem. J. 151, 337 (1975)

Strosberg, A.D., Hamers-Casterman, C., Van der Loo, W., Hamers, R.: J. Immunol. 113, 1313 (1974)

Wilkinson, J.M.: Nature (Lond.) 223, 616 (1969)

Wu, T.T., Kabat, E.A.: J. Exp. Med. 132, 211 (1970)

Yount, W.J., Hong, R., Seligman, M., Good, R., Kunkel, H.G.: J. Clin. Invest. 49, 1957 (1970)

Genetic Control of T and B Cell Receptor Specificity in the Mouse

K. Eichmann, C. Berek, G. Hämmerling, S. Black, and K. Rajewsky

Introduction

The function of the immune system is based on a large variety of receptor molecules whose combining sites can recognize virtually every
antigenic substance in the universe. This capacity is reflected in a
pronounced structural variability of the receptor molecules and particularly of their variable and hypervariable regions (1, 2). The
classical receptor, the antibody molecule, possesses a combining site
that is constructed from 6 hypervariable sequence regions, three of
which belong to the heavy and three to the light chain (1, 2). They
are fixed in their three-dimensional arrangement by the so-called
"framework stretches" of the variable regions.

The genetic problems connected with the generation of antibody variability have been amply discussed in this volume. One of the basic
questions concerned with this diversity problem is whether antibody
specificity can be encoded by germ line genes. Although there is no
doubt that antibody V genes are part of the genome of each individual
producing antibodies, it is not clear whether these genes directly
encode V regions or only their somatic variants. The demonstration of
genes that directly control antibody specificity would be a strong
indication for a direct coding of V genes for V regions.

A second question that can be asked in this connection is whether all
or only some of the variable regions of the immune system are encoded
by antibody V genes. This question can be divided into two parts, one
concerning only the classical antibody molecule, and the second concerning the antigen-receptor of T cells which does not appear to possess a classical immunoglobulin structure (3, 4).

Both these questions have been approached by a variety of laboratories
using two different kinds of genetic marker systems directly related
to antibody specificity. The first marker system is antibody specificity itself, using the observations that antibodies differ in cross-
reactivity with antigens structurally related to the immunizing antigen.
The most outstanding examples of these "fine specificity markers" are
the antibodies of inbred mice to the hapten NP, some of which have a
higher affinity to the related haptens NIP and NNP than to NP. Such
antibodies are termed "heteroclitic" and occur in certain strains of
mice, whereas other strains of mice produce normal, "non-heteroclitic"
antibodies (5). This difference is genetically controlled and inherited
as a single pair of Mendelian alleles which is linked to the genes
that encode the structure of the constant region of the mouse heavy
chains (C_H genes). Thus, the genes that control fine specificity of
anti-NP antibodies appear to be coding for variable regions of mouse
heavy chains (V_H genes) (6).

The other kind of genetic marker related to the specificity of an
antibody molecule is its idiotype. Idiotypes (Id) are antigenic determinants closely related to the binding site of antibody molecules,

Ig-1(2,3,4)	Immunoglobulin		Allotype
T 15	Myeloma Protein	Phosphorylcholine	Idiotype
J 558	Myeloma Protein	α1-3 Dextran	Idiotype
S 117	Myeloma Protein	A-CHO	Idiotype
A5A	Antibody	A-CHO	Idiotype
ARS	Antibody	Phenylarsonate	Idiotype
KLH	Antibody	KLH	Idiotype
NIP-NP	Antibody	NP	Fine Spec.
NBrP	Antibody	NBrP	Fine Spec.
Pre	Prealbumin		Electrophoresis

Fig. 1. Summary of genetic markers in mouse heavy chain linkage group(Ig-1 complex). Prototype marker is allotypic specificity (9), and any linkage analysis is performed in relation to Ig-1. For detailed review of these markers, their strain distribution, detection systems etc. see Ref. (8). KLH, Keyhole limpet hemocyanin; NP, nitrophenacetyl; NBrP, nitrobromophenacetyl

and antiidiotypic antibodies (anti-Id) can in principle distinguish any antibody molecule from all others (7). The use of idiotypes as V gene markers will be discussed in this article.

The Genetic Control of Idiotypes of Antibodies to Streptococcal Group A Carbohydrate in Mice

Antibodies to streptococcal carbohydrates are merely examples of a series of antibody molecules and myeloma proteins whose idiotypes are used for the study of the genetic control of antibody specificity. Figure 1 summarizes a variety of genetic markers, including idiotypes and fine specificity markers. All these genetic markers have been found linked to the allotype (Ig-1) locus which encodes the constant regions of the various classes of heavy chains in the mouse. Except for the prealbumin marker which is unrelated to the immunoglobulin system, all genetic markers in this linkage group are associated with antigen binding specificity (reviewed in 8).

In our laboratory, two different antibody molecules have been used as probes for the production of anti-Id. One of these antibodies is termed A5A (8, 10) and is the product of a clone of lymphocytes of mouse strain A/J. This clone is specific for N-acetyl-glucosamine, the major antigenic determinant of Group A streptococcal carbohydrate (A-CHO) (11, 12). The second antibody is a myeloma protein of Balb/c mice, termed S117, which also has specificity for N-acetyl-glucosamine (13).

The principle of the genetic work with idiotypes is to analyze antibodies induced in inbred strains of mice for their reactivity with anti-Id produced against the antibodies used as probes. In the case of the antibodies A5A and S117, both of which react with Group A streptococcal carbohydrate (10, 13), mice are immunized with Group A streptococcal vaccine (10, 14) and their antibodies are tested in radioimmune-inhibition assays for their capacity to inhibit the homologous idiotypic binding of radio-labeled A5A and S117 to their respective anti-idiotypic antisera. Three levels of idiotypic crossreactivity are observed in each of the two idiotypic systems. Antibodies of the strain from which the original idiotype is derived possess a certain fraction which is strongly inhibitory. These strains of mice and their antibodies are classified as A5A[+] or S117[+], respectively. Some other strains produce antibodies that are weakly inhibitory, suggesting some idiotypic crossreactivity. These strains are

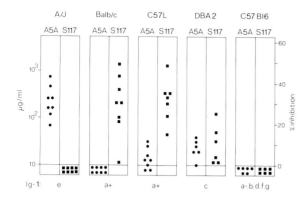

Fig. 2. Five different idiotypic phenotypes in inbred strains of mice. Each dot or square represents the A5A or S117 titer, respectively, in serum of a single mouse immunized with Group A streptococci. A5A titers indicated in μg/ml (10); S117 titers given as % inhibition by 1 μg of anti-A-CHO antibody (15)

	BREEDING	RI LINES	
Ig-1 / J558[+]	1/421	1/133	0.004
Ig-1 / S117[+]	5/154		0.032
Ig-1 / A5A[+]	11/344		0.032
Ig-1 / A5A^{cr}		1/72	0.014
Ig-1 / S117^{cr}		3/72	0.042
A5A^{cr}/ S117^{cr}		2/72	0.028

Fig. 3. Recombination frequencies observed between various idiotypic markers and Ig-1. Data partially obtained from backcross experiments (8) and partially from analysis of recombinant inbred strains (8, 15)

classified as A5A^{cr} or S117^{cr}, respectively. Most strains of mice produce antibodies which are non-crossreactive in either idiotypic system and these strains are classified A5A[-] or S117[-], respectively (10, 15). Thus, depending on which of the idiotypes they produce, the common laboratory strains are readily grouped into 5 distinct idiotypic phenotypes which are illustrated in Figure 2. These groups, except for the Ig-1^{a+} allotype which may be associated with either of two idiotypic phenotypes, correlate to the allotype groups indicated at the bottom.

Each of these idiotypic phenotypes is inherited as a single Mendelian factor linked to the Ig-1 locus. Extensive studies on the mode of inheritance of idiotypes have been published and will not be discussed in detail here (review in 8). It has become clear from these studies that a certain number of unexpected marker combinations were observed in both breeding experiments and surveys of congenic or recombinant inbred strains (review in 8). These unexpected marker combinations are interpreted as resulting from crossovers, and the frequencies by which some of these markers recombine are indicated in Figure 3. Here we compare recombination frequencies between various V_H markers and the Ig-1 locus, observed in breeding experiments and in recombinant inbred lines. It is apparent that data from both sources agree with each other. It is also clear that the absolute values merely represent orders of magnitude because of the low number of observations. Nevertheless it should be noted that these recombination frequencies are unexpectedly high, suggesting rather long distances between C_H and V_H genes.

68

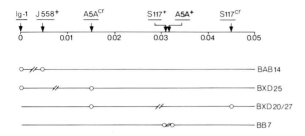

<u>Fig. 4.</u> Linear arrangement of V_H genetic loci as suggested by their recombination frequencies with Ig-1 (Fig. 3) and by several recombinant strains of mice. Strain BAB14 from cross between C57/Bl and Balb/c has all V_H genes from Balb/c in combination with a C57/Bl Ig-1 allele. Strains BXD20 and BXD27 from crosses between DBA/2 and C57/Bl showed crossovers at positions indicated. Strain BB7 resulted from cross between strain A/J and Balb/c; crossover recombined the A5A+ and S117+ genes on to one chromosome (15)

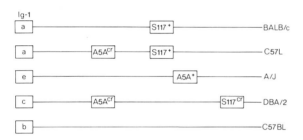

Fig. 5. Pseudo-allelic location of V_H genes in heavy chain linkage groups of strains whose idiotypic phenotypes are depicted in Fig. 2

Figure 4 shows a linear plot in which each of the V_H markers is assigned to a gene locus whose distance from the Ig-1 locus is defined by its recombination frequency (see Fig. 3). The order of gene loci thus established agrees surprisingly well with the order of gene loci suggested by the various recombinant Ig-1 complexes which are indicated below, except for the loci A5A+ and S117+ which have the same recombination frequency but are assigned to distinct loci by the recombinant BB7 (8, 15). For instance, the locus A5A^{cr} is by recombination rate analysis between Ig-1 and S117^{cr}, and the recombinant strains BXD25, BXD20 and BXD27 suggest the very same gene order.

As was indicated by the data in Figure 2, 4 of the 5 distinct idiotypic phenotypes are characterized by the expression of one or two of the 4 V_H loci that control antibody specificity to Group A streptococcal carbohydrate. Figure 5 therefore depicts the chromosomal location of these loci within the Ig-1 complexes of the reference strains of these idiotypic phenotypes. It is apparent that there are no functional alleles known for any of the V_H loci which are themselves nonallelic. The data suggest complex, pseudo-allelic relationships between different Ig-1 complexes of the mouse and an unusual evolutionary history for this part of the genome.

C_H	V_H II	V_H III (AB)	V_H III (C)
	J558	T15	
Balb/c	EVQLQESGPE	EVKLVESGGG	
		S117	
		—— L ——	
	ARS		A5A
A/J	—— Q—AG		—— E ——

Fig. 6. Tentative assignment of subgroup genes to chromosomal loci in Ig-1 complex. Amino-acid sequences represented in one-letter code (17); see Refs. 17, 18, 19. T15 gene allocated to position between A5A and J558 on basis of belonging to V_H III subgroup, not on genetic evidence. All other allocations based on genetic data

Relationships Between Primary Structure and Genetic Location of V Genes and Their Products

The relationship between the amino acid sequence of antibody V regions and the genes encoding them is particularly interesting with respect to the question of whether the total V region is encoded by one continuous gene or by a gene constructed by intragenic recombination of hypervariable episomes and less variable DNA coding for the framework (1, 16). Specificity-related genetic markers such as fine specificity or idiotype primarily represent hypervariable segments of the gene and its product, whereas the N-terminal sequence of the polypeptide chain is a relatively invariant framework segment (1, 2). If there is a single gene for each total V region one would expect those specificities that are primarily associated with a single subgroup (17) to be controlled by V genes that are tightly linked to each other, whereas specificities commonly associated with different subgroups are controlled by genes less tightly linked. On the other hand, if hypervariable and framework sequences were encoded by separate and freely combinable genes (1, 16), no such association should be observed.

Because of the paucity of data we are far from being able to decide between the above possibilities. Figure 6 shows an alignment of 10 N-terminal residues of the heavy chains from some of the antibodies whose idiotypes have been tentatively assigned to genetic loci within the Ig-1 complex (18). Three such assignments can be made for the Balb/c V_H genes and two for the A/J V_H genes. In the latter strain it is clear that the ARS gene whose product belongs to the V_H II subgroup (19) is closer to the C_H genes than the $A5A^+$ gene whose product belongs to the V_HIII subgroup (18). Furthermore, the T15 and $S117^+$ genes of Balb/c, whose products also belong to the V_HIII subgroup, are shown to be closer to C_H than $A5A^+$ of strain A/J. Thus, V_HIII coding genes are of at least two genetic loci and V_HII coding genes appear to be located between V_HIII and C_H genes. Many more data have to be accumulated to draw any conclusions as to intragenic recombination mechanisms from this type of analysis, but a picture of subgroup and hypervariable sequence determining genes begins to emerge.

Another conclusion to be drawn from amino acid sequence comparison concerns the pseudo-allelic nature of genes encoding V_H regions of antibodies against the same antigen. Figure 7 shows the 30 N-terminal residues of the A5A and S117 heavy chains, and within this 1/4 of the total sequence there are two amino acid replacements. This suggests non-allelism of the $A5A^+$ and $S117^+$ genes, as does the genetic mapping

Heavy Chains

```
              10              20              30
S117  E V K L L E S G G G L V Q P G G S L K L S C A A S G F D F S

A5A  _____  E  _____  M  _____
```

Fig. 7. 30 N-terminal amino-acid residues from heavy chains of antibodies A5A and S117. Data from Ref. 18

(see Fig. 5) (18). The V_H genes are thus comparable to the genes encoding histocompatibility antigens, whose amino acid sequences are equally difficult to reconcile with simple allelic relationships (L. Hood et al., this volume).

Genetic Control of T Cell Receptor Specificity

Thus far, this article has dealt only with the antibody molecule. Antibody molecules are thought to function as antigen receptors on the surface of B lymphocytes, and this notion is fully supported by experiments in which anti-Id was used to suppress idiotype production by injecting the anti-idiotypic antisera into mice prior to the injection of antigen (20, 21, 22). Mice treated in this way are temporarily or permanently unable to respond with their idiotype-positive precursor cells, suggesting that these cells have been blocked or inactivated by the interaction of the anti-Id with the antigen receptors on their surface.

More exciting were observations in which it was found that anti-Id antibody can also activate lymphocytes (23, 24). This activation was found to take place in both the B precursor and T helper cell compartments (24, 25, 26). Activated B and T cells were specific for the antigen to which the original idiotype was raised, namely Group A streptococcal carbohydrate (24, 25).

Figure 8 shows the experimental protocol which was used to analyze specific T helper and B precursor cell induction by anti-idiotypic antibody. The experiment is a typical adoptive transfer system (27) in which hapten-sensitized B cells and carrier-sensitized T helper cells cooperate to produce a secondary anti-hapten response. In our experiments, cells from mice sensitized with hapten NIP on an irrelevant carrier were mixed with cells from mice treated with anti-idiotypic antibody (anti-A5A or anti-S117) and injected into irradiated syngeneic hosts. The hosts were boosted with NIP coupled to streptococci and 10 days later the antibodies to NIP and to A-CHO were determined. In the presence of excess NIP-primed precursor cells, the anti-NIP-response of the host is entirely dependent on T cells primed to streptococcal antigens, and our experiments showed that these cells can be obtained from mice treated with anti-A5A or anti-S117 antiidiotypic antibodies. Furthermore, these mice had B memory cells specific for A-CHO. The activation of both types of lymphocytes in mice treated with anti-idiotypic antibody can be demonstrated by treatment of the cells with anti-Thy 1.2 serum and complement, or by passage of the cells through an anti-IgG immunoabsorbent column. The enriched B and T cell populations thus obtained were fully active as memory and as helper cells, respectively (24).

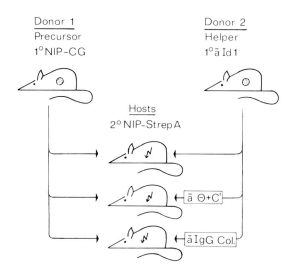

Donor 1
Precursor
1°NIP-CG

Donor 2
Helper
1°ā Id 1

Hosts
2° NIP-Strep A

ā Θ+C'

ā IgG Col.

Fig. 8. Adoptive transfer system to demonstrate B and T
helper cell activation by anti-
idiotypic antibody (24). NIP-CG,
nitroiido-phenacetyl-chicken-
gammaglobulin; āId1, anti-idiotypic antibody of the IgG1 class;
NIP-Strep A; NIP coupled to
Group A streptococci; āθ+C'
treatment of cells with anti
Thy 1.2 serum and complement to
remove T cells; āIgG Col., passage of cells through column
made from IgG-anti-IgG complexes
attached to Degalan beads, to
remove B cells

These experiments strongly suggested that anti-idiotypic antibody,
prepared against B cell-derived antibody molecules, not only reacts
with B cell surface receptors but also with that of T helper cells.
This was subsequently verified by in vitro experiments which showed
that T helper cells activated by anti-idiotypic antibody had specificity for A-CHO, and that their helper function could be inhibited by
anti-idiotypic antibody (25, 26).

As was mentioned before, the expression of the A5A and S117 idiotypes
on antibody molecules is a strain-specific character and shows genetic
polymorphism. With respect to the expression of these two idiotypes,
a number of idiotypic phenotypes can be clearly distinguished from
each other (see Fig. 2), and these phenotypes are controlled by V_H
genes in the Ig-1 complex (see Fig. 5).

The genetic polymorphism of these V_H genes allows us to study the
question of whether the idiotypes on T helper cells are controlled
by the same V_H genes that control B cell or antibody idiotypes. This
is done by recording the magnitude of the stimulatory effect of the
two different anti-idiotypic antibodies on T helper cells in various
strains whose antibody-associated idiotypes are known (8, 26). If
antibody-associated idiotypes and T cell responsiveness to anti-Id
show a parallel genetic polymorphism, both B and T cell idiotypes
are likely to be controlled by the same genes.

Table 1 shows the effect of helper cell stimulation with anti-A5A and
with anti-S117 in some of the strains whose idiotypic phenotypes are
depicted in Figure 2. It is clear from the data that those strains
that produce antibodies with a given idiotype also respond to the corresponding anti-idiotype with helper cell production, whereas those
strains whose antibodies do not bear a given idiotype do not respond
to the corresponding anti-idiotype. These data clearly show that the
genetic polymorphism of T and B cell idiotypes is the same (26).

Further analysis indicated that T cell response to anti-idiotype is
independent of the H-2 type of the strain. Table 2 shows an example
in which the H-2 complexes of responder strains, when inserted in the
genome of a non-responder strain, do not establish responsiveness.

Table 1

Strain	Antibody[a] Idiotypes	Responsiveness of helper cells to[b] anti-A5A	anti-S117
A/J	A5A[+] S117[-]	93	1.2
Balb	A5A[-] S117[+]	4.1	76
DBA/2	A5A[cr] S117[cr]	140	110
C57Bl10	A5A[-] S117[-]	7.5	5.9

[a] Data from Ref. 8.

[b] Data from Refs. 25, 26. Helper cell responses are determined as plaque forming cells to TNP in micro cultures of spleen cells from mice treated with anti-A5A and anti-S117, respectively, which were boosted in vitro with TNP-Strep A. Responses expressed as % of response of cells primed with Strep A.

Table 2

Strain[a]	Responsiveness of helper cells to[b] anti-A5A	anti-S117
A	263	15.5
DBA/2	140	110
B 10.A	5.2	17.5
B 10.D 2	5.1	6.1
C57Bl/10	7.5	5.9

[a] Strains B10.A and B10.D2 possess H-2 complexes of strains A and DBA/2, respectively, together with rest of genome from strain C57Bl/10.

[b] Data from Ref. 26; see footnote Table 1.

This is particularly important because of the well documented H-2-linked Ir gene control of T cell responsiveness to a number of antigens, which does not appear to apply to the response of T cells to anti-idiotypic antibody.

The data clearly establish that the H-2 complex in the mouse is irrelevant for the stimulation of helper T cells by anti-idiotypic antibody. Rather it appears that the Ig-1 complex which contains the genes that encode the antibody heavy chain is responsible for idiotypic T helper cell activation. We therefore conclude that the V_H genes control the variable region of the heavy chain of antibodies as well as that of a heavy chain analog of T cell receptors.

Conclusion

The data presented in this article may be summarized as follows:
1. Antibody specificity may be controlled by genes in the germ line.

73

These genes appear to encode the variable regions of the heavy chains
of certain antibody molecules, although it is not impossible that frequently occurring somatic variants of germ line genes are responsible
for the coding. In the latter case, the somatic variation of these
genes is such that their polymorphic character is maintained. The antibody molecules whose heavy chain V regions appear to be under the
germ line gene control are exceptions as brought out by the experience
of many laboratories that the idiotypes of only a few special antibodies or myeloma proteins show inheritance. Thus the data presented
here are not to be taken as in favor of a strict germ line theory of antibody diversity. Rather, the results suggest that several antibody
specificities are under close germ line control, whereas in general
it is merely the potential receptor repertoire that is restricted by
the germ line.

2. V and C genes are arranged discontinously in the chromosome. This
has long been suspected to be the case because of indirect evidence,
but the analysis of recombinant Ig-1 complexes clearly brings out that
third party V genes may be between a pair of V and C genes that together specifies a heavy chain. A picture of the subgroup-specific genetic loci emerges suggesting that the V_HII locus may be between the
C_H genes and the V_HIII locus. There are at least two distinct loci
containing genes encoding V_HIII sequences, whereas no such subdivision
within the V_HII genes has as yet been observed, and no information as
to the location of the V_HI locus is available.

3. V and C genes recombine surprisingly frequently. This has previously
been thought to be mainly due to mistyping of individual mice in breeding experiments, but recently the typing of recombinant inbred strains,
which avoids the uncertainty of individual typing, also showed a cross-
over rate of up to 3 %. Thus, if the rules of classical genetics are
valid for the Ig-1 complex it appears to occupy a rather extended chromosomal segment. Again, conclusions as to the number of genes that may
be fitted into this segment suffer from the uncertainty over the proportion of actually transcribed DNA in this segment.

4. V genes that control V_H regions also play a role in the coding of
the antigen receptor of T helper cells. Since the response of T helper
cells to anti-idiotypic acitivation shows the same genetic polymorphism
as do the V genes for the heavy chain, it appears an unescapable conclusion that the variable region of the antibody heavy chain is also
part of the T cell receptor. The implications of this finding for the
overall structure of the T cell receptor have been discussed in more
detail elsewhere (28), and it may suffice in this connection that the
data of many laboratories suggest a construction of this molecules out
of a heavy chain analog encoded by the Ig-1 complex and a light chain
analog encoded by the MHC complex. Alternatively, the product of the
MHC complex may be a separate recognition unit on the cell, and double
recognition is required for T cell activation.

In general, the analyis of genetic questions with antibodies to specificity-related determinants on receptor molecules has been useful in
that a clearer picture of the construction of the heavy chain linkage
group has emerged. Furthermore, some light is shed on the largely unknown second world of immunology, namely the T cell compartment. It
is particularly this second area in which anti-idiotypic antibodies
may be of further use as specific reagents for the isolation and biochemical characterization of the antigen recognition system of T cells.

References

1. Wu, T., Kabat, E.A.: J. Exp. Med. 132, 211 (1970)
2. Capra, J.D., Kehoe, J.M.: Advan. Immunol. 20, 1 (1975)
3. Vitetta, E.S., Uhr, J.W.: Science 189, 964 (1976)
4. Marchalonis, J.J.: In: Contemporary Topics in Molecular Immunology. In press 1976
5. Imanishi, T., Mäkelä, O.: Europ. J. Immunol. 2, 323 (1973)
6. Imanishi, T., Mäkelä, O.: J. Exp. Med. 140, 1498 (1974)
7. Hopper, J.E., Nisonoff, A.: Advan. Immunol. 13, 58 (1971)
8. Eichmann, K.: Immunogenetics 2, 491 (1975)
9. Herzenberg, L.A., McDevitt, H.O., Herzenberg, L.A.: Ann. Rev. Genet. 2, 209 (1968)
10. Eichmann, K.: J. Exp. Med. 137, 603 (1973)
11. McCarty, M.: J. Exp. Med. 108, 311 (1958)
12. Coligan, J.E., Schnute, W.C., Kindt, T.J.: J. Immunol. 114, 1654 (1975)
13. Vicari, G., Sher, A., Cohn, M., Kabat, E.A.: Immunochemistry 7, 829 (1970)
14. Krause, R.M.: Advan. Immunol. 12, 1 (1970)
15. Berek, C., Taylor, B., Eichmann, K.: J. Exp. Med. in press (1976)
16. Capra, J.D., Kindt, T.J.: Immunogenetics 1, 417 (1975)
17. Barstad, P., Weigert, M., Cohn, M., Hood, L.: Proc. Nat. Acad. Sci. 10, 4096 (1974)
18. Capra, J.D., Berek, C., Eichmann, K.: J. Immunol. 117, 7 (1976)
19. Capra, J.D., Tung, A.S., Nisonoff, A.: J. Immmunol. 115, 414 (1975)
20. Nisonoff, A., Bangasser, S.A.: Transpl. Rev. 27, 100 (1975)
21. Köhler, H.: Transpl. Rev. 27, 24 (1975)
22. Eichmann, K.: Europ. J. Immunol. 4, 296 (1974)
23. Eichmann, K.: Europ. J. Immunol. 5, 511 (1975)
24. Eichmann, K., Rajewsky, K.: Europ. J. Immunol. 5, 661 (1975)
25. Black, S.J., Hämmerling, G.J., Berek, C., Rajewsky, K., Eichmann, K.: J. Exp. Med. 143, 846 (1976)
26. Hämmerling, G.J., Black, S.J., Berek, C., Eichmann, K., Rajewsky, K.: J. Exp. Med. 143, 861 (1976)
27. Mitchison, N.A., Rajewsky, K., Taylor, R.B.: In: Developmental Aspects of Antibody Formation and Structure. Sterzl, J., Riha, I. (eds.). Prague: Czechoslovak Academy of Science 1970, Vol. II, p. 547
28. Rajewsky, K., Eichmann, K.: In: Contemporary Topics in Immunobiology. In press 1976

Immunoglobulin mRNA and Immunoglobulin Genes

C. Milstein, G. G. Brownlee, C. C. Cheng, P. H. Hamlyn, N. J. Proudfoot, and T. H. Rabbitts

The genetic origin of antibody diversity has been a subject of much thought and controversy over the past 20 years. Extensive studies of immunoglobulins, particularly at the amino acid sequence level, have provided us with a general scheme of the genetic loci which encode antibody molecules (Fig. 1). But many problems remain controversial, in particular the number of V genes and the mechanisms whereby two separate genes (V and C) encode a single, continuous polypeptide chain.

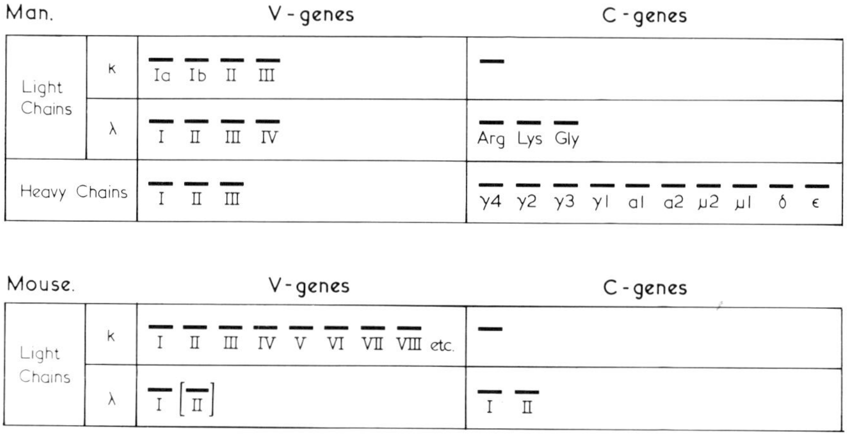

Fig. 1. Possible arrangement for a minimum of genes for human Ig and mouse light chains. Scheme taken from Ref. 13. The three unlinked sets of genes coding for the different chains thought to occur on different chromosomes. A single polypeptide chain results from expression of a V and C gene from same set. Number of V genes depicted lower than present estimates of minimum derived by protein sequences. Second λ V gene uncertain

The succesful, cell-free translation of immunoglobulin mRNA (3, 20) opened the way for the purification of the mRNA. Consequently a fresh look at the genetic bases of antibody specificity became possible at a level much closer to the genes themselves than their translation products. This approach depends directly or indirectly on the mRNA structure for the interpretation of the results.

Structural Studies of Immunoglobulin mRNA

Molecular hybridisation experiments are based on the use of labelled mRNA or complementary DNA (cDNA) prepared from it. The techniques rely on the chemical nature of the material used rather than on its biological properties. We are currently investigating the nucleotide sequence of the mRNA coding for the immunoglobulin secreted by the mouse myeloma MOPC 21. The amino acid sequences of both light and heavy chains have been established (11, 22).

Our first approach to the nucleotide sequence of both light (2, 12) and heavy chains (6) mRNA was to employ the established methods for ^{32}P-RNA which we were able to prepare using tissue culture techniques. The labelled mRNA was digested with the enzyme T_1 ribonuclease which cleaves at all G residues. Large products can then be purified by two-dimensional fractionation procedures. These products can be further analysed by digestion with pancreatic ribonuclease or U_2 ribonuclease. In general this procedure does not give an unambiguous sequence, but defines a restricted number of alternatives. When these alternatives are compared with the amino acid sequence, a coincidence between one of the oligonucleotide alternatives and the protein sequence is often observed. An assignment is thus made for a certain oligonucleotide to represent the amino acid sequence of a certain portion of the chain. Since the assignment is statistically based, the validity of each individual assignment is questionable, but experience has shown the procedure to be quite reliable on oligonucleotides longer than eight to ten bases. This will be clearly illustrated as the details of Figure 2 are discussed. Oligonucleotides so far assigned to the C region of MOPC 21 light chain sequence are shown in Figure 2 (which includes some unpublished recent data). A similar approach to the MOPC 21 heavy chain has been made, and oligonucleotide assignments for considerable sections of the peptide chain have been published (6).

The procedure, however, has serious limitations, and the critical one is the amount of radioactivity available. This drawback motivated the development of new sequencing techniques not only using light chain mRNA but also globin mRNA and ovalbumin mRNA (4, 14). Nonetheless, fingerprint analysis of ^{32}P-mRNA has remained a method of choice for a simple characterisation of the mRNA or of its fragments. Examples of such applications are the detection of inactive mRNA in a mutant line (11), and the demonstration that light chain mRNA hybridises with unique genes (16). A critical point in molecular hybridisation studies is the chemical purity. Fingerprints are reliable in giving an assessment of sequence complexity, but we have devised a more accurate approach than visual estimation. It consists of calculating purity from the molar yield of specific oligonucleotides isolated from fingerprints (16).

Complementary transcripts of mRNA can be made by the enzymes reverse transcriptase and DNA polymerase. They require a primer to initiate transcription. A commonly used primer is oligo(dT) which is complementary to the poly(A) tract present at the 3' end of the mRNA. Transcription therefore starts at variable points along the poly(A) tract and proceeds towards the 5' end of the molecule. This complementary copy (cDNA) is the most reliable probe for molecular hybridisation studies. Methods for sequencing DNA are at present under active development in our laboratory (19). It was therefore obvious to attempt sequencing the cDNA of the light chain, (a) because it was being used as a hybridisation tool, and (b) to test the suitability of this indirect approach for the sequence analysis of the mRNA.

Radioactive cDNA was made under conditions in which short copies of cDNA were synthesized (14) and then digested with endonuclease IV which splits at C residues. Unfortunately the splits are only partial and many C bonds are not split at all. Fingerprints are therefore very variable. On the other hand the sequence of a particular oligonucleotide can be deduced with partial accurarcy by the very simple procedure of a two-dimensional fractionation of all the intermediate products of digestion with venom 3' phosphodiesterase (9). For greater accuracy more elaborate analyses are required (nearest neighbour and depurination analysis). In this way a series of sequences, some over-

77

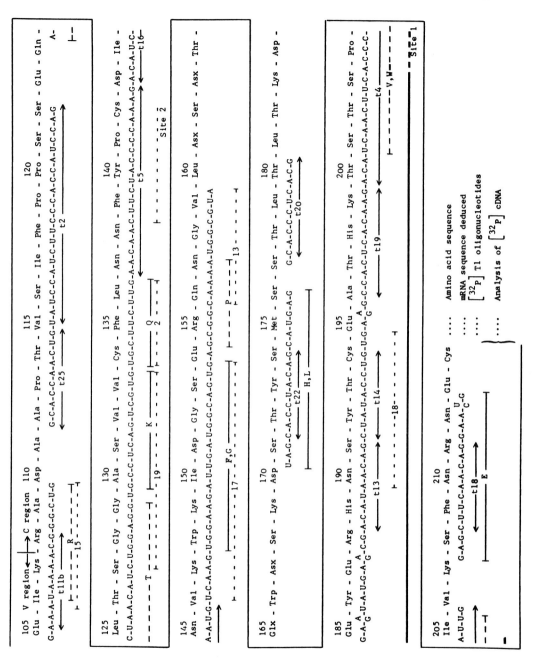

Fig. 2. Oligonucleotide sequences coding for C region of light chain mRNA of MOPC
21. Data only preliminary since depurination and nearest neighbour analysis are often
lacking. Amino acid sequence from Svasti and Milstein (22). All T1 oligonucleotides
analysed by pancreatic and U$_2$-ribonuclease digests. Preparations of cDNA and analysis
of products as follows: Capital letters (E, P, R, etc.) indicate endonuclease IV
digestion products of cDNA primed with oligo(dT); numbers (18, 17, 15, etc.) are
similar products of cDNA primed at sites 1 and 2. Thick line (amino acid residues
185-205): sequence analysis of oligonucleotides made by specifically restricting
length of cDNA copy and obviating necessity of endonuclease IV digestion. These products were subjected to nearest neighbour and depurination analysis

lapping and some not, were obtained. By varying the length of the transcript, it was possible to obtain sequences further away from the primer, although here the greater complexity of the cDNA made analysis more difficult.

This approach appeared at first very promising, but recently we discovered some shortcomings. It has become clear that large fragments of the cDNA digest do not show up in the two-dimensional fractionation procedure. This difficulty is reminiscent of the classic protein chemistry problem of the "insoluble core", so that when all the products from a given cDNA digest are analyzed they do not correspond to a continuous stretch. The possibility of wrong overlaps becomes significant and some of the advantages of endonuclease IV as a sequencing tool are lost. But the procedure has given a considerable amount of data (Fig. 2) and longer regions of continuous sequence have been established than was possible when using uniformly ^{32}P-labelled mRNA (see above).

The use of alternative primers has opened up new possibilities. Instead of using oligo(dT) as primer (complementary to poly(A)), specific oligonucleotides synthesized chemically and complementary to known stretches of the mRNA have been used. Data collected by this method using a specific primer which recognises two sites in the C region are also shown in Figure 2.

Unlike oligo(dT), which starts somewhere along the poly(A) tract, specific primers have defined starting points. Therefore, by restricting the length of the cDNA, specific oligonucleotides can be purified without using endonucleases. This has proved a major advantage which overcomes the oligonucleotides core problem. Such short cDNA fragments can be analyzed by their venom phosphodiesterase digestion products, depurination fragments and nearest neighbour analysis. The succesful use of this method has permitted the continuous sequence of about 50 residues in the C region of the light chain mRNA (Fig. 2, thick line starting at site 1). A sequence of particular interest in Figure 2 is the one which overlaps the V and C regions: this sequence is somwhat longer than previously published.

It is clear from this brief summary that no single method has proved totally satisfactory. Further technical developments will be required for succesful completion of the mRNA sequence.

As can be seen in Figure 2, at present we have data which account for about 80 % of the C region. The data are only preliminary and confirmation by depurination and nearest neighbour analysis is often lacking. In fact, we have already discovered an error in our 3' terminal sequence, which now reads -C-U-U-G-poly(A)$_{OH}$ and not -C-U-U-G-C-poly(A)$_{OH}$ as published (12).

In addition to the data presented in Figure 2, we have accumulated a considerable amount of information on other regions of the mRNA, particularly the 3' untranslated region (Fig. 3). In this region the data are more difficult to rationalise, as there is no amino acid sequence to help establish the order of the fragments.

Figure 3 shows a schematic representation of the MOPC 21 light chain mRNA. The possibility of integration during or after translation seems now quite untenable. Additional evidence, derived from studies of hybrid cells (5, 10) strongly suggests the integration occurs during or prior to transcription, most likely during lymphocyte differentiation.

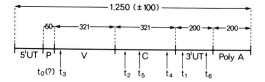

Fig. 3. Scheme of MOPC 21 light chain mRNA. Location of the oligo-nucleotides (t_0-t_6 derived by Ty_1 digestion and product of reverse transcription using oligo(dT) as primer are indicated

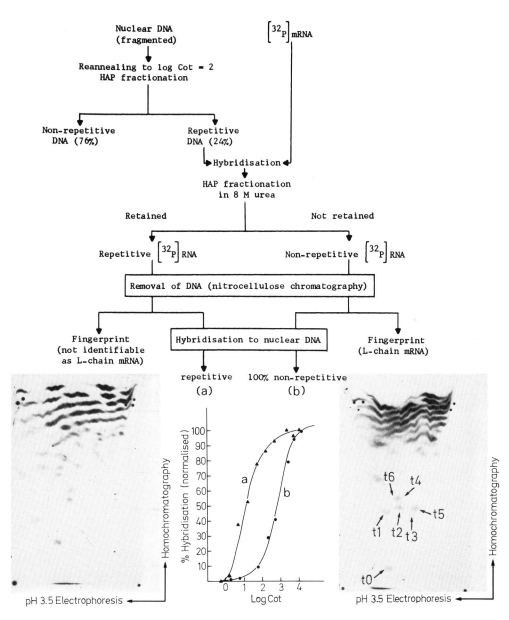

Fig. 4. Demonstration that light chain mRNA of MOPC 21 hybridises with non-respective DNA

Molecular Hybridisation and Antibody Diversity

mRNA as Probe

Initial hybridisation experiments with mRNA from MOPC 21 yielded bi-
phasic kinetics (15). Several other workers also reported the detection by other light chain mRNA preparation of a gene repeated very
few times in addition to a family of similar or identical genes repeated over 200 times (2, 21, 24). The interpretations of these bi-
phasic kinetic curves were various, and this was not surprising in
view of the number of variables. (a) The fingerprint analysis indicated
the presence of RNA contaminants not tranlated by cell-free synthesis.
(b) The first molecular weight determinations in formamide gels (2)
clearly established that in addition to the RNA sequences required to
translate V and C regions plus the precursor region, there was a long
stretch which was apparently untranslated (Fig. 3).

The experiment described in Figure 4 seems to settle the problem of
the repetitive component present in the mRNA. This component appears
to be due to a contaminant RNA species. The experiment consisted of
the preparation of repetitive DNA (i.e. the DNA fraction which contains
the genes repeated many times) and the use of the fraction to separate
the complementary components of the mRNA. The "absorbed" and "unabsorbed"
fractions were then analysed by fingerprinting. The light chain was unabsorbed, i.e. none of the mRNA is represented in the fraction of the
DNA which contains repetitive genes. Moreover, the nonabsorbed fraction
was totally free of a repetitive component, while still containing the
sequences which characterise the V region (16). Other workers have now
reached a similar conclusion using other mRNA species (8, 23). It is
our opinion that in future any reports indicating the presence of a
repetitive component can only be considered significant if it is shown
to be an integral part of the immunoglobulin mRNA molecule.

cDNA as Probe

In general cDNA is not long enough to include a significant portion
of the V region. By a suitable fractionation procedure, we have been
able to separate a very small fraction of cDNA which includes a significant portion of the V region. Such cDNA does not contain a repetitive component (17). Figure 5 shows a more recent repeat of such an
experiment and a comparison with a similarly prepared cDNA of mouse
globin. The results are in full agreement with our previous conclusions
and indicate the presence of one to eight V genes detected by the MOPC
21 light chain cDNA.

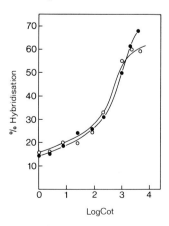

Fig. 5. Hybridisation of MOPC 21 light chain cDNA
to liver nuclear DNA analysed by HAP. cDNA for
light chain (-●-●-) and globin (-o-o-) 600-900
bases long. Experiment similar to one described
in Ref. 21

Table 1. Differences between V and C regions

(a) Mouse C_K region and C_K regions of:

Rat	31
Human	50
Rabbit	69

(b) Individual mouse V regions

MOPC 21	61					
MOPC 41	38	55				
MOPC 70	52	58	57			
MOPC 321	52	48	56	26		
MOPC 63	50	46	54	24	8	
T 124	50	50	54	25	4	9
	MOPC 173	MOPC 21	MOPC 41	MOPC 70	MOPC 321	MOPC 63

(c) Range of variation between human V regions of same and different subgroups

$V_{\kappa Ia}$	16-27		
$V_{\kappa II}$	56-66	28	
$V_{\kappa III}$	44-58	44-49	13-22
	$V_{\kappa Ia}$	$V_{\kappa II}$	$V_{\kappa III}$

Numbers given refer to minimum base changes per 100 amino acid residues (i.e. 300 bases)

The results with cDNA giving single hybridisation kinetics and mRNA indicating that the repetitive component is a separate molecule are contradictory unless the repetitive impurities of the mRNA are not transcribed and therefore absent from the cDNA. The transcription of mRNA appears indeed to be selective, since the repetitive component of the mRNA does not hybridise with cDNA (17). There is, in addition, another repetitive component detected by the longest cDNA molecules (comprising less than 1 % of the cDNA transcript), and this must be different from the repetitive component detected with mRNA, since it is not observed on short cDNA copies. In addition this other component can only be detected by fractionation with hydroxyapatite (HAP) but not with S_1 nuclease (S_1 nuclease discriminates against mismatched segments and unhybridised tails while HAP fractionation does not). Our interpretation of this observation was that there exists a segment of RNA located towards the extreme 5' which is recognised by a population of perhaps 200 genes. But it remains uncertain whether this segment is part of the light chain mRNA or of a contaminant RNA species.

How Many Genes

There is thus increasing evidence that the number of V genes detected by specific probes is very small and very similar to the number of C genes.

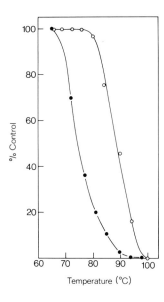

"Melting" curves of rat DNA and of hybrids between mouse MOPC 21 myeloma light chain cDNA with rat DNA. o, reannealed rat DNA; ●, mouse cDNA-rat DNA hybrid. Experiment taken from Ref. 25

Normally hybridisation experiments are carried out under conditions which minimise mismatching and infidelity of hybrid formation. How much will closely related V gene sequences be able to crosshybridise? This varies with hybridisation conditions and with the nature of the probe. To gain information on this problem we have measured the hybridisation of mouse cDNA (copies not longer than the C region) with DNA from rat, rabbit and human.

Table 1 shows the homologies between the C regions of these species in comparison with the V regions within humans and mouse. Proteins of the same subgroup (mouse or human) are more homologous than any of the C regions. On the other hand the mouse-rat C region is more homologous than two V regions of different subgroups. The hybridisation properties of the C regions provide a model for the V region cross hybridisations. The protein homologies, however, do not tell us unequivocally how similar the DNA sequences will be. Some nucleotide differences do not lead to amino acid sequence differences (wobble positions and untranslated segments). Therefore an independent measure of DNA sequence similarity was made by measuring the melting temperature of reannealed rat DNA compared to the hybridised heterologous rat DNA mouse cDNA (Fig. 6). The difference in melting temperature can be correlated with the level of mismatching (1, 25).

We conclude that the mismatching between the mouse cDNA and rat DNA is of the order of 8-15 %. Table 2 shows that the assay of hybridisation by HAP is unable to differentiate such a level of mismatching, while nuclease analysis can reveal it quite clearly. In terms of V region sequence it means that our analysis by HAP should have recognised sequences differing perhaps by as much as 30 bases/100 amino acid residues (10 % of bases). This value loosely means proteins within one subgroup. Nuclease digestion is on the other hand much more restricted and may not include all members of a subgroup.

Proteins from different subgroups may be expected to vary by as much as rabbit and human C region differ from mouse. Cross-hybridisation, although still detectable by HAP fractionation, is very much reduced

Table 2. Hybridisation of mouse C_K-DNA with DNA of different species

Hybridisation conditions		Source of DNA	% hybridisation (high Cot)	
Temp. ^{O}C	Salt		HAP	S_1
70°	0.24 M PB	Mouse	65	56
		Rat	61	20
		Human	13	4
		Rabbit	21	4
65°	0.48 M PB	Mouse	69	
		Human	28	
		Rabbit	32	
60°	0.48 M PB	Mouse	62	
		Human	36	
55°	0.48 M PB	Mouse	77	
		Human	45	
		Rabbit	56	

The cDNA was a fraction prepared by alkaline sucrose gradient and shown by acrylamide gels to be 100-600 bases long (average 500).
PB: sodium phosphate buffer, pH 6.8. Cot = 10,000

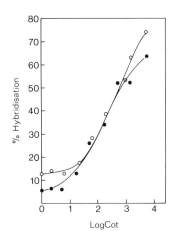

Fig. 7. Kinetics of hybridisation of MOPC 21 light chain cDNA to vast excess of mouse and rat DNA. cDNA 100-500 bases long and does not include V region sequences. ● rat and o mouse liver DNA. Hybridisation at 70° in 0.24 M phosphate buffer

(Table 2). Lowering the stringency (lower temperature and higher salt concentration) results in a dramatic increase in cross-hybridisation. This has obvious potential in our goal to estimate the total number of V_K genes by cross-hybridisation methods.

The effect of mismatching is to lower the rate of hybridisation and consequently may lead to underestimation of pool size. We decided therefore to study the effect of mismatching on hybridisation rates under our experimental conditions. Figure 7 shows a comparison of the

kinetics of hybridisation of cDNA to a vast excess of mouse and rat
DNA. Under our standard hybridisation conditions no difference was
detected, indicating that the differences expected between proteins
of the same V-region subgroups are unlikely to affect seriously our
numerical conclusions.

These data, especially when taken together with the results of other
laboratories on other light chain mRNAs, strongly suggest that the
number of V-region sequences detected by hybridisation kinetics is
lower than the number of V-region sequences predicted by the amino
acid sequence analysis of myeloma proteins. This conclusion, although
not yet totally free of uncertainties and difficulties, implies that
somatic processes contribute to antibody diversity. The extent of this
contribution remains undecided, although we would suggest it is appreciable by the very fact that it can be detected.

The required contribution of somatic processes will depend on the gap
between the number of germ-line V genes and the number of immunoglobulin chains necessary for optimum antibody diversity. It has been pointed
out that this gap depends on the number of V genes, but also on the
time available to the embryo to develop a functional system, the presence of mutational "hot spots" and the mechanisms for selection of
suitable somatic mutants (13). It is likely that the balance of these
variables is not the same in each species and that it is under continuous evolutionary change. To develop a comprehensive model of the antibody diversity, it is necessary to define these variables. Molecular
hybridisation techniques may be developed along the lines which we
have described above, to allow a measure of the total number of V genes
in a given species.

At present, however, there is no reason to believe that the total number of genes is going to be significantly larger than the maximum number of subgroups as defined by protein sequence studies. If this is
so, it is expected that, as with the subgroups, the total number of
V genes coding for the heavy chains of mouse or for the κ chains of
humans will be lower than the total number of V_{κ} genes of mouse.

References

1. Bonner, T.I., Brenner, D.J., Neufeld, B.R., Britten, R.J.: J. Mol.
   Biol. 81, 123-135 (1973)
2. Brownlee, G.G., Cartwright, E.M., Cowan, N.J., Jarvis, J.M.,
   Milstein, C.: Nature (New Biol.) 244, 236-240 (1973)
3. Brownlee, G.G., Harrison, T.M., Mathews, M.B., Milstein, C.:
   FEBS Lett. 23, 244-247 (1972)
4. Cheng, C.C., Brownlee, G.G., Carey, N.H., Doel, M.T., Gillam, S.,
   Smith, M. (1976), in press.
5. Cotton, R.G.H., Milstein, C.: Nature (Lond.) 244, 42-43 (1973)
6. Cowan, N.J., Secher, D.S., Milstein, C.: Europ. J. Biochem. 61,
   355-368 (1976)
7. Delovitch, T., Baglioni, C.: Cold Spring Harbor Symp. Quant. Biol.
   38, 739-752 (1973)
8. Farace, M.G., Aellen, M.F., Briand, P.A., Faust, C.H., Vassalli,
   P., Mach, B.: Proc. Nat. Acad. Sci. 73, 727-731 (1976)
9. Galibert, F., Sedat, J., Ziff, E.: J. Mol. Biol. 87, 377-407 (1974)
10. Kohler, G., Milstein, C.: Nature (Lond.) 256, 495-497 (1975)
11. Milstein, C., Adetugbo, K., Cowan, N.J., Secher, D.S.: In: Progress
    in Immunology. Brent, L., Holborrow, J. (eds.). Amsterdam: North-
    Holland Press 1974, Vol. I, p. 157-168

12. Milstein, C., Brownlee, G.G., Cartwright, E.M., Jarvis, J.M., Proudfoot, N.J.: Nature (Lond.) 252, 354-359 (1974)
13. Milstein, C., Munro, A.J.: In: Defence and Recognition. Porter, R.R. (ed.). London: Butterworths 1973, pp. 199-228
14. Proudfoot, N.J., Brownlee, G.G.: Nature (Lond.) 252, 359-362 (1974)
15. Rabbitts, T.H., Bishop, J.O., Milstein, C., Brownlee, G.G.: FEBS Lett. 40, 157-160 (1974)
16. Rabbitts, T.H., Jarivs, J.M., Milstein, C.: Cell 6, 5-12 (1975)
17. Rabbitts, T.H., Milstein, C.: Europ. J. Biochem. 52, 125-133 (1975)
18. Rabbitts, T.H., Milstein, C.: Trans. Biochem. Soc. 3, 870-872 (1975)
19. Sanger, F.: Proc. Roy. Soc. Ser. B 191, 313-333 (1975)
20. Stavnezer, J., Huang, R.C.C.: Nature (New Biol.) 230, 172-175 (1971)
21. Storb, U.: Biochem. Biophys. Res. Commun. 57, 31-38 (1974)
22. Svasti, J., Milstein, C.: Biochem. J. 128, 427-444 (1972)
23. Tonegawa, S.: Proc. Nat. Acad. Sci. 73, 203-207 (1976)
24. Tonegawa, S., Bernardini, A., Weimann, B.J., Steinberg, C.: FEBS Lett. 40, 92-96 (1974)
25. Ullmann, J.S., McCarthy, B.J.: Biochim. Biophys. Acta 294, 416-424 (1973)

Differentiation of Immunoglobulin Genes

S. Tonegawa and N. Hozumi

Introduction

Recent advances in the technology of nucleic acid biochemistry, in particular the availability of highly purified, light and heavy chain mRNAs from mouse plasmacytomas, have enabled several groups of researchers to study immunoglobulin genes at the molecular level. In this article we will summarize our recent efforts directed to the following two questions:

What is the genetic origin of the enormous diversity of antibody molecules?

How is the genetic information encoded in two seemingly separate segments of DNA, V and C, integrated to generate a contiguous polypeptide chain?

The Problem of the Genetic Origin of Antibody Diversity

One of the most intriguing features of the immune system is the vastness of the diversity of its components. On the molecular level this diversity is seen in the primary structure of antibody molecules. On the cellular level it is seen in the clones of antibody producing cells. The two kinds of diversity are connected by the well-established clonal selection theory (Burnet, 1957; Jerne, 1955). The genetic origin of such diversity has been one of the central issues in immunology. The issue is whether or not structural genes for the full repertoire of antibody molecules are contained as such in energy cell of an organism. The alternative is that a large proportion of the diversity is generated by somatic process, such as somatic mutation. The immense complexity of the vertebrate genome has hampered direct approaches to this question. Inevitably, arguments for and against these contrasting views have been based on indirect experimental observations (primarily of amino acid sequences of immunoglobulin chains) and general principles of genetics or evolution. (See for example Capra and Kihoe, 1974; Wigzell, 1973).

Availability of light and heavy chain mRNAs of a purity sufficient for detailed nucleic acid hybridization studies has opened a way for direct experimental approaches to this controversial problem.

The Principles of the Approach

Given purified preparations of mRNAs coding for immunoglobulin chains, several approaches to the present issue can be envisaged. Among these the ultimate appraoch is probably to isolate, perhaps making use of nucleotide sequence complementarity, the immunoglobulin genes both

from germ line and plasma cells, and determine whether or not the nucleotide sequences in the corresponding genes from the two sources are
different. Such an approach, in spite of the enormous complexity of
eucaryotic genomes, now seems quite feasible thanks to the recently
developed techniques for cloning recombinant DNA molecules constructed
in vitro (Morrow et al., 1974; Thomas et al., 1974). Another approach
which we have employed in the past few years is to count the number
of immunoglobulin genes by RNA-DNA hybridization. In this approach the
experimental steps and logic listed below were followed:

1. Purify mRNAs coding for particular light of heavy chains from mouse
myeloma cells.

2. Radioiodinate the purified mRNA to a specific activity higher than
10 million cpm per microgram.

3. Determine the reiteration frequency of the DNA sequences complementary to the mRNA. This is done by annealing under proper conditions
the radioiodinated RNA with denatured, sonicated liver DNA (liver is
in this case a surrogate of germ line cells), and then determining, as
a function of time, the fraction of RNA in the hybrid by measuring the
resistance of hybridized RNA to RNase. Under a given set of annealing
conditions, the fraction of hybridized RNA (f) is dependent upon the
RNA concentration (R_O), the DNA concentration (C_O) and the time of incubation (t). If DNA is present in large excess, f is independent of
R_O, and depends only on the product of C and t ($C_O t$). The value of
$C_O t$ necessary to achieve 50 % hybridization, $C_O t_{1/2}$, will depend upon
the fraction of the DNA which is complementary to the RNA. The reiteration frequency (F) of the gene in question is calculated by the
following formula

$$F = F* \frac{C_O t_{1/2} *}{C_O t_{1/2}} \frac{C}{C*}$$

where C designates the genome complexity and * designates a gene used
for standardization and whose reiteration frequency F* is known by an
independent method (Melli et al., 1971).

4. Carry out competition hybridization experiments by annealing the
radioiodinated RNA used in the $C_O t$ curve experiment in the presence
of varying amounts of unlabeled mRNA coding for another V region. The
purpose of the competition experiments is to determine the degree of
base sequence homology between the two mRNAs, and consequently between
the two corresponding V genes. The results will tell us whether and
how much the radioiodinated mRNA used in the $C_O t$ curve experiment would
cross-hybridize with other V genes if they existed as separate germ
line genes.

5. Estimate, on the basis of such competition experiments and on the
available amino acid sequence data, the number of different V regions
whose genes would have cross-hybridized with the particular mRNA used
in the $C_O t$ curve experiment. If this number is significantly larger
than the experimentally determined reiteration frequency, it constitutes formal evidence for somatic generation of antibody diversity.

Evidence for Somatic Generation of Antibody Diversity

We have applied the approach described in the last section both to
mouse κ and λ chains (Tonegawa et al., 1974; Tonegawa, 1976). Partial

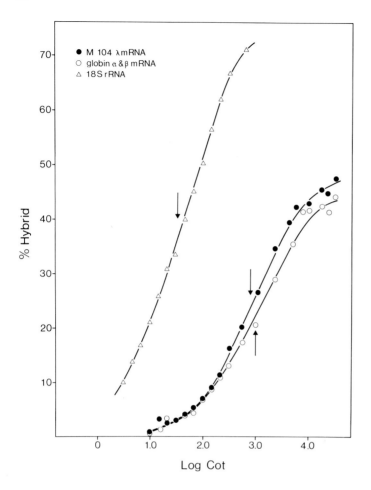

Fig. 1. Hybridization kinetics of mouse RNAs with mouse liver DNA (C_Ot curve).
Arrows: positions of $C_O t_{1/2}$

application was also made to heavy chains (Bernardini and Tonegawa, 1974; Schuller, 1976). Because they are simple and straight-forward we will describe here the experimental results obtained with λ chains in some detail.

In the λ chain studies, two mRNAs coding for two different V_λ regions were prepared to a purity of 90 % or higher. The sources of the mRNAs were the Balb/s plasmacytomas MOPC 104E and HOPC 2020. The two V regions differ only in two amino acids (see Fig. 3). Each of the two pairs of amino acids is related by a single base substitution in the corresponding triplet codons. The C_Ot curve of M 104E mRNA is illustrated in Figure 1. A mixture of mouse α and β globin mRNAs as well as 18S ribosomal RNA were used as kinetic standards. The gene reiteration frequencies of these RNAs have previously been determined at less than 3 and at 200-250 respectively (Harrison et al., 1972; Melli et al., 1971). The reiteration frequency of M 104E λ chain gene is calculated to be 2 to 3 from the results represented in Figure 1.

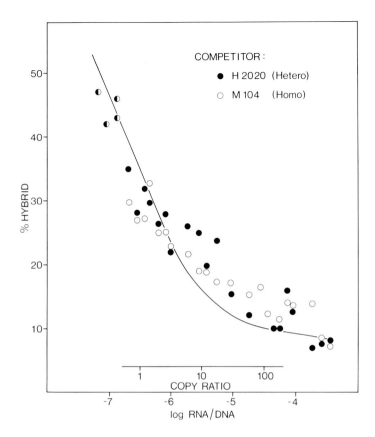

Fig. 2. Competition hybridization of λ mRNAs. Theoretical curve derived by computer program and represents homologous competition of hypothetical RNA mixture containing 85 % unique, major component and 15 % impurities, assumed to consist of 1500 different equimolar species

In the competition hybridization experiments illustrated in Figure 2, a fixed amount of radioactive M 104E mRNA and an excess of denatured liver DNA were annealed to a high C_0t value (C_0t = 10,000) in the presence of varying amounts of unlabeled H 2020 mRNA (heterologous competition) or M 104E mRNA (homologous competition). Radioactive M 104E mRNA in the hybrid is plotted as a function of the ratio of total RNA to DNA. The fact that the two competition curves are indistinguishable indicates that there is extensive nucleotide sequence homology between the two mRNAs and hence between the two V_λ genes. Is such an extensive sequence homology unique to this particular pair of λ chains? That this is not the case is strongly suggested in the amino acid sequence data of myeloma λ chains summarized in Figure 3. As discussed elsewhere (Tonegawa and Steinberg, 1976), a simple statistical calculation applied to the data in Figure 3 suggests that there must be at least 20 to 30 different V_λ regions in the pool of Balb/c mouse myeloma proteins. Each of these V_λ regions is expected to be as homologous to M 104E in sequence as is H 2020. Thus if there were a germ line V gene for each of these sequences, M 104E mRNA would cross-hybridize with every one of them. Yet the number of germ line V_λ genes was determined to be no more than 2 to 3 by the C_0t curve experiment.

90

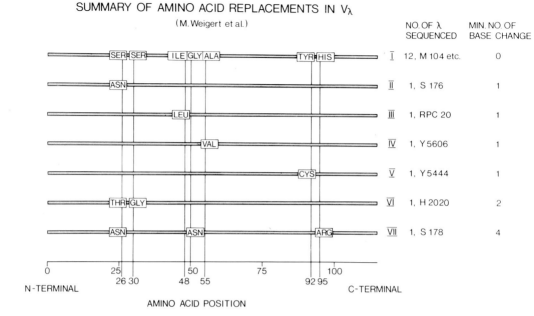

SUMMARY OF AMINO ACID REPLACEMENTS IN V$_\lambda$

(M. Weigert et al.)

	NO.OF λ SEQUENCED	MIN. NO. OF BASE CHANGE
I	12, M 104 etc.	0
II	1, S 176	1
III	1, RPC 20	1
IV	1, Y 5606	1
V	1, Y 5444	1
VI	1, H 2020	2
VII	1, S 178	4

Fig. 3. Schematic representation of differences among λ chains. (After Weigert et al., 1970)

Sequence variability is considerably greater in mouse κ chains than in λ chains. This makes it more difficult to apply the kind of approach described above to κ chains. Nevertheless, we could define a group of κ chains, comprising over 10 % of all κ chains for which sequence information was available, within which there was over 80 % amino acid homology. For this group of κ chains we have shown that the numbers of germ line V genes are no more than 1 to 2 (Tonegawa et al., 1974).

Hence, the hybridization evidence indicates that the numbers of germ line V genes are too small to account for the observed diversity in antibody molecules.

The Problem of Integration of V and C Gene Information

Uniqueness (one gene per haploid genome) of C genes has been suspected from normal Mendelian segregation of allotypic markers (Milstein and Munro, 1972). The hybridization studies just described confirm this notion. The same hybridization studies demonstrate that a group of closely related V regions are somatically generated from a few, possibly a single, germ line V gene(s). They do not, however, give us any reliable estimate of total number of germ line V genes. Given the enormous diversity of V regions, however, the existence of multiple germ line V genes is almost inescapable. This raises a second intriguing problem, one to which Dreyer and Bennett addressed themselves more than a decade ago (Dreyer and Bennett, 1965). How is the information in the V and C genes integrated? Since V and C gene sequences exist in a single mRNA molecule as a contiguous stretch (Milstein et al, 1974), such integration must take place either directly at the

SPECIFICITY OF RESTRICTION ENZYMES

Fig. 4. Base sequence specificities of 2 restriction endonuclease

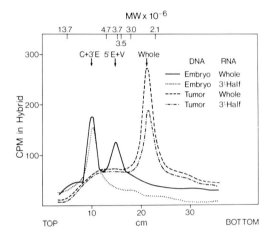

E.coli R-I   —GAATTC—
              —CTTAAG—

B.am H-I   —GGATCC—
              —CCTAGG—

Fig. 5. Hybridization patterns of Bam H-I digests of mouse DNA with MOPC 321 κ chain mRNA and its fragments

DNA level or during transcription, or after transcription by joining of two RNA molecules.

We will describe experiments which provide evidence for joining of V and C sequences at the DNA level.

Evidence for Somatic Joining of V and C Genes

Bacterial restriction endonucleases recognize and cleave specific sequences of base-pairs within a DNA duplex. The specificity of two such enzymes, *Escherichia coli* R-I (Eco R-I) and *Bacillus amyloliquefaciens* H-I (Bam H-I) is shown in Figure 4. We have purified these enzymes in a large quantity and digested high molecular weight mouse DNA from various sources. The resulting DNA fragments were fractionated according to size in a preparative agarose gel electrophoresis. The gel was cut into slices, and DNA extracted from each slice was assayed for V and C gene sequences by hybridization with radioiodinated whole light chain mRNA or its 3'-end-containing fragments. The assay is based on the fact that the sequences corresponding to V and C genes are on the 5'-end and 3'-end halves of the mRNA molecules respectively. The details and justification of such assay procedures are described elsewhere (Hozumi and Tonegawa, submitted).

Figure 5 illustrates such an experiment carried out for a κ chain. DNA from 12-day-old Balb/c mouse embryos or from MOPC 321 tumors was

digested with the Bam H-I enzyme. Assay was carried out with MOPC 321
κ-mRNA. Results obtained from two separate gels, one of embryo DNA and
the other of tumor DNA, are superimposed. Since both electrophoresis
and hybridization were performed under identical conditions, profiles
from the two gels are comparable. With the embryo DNA, two DNA components of 6.0 and 3.9 million MW hybridized with the whole κ-mRNA, whereas only the former hybridized with the 3'-end half. Hence the C_K gene
is in the 6.0 million component and the V_K gene should be in the 3.9
million component. The fact that the extent of hybridization in the
larger component is nearly identical with the two RNA probes supports
this interpretation. Since the two genes are in separate DNA fragments,
they are probably some distance away from each other in the embryo
genome. The possibility that the enzyme cleaved contiguously arranged
V and C genes near the boundary is not entirely eliminated. Two candidates for such cleavage sites are represented by the amino acids
in the V region at positions 93-95 and 97-98. However, the probability
that either of these amino acids provides the exact nucleotide sequence
required is low.

The pattern of hybridization is completely different in the DNA from
the homologous tumor. Both RNA probes hybridized with a new DNA component of 2.4 million MW. There is no indication that either of these
RNA probes hybridizes with other DNA components above the general background level. These results strongly suggest that both V and C gene
sequences are contained in the 2.4 million MW component of this tumor
DNA. The whole RNA hybridizes with this component nearly twice as well
as does the 3'-end half, thereby supporting this notion. Hence the V_K
and C_K genes, which are most likely some distance away from each other
in the embryo genome, are brought together in the plasma cells presumably to form a contiguous nucleotide stretch. Such rearrangement of
immunoglobulin genes takes place in both the homologous chromosomes.
An alternative explanation, namely that fortuitous mutations by loss
and gain of Bam H-I sites generated the observed pattern difference, is
not impossible. It is, however, extremely unlikely because of the multiplicity of events which must be assumed. Furthermore such pattern
changes are not unique to this particular combination of enyzme and
DNA. With *E. coli* R-I enzyme, and embryo and H 2020 DNAs we have obtained
results which lead us to a similar conclusion (Hozumu and Tonegawa,
manuscript in preparation).

Models for V - C Gene Joining

Various models have been suggested for integration of V and C information at the DNA level. Some of these models are schematically illustrated in Figure 6. The "copy-insertion" model assumes that a specific
V gene is duplicated and the copy is inserted at a site adjacent to
the C gene (Dreyer et al., 1967). The "lateral duplication" model en-
visages multiple V genes which are arranged in parallel, each adjacent
to the C gene (Smithies, 1970). According to either of these two models,
lymphocytes expressing a particular V gene should retain embryonic
context as far as this V gene is concerned. Our results are clearly
incompatible with this prediction. The "excision insertion" model
suggests that a specific V gene is excised to form an episome-like
structure, which in turn is integrated adjacent to the C gene (Gally
and Edelman, 1970). The "deletion" model originally presented for the
globin gene system can be applied to the immunoglobulin genes. According to this model DNA intervening a specific V gene and the C gene
lopps out and is lost upon subsequent cell multiplication (Kabat, 1972).
The latter two models are both consistent with the present results.

A copy insertion

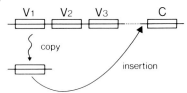

Fig. 6. Models for V-C gene
joining

B lateral gene
   duplication.

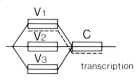

C excision insertion

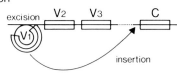

D deletion

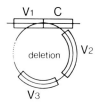

One may note, however, that the excision insertion model requires at least two recombinations, whereas the deletion model requires only one. Further experiments are required to distinguish between the two models.

Concluding Remarks

The hybridization experiments described in this article demonstrated that both content and context of immunoglobulin genes are altered during differentiation of lymphocytes. We know, however, very little about the molecular mechanisms underlying these phenomena. The gene isolation experiments now underway in several laboratories are expected to help in understanding these mechanisms as well as other problems associated with immunoglobulin genes.

Are alterations in genetic content and context restricted to the immune system? As mentioned above, the "deletion" model was originally proposed for the globin system. We feel that this mechanism will turn out to be important in the differentiation of other eucaryotic systems as well.

Acknowledgment. We acknowledge the expert technical assistance offered by Mr. G.R. Dastoornikoo, Ms. S. Stutz and Ms. R. Schuller.

References

Bernardini, A., Tonegawa, S.: FEBS Lett. 40, 73 (1974)
Burnet, F.M.: Australian J. Sci. 20, 67 (1957)
Capra, J.D., Kehoe, J.M.: Scand. J. Immunol. 3, 1 (1974)
Cohn, M., Blomberg, B., Geckeler, W., Raschke, W., Riblet, R., Weigert, M.: In: The Immunse System: Genes, Receptors, Signals. Sercarz, E.E., Williamson, A.R., Fox, C.R. (eds.). New York, London: Academic Press 1974, p. 89
Dreyer, W.J., Bennett, J.C.: Proc. Nat. Acad. Sci. 54, 864 (1965)
Dreyer, W.J., Gray, W.R., Hood, L.: Cold Spring Harbor Symp. Quant. Biol. 32, 353 (1967)
Gally, J.A., Edelman, G.M.: Nature (Lond.) 227, 341 (1970)
Harrison, R.P., Hell, A., Birnie, G.D., Paul, J.: Nature (Lond.) 239, 219 (1972)
Jerne, N.K.: Proc. Nat. Acad. Sci. 41, 849 (1955)
Kabat, D.: Science 175, 134 (1972)
Melli, M., Whitfield, C., Rao, K.V., Richardson, M., Bishop, J.O.: Nature (New Biol.) 231, 8 (1971)
Milstein, C., Brownlee, G.G., Cartwright, E.M., Jarvis, J.M., Proudfoot, N.J.: Nature (Lond.) 252, 354 (1974)
Milstein, C., Munro, A.J.: Ann. Rev. Microbiol. 24, 335 (1972)
Morrow, J.F., Cohen, S.N., Chang, A.C.Y., Boyer, H.W., Goodman, H.M., Helling, R.B.: Proc. Nat. Acad. Sci. 71, 1743 (1974)
Schuller, R.: M.Sc. thesis, Univ. Alberta (1976)
Smithies, O.: Cold Spring Harbor Symp. Quant. Biol. 38, 725 (1970)
Thomas, M., Cameron, J.R., Davies, R.W.: Proc. Nat. Acad. Sci. 71, 4579 (1974)
Tonegawa, S.: Proc. Nat. Acad. Sci. 73, 203 (1976)
Tonegawa, S., Steinberg, C.: In: The Generation of Antibody Diversity. Cunningham, A.J. (ed.). London, New York: Academic Press 1976, pp. 175-182
Tonegawa, S., Steinberg, C., Dube, S., Bernardini, A.: Proc. Nat. Acad. Sci. 71, 4027 (1974)
Wigzell, H.: Scand. J. Immunol. 2, 199 (1973)

Discussion

Dr. Hood: I think the two last papers are very exciting in the sense that (1) they give us an overview of nucleic acid chemistry with a clear picture of the messenger for antibody molecules that has resulted from it, (2) the results reported by Tonegawa suggest rather strongly that the joining mechanism of V and C regions of antibody molecules that one has talked about for the past ten years is indeed a legitimate mechanism that occurs at the level of DNA and as such is almost certainly going to be a fundamental mechanism for differentiating the immune system and perhaps even other complicated kinds of system with a large information-handling problem, and (3) the hybridization data reported both by Drs. Milstein and Tonegawa has suggested to them very strongly that each subgroup of κ chains on one hand, or λ chains on the other, contains gene numbers that are lower than are consistent with amino acid sequences that are available today. Let me again parenthetically add that, in my own mind, I still have great reservations about this latter point in the sense that we do not really understand the limits of the error bars. In fact I was amused to see Dr. Tonegawa use hemoglobin as his standard when in the case of human hemoglobins one can point out that some years ago Eric Davidson did some hybridization studies and said the human hemoglobulins were consistent with single cot characteristics and it is only more recently that a very careful cot analysis of the α chain of human hemoglobulins suggests that there are at least four closely related genes and this is consistent with the sequence data from which we know that at least in certain individuals there are probably three different closely related genes. So I think the reservations that we should keep in mind are how big are the error bars, that is, is it possible still to have 5-10 genes per subgroup. Now, if that is a possibility, and I just raise it for your consideration, the next really important question one has to consider is, what is the total repertoire of antibody molecules. Clearly, if it is 10^7 or 10^8, there are too many antibody molecules to be accounted for by a germ-line model. If there are 10^5 or 10^6 and in addition other mechanism such as combinatorial association of ^{V}H and ^{V}L, then perhaps a germ-line contribution would still end up by being freely a major contribution to antibody genes. But with that brief summary, I would like to open these two fine papers for questions from the audience.

Dr. Rajewsky: I should like to ask the speakers to answer the Chairman's challenge. Cesar Milstein was clearly saying that he wanted to make a point on the basis of his data, namely that he concludes that the number of germ-line genes is lower than would be required for the expression of antibody variability. Dr. Tonegawa made the same point. The Chairman was saying that he was not convinced. Wouldn't you like to comment on this?

Dr. Milstein: In this story of number of genes and diversity there have been all sorts of ups and downs and the fashion generally accepted has dramatically changed in a short number of years. There have always been a number of stubborn people who kept saying "I am not convinced," who eventually changed the fashion by some new experiments. I think it is very healthy that there are some stubborn people who will look for small flaws which do occur in these experiments. The experiments point in one direction but they are not absolutely cut and dried. If one wants to look for uncertainties, one could say, we are not really absolutely sure, and, in fact, he is right in raising the point. We are not absolutely sure and we could certainly destroy each individual experiment. I think, what to me is particularly convincing (and I would like to perhaps address now the Chairman and explain to him why I am convinced, or *fairly* convinced, anyway) is the accumulation of data. There are several cases and not one which is well done, there are several which are well done and that is what convinces me.

Dr. Hood: Yes, I have to agree that the bulk of the nucleic acid hybridization data certainly points to a smaller number of antibody genes; the issue, however, is open whether you need to have somatic mutation as an important mechanism. However, one can look at other experiments, too, such as the very beautiful ones reported by Dr. Eichmann which suggest that rather closely related genes the products of which all bind with a particular kind of antigen are pseudo-alleles, that is, separate gene products that can be mapped into the genome. I think this kind of data raises very intriguing and important questions. It is not as elegant nor as satisfying as the nucleic acid data will eventually become, because we are one step removed from the direct source of information, the genes themselves, but to me this argues in a very compelling way that the final number of antibody V genes, whether or not somatic mutation is important, is going to be rather large.

Dr. Tonegawa: Well, first of all one technical point, whether one can distinguish one to two copies from say five copies, and I completely agree with the Chairman that this would be very difficult. Now when you start talking about distinguishing 20 to 30 genes from one to five copies I feel quite comfortable. Dr. Hood said that we were using hemoglobin messenger as a standard, in fact I thought I explained that we did use other RNA which is known to be unique as a standard like *E. coli* RNA made *in vitro*. The data is such that the cot-curve of reiteration frequency of lambda messenger RNA cannot correspond to more than a few genes. My second point is that the total number of germ-line V genes and the relative number of germ-line V genes as compared to the size of the V-region repertoire in an adult animal are two completely separate issues. The issue which we addressed ourselves to is the second one. We still do not know what the total number of V genes is.

Dr. Starlinger: Regarding your translocation mechanisms, if I understood you correctly, you discuss either a direct translocation, that means excision and translocation or, as an alternative, you discuss the removal of intervening material by a Campbell-type mechanism. For the latter I think you could have a test because, if I understood correctly, people are discussing that in the course of the maturation of a specific clone, you have first a specific V gene attached to a μ chain and, then later, you have the same V gene attached to a γ chain. Now, if the excision-type mechanism were correct, then you would expect that the C region for the μ chain should be removed upon putting the V gene to the γ chain and this you could test.

Dr. Tonegawa: I don't want to generalize at this time that this V-C gene joining in the light chain is a similar situation to that existing in the so-called μ-γ- or μ-α-switch. I like the unifying model but we don't know it, and we should be careful about it. There are ways of testing this model but it has not been done.

Dr. Hood: Let me add a postscript to that - I will be very surprised if either of the models you think of are most likely correct because of what we have seen with heavy chains. The pertinent observation there, although not completely verified, it that, on rare occasions, a single cell can make two different classes of heavy chains which apparently have precisely the same V gene by idiotypic analysis. That is the important reservation. If those V-region sequences from both heavy chains are identical, then it means that the mechanism for heavy chain VC joining must be one in which a single V is placed adjacent to many Cs that a single cell can simultaneously secrete both of these gene products. However, the critical analysis that is in progress now, is whether or not those molecules of different class that are idiotypically identical have V regions that are identical.

Dr. Poljak: I would like to ask a general question as a naive member of the audience here and this is, we have heard different proposals about the number of V genes in the germ line, and about somatic mutations giving rise to the diversification. I think positions have in later years tended to come together; also we do know that the maximum number of structural genes that we can have in a vertebrate genome is only about 40,000. Now I would like to ask you as a proponent of the germ-line theory how many genes do you need, and then I would like to ask the people who propose somatic mutations how many genes do they need? and see if the number is not in fact fairly close.

Dr. Hood: Well, there are two points. I suspect there are people who might argue with the 40,000 gene number in either direction. My own view is, I can't answer that question because we have no idea about the total repertoire. Obviously, in a sense, if one believes in a straight germ-line theory you need the square root of the number of genes that are required in the repertoire if you can put all light and heavy chains together in this combinatorial fashion. I would point out that a thousand light and a thousand heavy chain genes can generate 10^6 different specificities to antibody molecules and, in my view, that is more than enough. The whole issue of how big the repertoire is and notwithstanding some recent estimates that are very high, I think it is entirely unsettled and extremely intriguing and important. Suppose the number is 10^5, then I think there would be no problem whatsoever with the number of germ-line genes. There will be people that will jump up and say that can't possibly be right, but I don't think there is compelling data one way or the other.

Dr. Milstein: I said that actually. I said, if you ask me how many genes there are, the first thing I will ask you is, in which species and of which chain. One of the main points I am trying to make is that the number is going to vary. There are other variables which are, in my mind, very important, which are the ones that are going to diminish the load of the number of genes because of the facility of production of mutants which will be useable at the time the animal needs it, which is when it is born. I will say that, as a general rule of thumb, the number of genes is roughly of the order of the number of subgroups which are defined by protein chemistry data. Does that answer your question?

Immunocompetent Cells

Phylogenesis of the Vertebrate Immune System

L. Du Pasquier

Introduction

Our current knowledge of the phylogenesis of immunity is exclusively
based on studies of comparative immunity. We know, for example, that
some functional components of the vertebrate immune system such as
allogeneic recognition and allograft rejection are found in living
representatives of primitive diploblastic invertebrates, whereas others,
such as specific immunological memory, appear only in more advanced
invertebrates where, as in vertebrates, they are associated with the
presence of lymphocyte-like cells (reviewed in Ref. 43). We also know
that the vector of antibody activity, the multichain immunoglobulin
(Ig) molecule, has been detected so far in vertebrates only (reviewed
in 61). If these comparative studies have provided a considerable har-
vest of facts and a reasonable picture of the possible evolution of
the immunse system, they have also left us with the disquieting feeling
that they may not deal with evolution itself but rather with the present immunological status of living highly evolved representatives of
extinct species. This is why, after reviewing the expression of some
elements of the mammalian immune system in a variety of primitive species (Ig molecules, lymphocyte diversity, and major histocompatibility
complex), it seemed worthwhile to present a new approach to the study
of the evolution of the immune system. This approach was made possible
by the existence of species where the role of gene duplication in evolution (63) can be more or less mimicked via the production of "artificial" polyploid individuals. An example will be given in Xenopus, the
South African frog, where the genetic control of acute graft rejection
and the mixed leukocyte reaction in "artificially" produced polyploid
individuals and an naturally occurring polyploid species have been
compared.

The Immune System in Various Animal Species

The following three features of the immunse system will be considered:
the immunoglobulin molecule, the diversity of lymphocytes, and the
major histocompatibility complex (MHC). Since there are several reviews
on each of these subjects (10-13, 15, 16, 20, 22, 42, 43, 61), this
review will concentrate on recent findings which may modify our concept in each of these three areas of interest.

A general picture of the evolution of the immune system is presented
in Figure 1. This figure details additional elements of the immune
system (complement (reviewed in Ref. 35) and the organization of lymphoid tissue (reviewed in Ref. 13)) which, because of the topics of
this symposium, will not be discussed in the text.

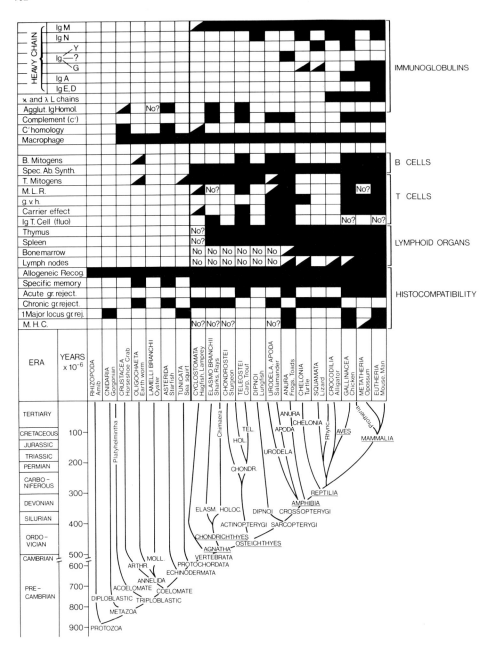

Fig. 1. Evolution of the immune system. (Abbreviations see p. 103.)

Data from invertebrates are restricted to components showing homology with vertebrate immune system. Data used in table are referred to in text and in reviews
quoted in Sect. A. For Protozoa see Ref. 36. Lack of MLR in Opossum reported by
D.H. Fox, D.T. Rowlands Jr., and D.E. Wilson in: Transplantation 21, 164-168 (1976)
and justifies partial shading in MHC compartment

The Immunoglobulin Molecule

Most theories of immunoglobulin phylogenesis place at the origin an
ancestral gene which codes for the equivalent of one domain (i.e. a
polypeptide of 110 amino acids) of the Ig molecule. It has been proposed that this gene has undergone duplication to give rise to variable part genes (V genes) and constant part genes (C genes). These
genes probably continued to duplicate, thereby creating genes for V
region subgroups and C region classes. A survey of animal species suggests when and where these hypothetical events might have occurred.

The IgM Molecule. Living representative of practically all vertebrate
classes possess high molecular weight immunoglobulins which resemble
the IgM of eutherian mammals. In all species the light (L) and heavy
(H) chains, have similar molecular weights (MW, H = 70,000, L =
23,000) and show their relatedness by virtue of their amino acid composition (59, 61). However, in the most primitive class of vertebrates,
the Agnatha, the high molecular weight Ig does not "fit" well with the
classical IgM model. For example, although the H chain of the lamprey
(Cyclostomata) Ig is similar in its amino acid composition to the µ chain
of higher vertebrates (59), there is no evidence for a typical light
chain. In addition, the subunits of the molecule are not disulfide linked
(reviews 3, 54). The case of the hagfish, another cyclostome, is even
more puzzling. The molecular weight determination of hagfish Ig fractions is consistent with, but does not demonstrate the existence of, a
"multimer of low molecular weight polypeptides consisting predominantly
of a light chain-like peptide" (18).

If conservatism can be demonstrated at the level of the IgM subunit
from Chondrichthyes and Osteichthyes upward, the degree of polymerization of the molecule varies from family to family. Free subunits can
be found in the serum of fish (Chondrichthyes and Actinopterygi).
Pentamers of IgM (thus resembling the mammalian type) can be found
in Chondrichthyes, Dipnoi, some Amphibia, Reptilia, Aves and Mammalia.
Tetrameric IgM seems restricted to the actinopterygian fish (review
54), while Xenopus IgM is a hexamer (39, 64). In this latter species,
the J chain presents crossreacting determinants with the human J chain
(64).

The flexibility of the Ig molecule also shows some changes during
phylogenesis. Studies by fluorescence depolarization suggest that the
sites of flexibility and the relative amplitudes of motions are not
the same in mammals and sharks (5).

Low Molecular Weight Classes of Ig. The generation of Ig heavy chains
distinct from the µ chain of IgM apparently occurred in the Sarcopterygi,
whose modern descendants are the lungfish. This first non-IgM molecule,
the IgN, is composed of a typical light chain and of a heavy chain

Abbreviations to Fig. 1

M.L.R.	:	Mixed leukocyte reaction	Moll.	:	Mollusca
g.v.h.	:	Graft versus host reaction	Elasm.	:	Elasmobranchii
fluo	:	Detection of surface Ig by immunofluorescence	Holoc.	:	Holocephali
Recog.	:	Recognition	Chondr.	:	Chondrostei
gr. reject.	:	Allograft rejection	Hol.	:	Holostei
M.H.C.	:	Major histocompatibility complex	Tel.	:	Teleostei
Arthr.	:	Arthropoda	Rhync.	:	Rhynchocephalia

Code: No: recognized absence; No?: So far, evidence for absence; Complete shading:
presence; Half-shading: partial evidence for presence or homology

whose MW is only 38,000. This type of heavy chain (called nu) is also found in reptiles, birds and mammals (61). Further diversification of low molecular weight Ig is found in anuran amphibians where the 7S Ig has an as yet unnamed (question mark in Figure 1) heavy chain of 64,000 MW (38, 59. This MW is distinct from nu and γ heavy chain but it is not very different from mammalian α or δ chains. A second non-μ heavy chain, antigenically distinct, has recently been identified in anurans (77). Low molecular weight Ig with 64,000 MW heavy chain is also found in birds but its amino acid composition differs from that of its amphibian equivalent (59). Among reptiles, turtles (Chelydra serpentina) lack this type of heavy chain and thus resemble primitive Dipnoi (59). On the other hand tortoises (Agryonemys horsfieldi) produce three types of Ig in the course of their anti-hapten response: IgM, an Ig 5S (probably an IgN homologue) and an Ig 7S which is thought to be the reptilian IgG (1). The Sphenodon, an interesting living fossil, of the Rhynchocephalia (i.e. a reptile branch close to that leading to birds) also has a non-μ heavy chain which resembles the γ type more than the nu (61).

The reptiles are also at a crucial point in the evolution of light chains. They are the most primitive vertebrates known to possess two types of light chains (κ and λ equivalents) (72). The appearance of α chains seems to be restricted to birds and mammals. However, the negative outcome of the search for IgA in the tortoise need not be definitive (85) and the Ig heterogeneity of the Sphenodon, which belongs to the same familiy Diapsida that gave rise to birds which express the α chain, has not yet been systematically investigated.

It is not until the descendants of therapsid reptiles that the γ chain (MW 50,000) can be detected, both in eutherian, protherian and metatherian mammals. Further diversification occured in mammals where there are as many as 4 non-μ classes (α, γ, δ, ϵ) (59). The expression of more than one class of heavy chain per cell has been demonstrated in mammals (65) but so far the only comparative immunological data on this subject have failed to reveal a similar phenomenon in anuran amphibians (47).

Immunoglobulin Equivalent in Invertebrates? The most primitive vertebrate Ig is already a multichain molecule; no homologous molecule has been found in invertebrates. The tertiary and quaternary structures of those molecules involved in recognition of foreignness in invertebrates are very different from Ig. This does not exclude some homology at the level of the primary structure. Indeed, Carton's comparison of amino acid composition by the method of Marchalonis and Weltmann (58) suggests that some relationship exists between the agglutinins of some echinoderms and the Ig of vertebrates. Moreover, homology has also been detected between vertebrate Ig and natural agglutinins of fish (4). Even though further studies are needed to confirm this hypothesis (which is only based on amino acid composition comparison) it is remarkable that the class whose agglutinins show the strongest homology with Ig is the Echinodermata. These animals are thought to be among the direct invertebrate ancestors of the vertebrates. In comparison, both structural data and amino acid composition comparison (4, 50) for the horseshoe crab (Arthropoda) agglutinin showed no or little homology with vertebrate Ig. Similarly, no homology was found between a molluscan agglutinin and Ig. Neither of these two phyla, Arthropoda and Mollusca, is on the main evolutionary line to vertebrates.

Several hypotheses have been proposed to link Ig ancestral genes with other genetic regions involved in the immune system. For instance, Urbain (83) suggested that primitive recognition mechanisms are under the control of a set of duplicated genes coding for proteins interacting at the surface of cells by means of isologous and heterologous association. Some of these genes could be the ancestor of the major histocompatibility complex. The first immunoglobulin genes are supposed to have arisen by duplication of these genes. In fact, amino acid composition comparison suggests homology between strong histocompatibility antigens of man and IgM or IgD (78). However, partial structural data (75) on H_2K and D products in the mouse do not support the existence of a primary structure homology between these proteins and Ig. Still, they do not exclude a possible homology for residues determining the three-dimensional structure of the molecule (75). Another interesting hypothesis (51) tries to link together Ig, complement (C'), and MHC genes. It has been shown recently that some genes of the MHC code for or control the expression of some C' components (32, 33, 86). Moreover it has been suggested that the complement component C_3a, showed homology with Ig variable region, which may suggest at the same time a link between C' and MHC and a common origin of some complement and Ig genes (51). Indeed the existence of a C_3 proactivator (cofactor) in the starfish (Echinodermata) (17) and of cofactors plus late acting components in the horseshoe crab (Arthropoda) and in sipunculid worms (Annelida), may indicate that some complement genes are of very ancient acquisition and may have evolved before Ig genes. It would be interesting to know whether the agglutinin and the C_3 proactivator of echinoderms are genetically or functionally related to each other.

Lymphocyte Diversity

In birds and mammals, lymphocytes fall into two categories: (1) the T cells, which originate from the thymus and are involved in cell-mediated immunity, helper activity, and production of non-specific lymphokines, (2) the B cells which produce antibody. These two populations are distinguishable from each other by morphological and functional criteria. For instance, in mammals, immunofluorescence methods fail to detect Ig at the surface of T cells whereas they detect it easily on B cells. T cells, which exhibit a specific surface marker (Thy 1), respond by proliferation in the mixed leukocyte reaction (MLR) assay and to certain mitogenic substances such as phytohemagglutin (PHA) or Concanavalin A (Con A). B cells respond by proliferation or Ig synthesis to another set of mitogenic substances including lipopolysaccharide (LPS) and other endotoxins from gram-negative bacteria and sugars (dextran sulfate). Within each cell class, lymphocytes can be further divided into subclasses, according to surface markers such as Ly for T cells (73) or immunoglobulin class for B cells. All these categories reveal that lymphocyte population are very heterogeneous. The purpose of this section is to discuss the phylogenetic origins of this lymphocyte diversity.

The Genesis of B and T Lymphocytes. On the basis of their involvement in cell-mediated transplantation immunity in annelids, the lymphocyte-like cells which appear first in coelomates have been considered to represent primitive T cells (43). In the same class of animals, however, stimulation of coelomocytes by Con A and LPS showed that these cells exhibit functional similarities with both T and B mammalian lymphocytes (68). So far, studies in invertebrates other than annelids are still very preliminary and the full range of techniques for assaying lymphocytes has not yet been applied to the most interesting Echinodermata and Protochordata. We know, however, that these animals possess lymphocyte-like cells which participate in graft rejection (49) and that in prochordates these cells appear to respond to PHA (43). Hopefully, the coming years will see extensive research in this areas as well as toward the elucidation of the nature of receptors present at the surface of lymphocyte-like cells of invertebrates.

Heterogeneity of lymphocyte populations has not yet been extensively
studied in cyclostomes (Agnatha). Indeed it was only recently that the
existence of lymphocytes in hagfish was demonstrated (53). Despite the
absence of a thymus, some hagfish and lamprey lymphocytes express
functional properties of T cells such as PHA-induced blast transformation and a kind of MLR reactivity in lamprey and hagfish (reviewed
in 13). The same animals obviously possess B cell equivalents, since
they can make antibody (reviewed in 13). Given the very high percentage (about 90 %) of lymphocytes responding to PHA in the lamprey it
seems likely that there is an overlap between PHA-stimulable cells
and potential antibody producing B cells. If so it may be that the
primitive lymphocyte of these vertebrates lacks specialization and
express chracteristics of both B and T lymphocytes. In elasmobranchs,
responsiveness to T cell mitogens has been demonstrated (74) but no
MLR could be initiated. No information is available for B cell mitogens.
These vertebrates are the first to possess a thymus and a spleen, but
the occurrence of a few plasma cells in the thymus of sharks following
immunization (13) suggests that if a certain level of heterogeneity
exists, the division into functional compartments may not be complete.
Thymus lymphocytes of the skate, another elasmobranch differ from
other skate lymphocytes at the level of expression of surface Ig. Although the majority of thymus and spleen cells exhibit endogenous Ig,
it seems that thymus cells possess less Ig on their surface than other
Ig-positive lymphocytes (29).

The use of specific mitogens, surface Ig markers, thymectomy, and
hapten-carrier immunization protocols has yielded more information in
the more advanced Teleostei (bony fish). For instance, lymphocyte heterogeneity resembling that of mammals exists in the trout. Thymus
cells respond exclusively to T mitogens (Con A), lymphocytes in the
kidney exclusively to B mitogens (LPS) and spleen cells to both (31).
Peripheral blood leukocytes can be activated in MLR in the trout (31)
but apparently not in another teleost species, the snapper (13). Further
experiments demonstrating the existence of a carrier effect in anti-
hapten response of sea robins also suggest the existence of two populations of lymphocytes cooperating during the immune response (89).
Finally, removal of the thymus gland during early life of the mouth
breeder (71) impairs allograft and antibody response. All lymphocytes
are stainable by fluorescent anti-Ig treatment in the trout (31), and
in the carp about 60-70 % of thymus cells can be stained (30). This
again suggests that at this level of the phylogenesis, the mammalian
distinction between B and T cells is not very sharp. Indeed the detection of surface Ig by immunofluorescence is considered as a criterion
for identifying B cells in warm-blooded vertebrates.

There are several reasons to believe that real functional B-T heterogeneity is achieved at the level of the anuran amphibian. In vitro
responses of their lymphocytes to T and B mitogens are good, even by
mammalian standards (37). Early thymectomy either abolishes or decreases
the allograft rejection capacity, as well as MLR and PHA responsiveness.
It also abolishes IgG-like synthesis and antibody responses to thymus-
dependent antigen (56) or to haptens in a hapten carrier immunization
(Du Pasquier and Horton, unpublished). However it does not abolish in
vivo (56) or in vitro (M.J. Manning and N. Donnelly, pers. comm.)
responses to thymus-independent antigens and mitogens. Moreover, heterogeneity can now be detected by immunofluorescence techniques. Although
larval anurans express easily detectable endogenous Ig on the surface
of about 90 % of their thymus cells, the amount of this Ig is distin-
guishably less than that found on the Ig positive cells of the spleen.
This organ, during development, always contains Ig negative lymphocytes. The difference between T and B populations becomes more apparent after metamorphosis, since Ig molecules are difficult to detect

(but they exist) on the surface of thymocytes from adults. Thymectomy
removes the population of weakly and apparently Ig-negative lymphocytes from the spleen (21, 47, 88). The existence of subpopulations
of T cells or of a sequential maturation of various T cell functions
is suggested, if not proven, by ontogenetic studies of MLR reactivity
and antibody production and by the effect of thymectomies performed at
various developmental stages (44, 56). First, MLR reactivity matures
before another thymus-dependent function, the ability to mount good
IgG synthesis in primary antibody responses (21, 28). In the intact
Xenopus or Rana catesbeiana this occurs only after metamorphosis, although the capacity of IgG-like synthesis does exist in larvae as evidenced by its presence in the serum of tadpole of several species (34,
23). Second, thymectomy at 7 days of age delays allograft rejection
and abrogates antibody response to thymus-dependent antigens whereas,
when it is effected later in larval life, the same operation abrogates
only the antibody response to at least some thymus-dependent antigens
(44, 56). This late operation apparently prevents the differentiation
of the T helper cell population. Conclusive demonstration of a existence of a T helper cell must await the results of reconstitution experiments in thymectomized individuals. Such experiments are now possible in amphibians because of the availability of histocompatible
strains (6, 19) and of large clones of genetically identical animals
(52).

Thus amphibians represent a key step in the achievement of T and B cell
heterogeneity. They are rendered even more interesting because one order, the Urodela (salamanders) is, immunologically speaking, profoundly
different from the anurans (frogs). In urodeles, thymocytes express
easily detectable Ig even at the adult stage (7). The response of their
lymphocytes to B mitogens is excellent, whereas that to T cell mitogens
is quite poor. The MLR response is also poor, and finally their antibodies are exclusively of the IgM class (14, 60, 45). Thymectomy abrogates allograft rejection (9, 81) but does not seem to sharply affect
serum antibody synthesis to any antigen (80). Indirect evidence for
cell cooperation in the newt (69) has also been theoretically explained
and interpretated without invoking the existence of a population of
helper T cells (71). Immunologically, urodeles seem more closely related to fish than to anurans, in that their lymphocytes seem to have
retained more B cell characteristics (as evidenced by surface Ig) than
those of anurans. So far all these data are compatible with the hypothesis that the primordial lymphocyte displays a composite of B and T
cell characteristics rather than T cell characteristics alone. This
conclusion is further supported by experiments performed in frog embryos. The replacement of thymus anlage of a young (tailbud) embryo
by isochronic thymus anlage of a donor animal bearing a cellular marker
results in an individual whose leukocytes are almost all of donor origin. This finding implies that both B and T cells can differentiate
from progeny of the thymus primordium (82). More refined experiments
have recently proved that donor-peripheralized lymphoid cells originated only from the thymus and their precursors were located in the
endoderm of the anlage (83).

Little information on heterogeneity of lymphocytes in reptiles is
available. That T-B heterogeneity exists is suggested by experiments
involving the use of anti-thymocyte antiserum, which resulted in alteration of some T cell-dependent functions (55, 67). Even though much
more is known about lymphocyte heterogeneity in birds and mammals, it
does not seem that a particularly significant step has been achieved
since anuran amphibians, except that a difference may exist at the
level of the T cell receptor. Although the nature of this receptor has
not yet been conclusively elucidated in mammals, phylogenetic data

argue that this receptor is either an Ig, a fragment of Ig, or is
strongly homologous with Ig.

Diversity at the B Cell Level = Antibody Diversity. Diversity of B
lymphocytes is reflected by the diversity of their gene products, Ig
of different specificities. In this respect one may wonder whether
primitive fish can make as many antibodies specificities as man. Immune response of higher vertebrates and shark do not seem to differ
markedly from the point of view of specificity and heterogeneity of
the antibody response (8). In other words, the "library" of shark an-
ti-carbohydrate antibodies is large, and the response can be very heterogenous in individual elasmobranchs. In fact the ability to mount a
restricted or a heterogenous response to one antigen may be encountered
within each class of vertebrate. For instance, in the mouse dinitrophe-
nol elicits very heterogenous responses (66) whereas in the guinea pig
it does not (40). Similarly, the same antigen promotes a very heterogeneous response in the bullfrog, even during larval stages, but not in
Xenopus (26). Heterogeneity does not depend on the number of cells,
since bullfrog tadpoles with only about 2-5×10^6 lymphocytes give a heterogenous response, while adult Xenopus which give more restricted response have about 1×10^8 lymphocytes. As far as the origin of antibody
diversity is concerned, DNA-RNA hybridization studies in the mouse (79)
tend to prove that the initial number of germ line V genes in mammals
is small, and would permit perhaps 10,000 antibody specificities. Provided that 10^4 different antibodies are not sufficient to explain the
diversity of mammalian immune response, this suggests that a somatic
process takes place to increase antibody diversity. However, data obtained in isogenetic amphibians (Xenopus) showed that these animals can
give identical antibody responses (i.e. identical isoelectric focusing
patterns) to the same antigen (26). This finding can easily be recon-
ciled with the possibility that at least lower vertebrates essentially
exploit their germ line genes, and that they have enough of them to
deal with most of the antigens present in the environment.

B cell heterogeneity can also be revealed by studies of the maturation
of the immune response (i.e. increase in binding affinity), which is
generally believed to represent the sequential expression of various
B cell clones (76). It is usually thought that IgM and IgG antibody
response are under the control of different regulatory mechanisms,
since the affinity of IgM shows little or no increase, whereas that
of IgG increases considerably. This phenomenon holds true in lower
vertebrates (1) although modest but significant (up to 10-fold) increases in IgM antibody affinity have now been reported in lower vertebrates (23) as well as in mammals (87). Thus, there do not seem to
be major differences from this point of view throughout the vertebrates.

Another aspect of B cell heterogeneity is the IgM-IgG switch observed
in mammals. In lower vertebrates, although IgG equivalent antibodies
appear late in primary responses and early in secondary or anamnestic
response, IgM antibodies are always present or at least easily detectable (38). This implies some differences at the regulation level that
may be quite important. For instance, if the switch in mammals is accompanied by the deletion of the μ chain gene, this would mean that
such a mechanism either does not occur in lower vertebrates or that
it occurs only in some cells.

The Major Histocompatibility Complex (MHC)

In mammals, acute graft rejection, in vitro manifestations of graft
reaction, strong histocompatibility antigens, control of immune re-

sponse, and control of some complement components are under the control of a single cluster of genes called the major histocompatibility complex. The biological role of this region is not known, and understanding its evolution could help in understanding its function or at least its biological importance. There are at least three models which could explain the MHC architecture of mammals (24).

1. It may represent a "frozen accident", and be preserved as it is with minor modifications since it first appeared.

2. It may have evolved from one or a few ancestral genes by a process involving gene duplications and translocations. Gene duplication alone could be at the origin of the whole complex, and then its components should all show some degree of homology. On the other hand, gene duplication may also have operated on genes which were already clustered in a complex due to a favorable translocation. In this model of cumulative acquisition, the MHC would appear more and more sophisticated from phylum to phylum, order to order. Each new acquisition would be automatically introduced in the MHC of the new species.

3. It may represent an example of convergent evolution and have differentiated in all classes of vertebrates (by duplication?) from an ancestor gene maintained in primitive species of each class of vertebrate (12). The complexes evolving independently from that homologous ancestor would show a certain degree of homology because of similar selective pressure.

The presence or the absence of an MHC in any vertebrate or invertebrate species and its degree of complexity should influence our choice of the mechanisms by which it may have evolved. For instance, if the MHC architecture in fish, amphibians, reptiles, birds, and mammals is found to be the same then the concept of convergent evolution will be difficult to maintain, since the MHC is not a single structure, and therefore evolution of a multigenic complex involving many gene duplications is unlikely to produce an identical MHC in all species.

So far our actual knowledge of the phylogeny of the MHC is very limited (12). The only component of the MHC which can be studied from the genetic point of view in primitive invertebrates is the control of histocompatibility reactions in general. It is of interest that in Hydractinia, a species of diploblastic Metazoa (cnidarian), histocompatibility reactions are apparently under the control of several genes (46). Only one of them is predominant, so much so that early investigators only detected this one (41). In far more advanced invertebrates (tunicates, protochordates), the situation is very similar: one major locus seems to control histocompatibility reactions (62). These loci have some characteristics of operationally weak H loci of higher vertebrates. Indeed, the apparently surprizing segregation pattern of rejection in F_1 sibs (25 % rejection, 75 % tolerance) found in these two phyla very much resembles the situation occurring in vertebrates in a species possessing the equivalent of an MHC (the South African frog), (24) when it is naturally immunosuppressed (i.e. during metamorphosis) (25). Thus the genetic control of histocompatibility reactions in invertebrates and vertebrates may not be fundamentally different, and the major histocompatibility locus (not complex) of Hydractinia may have served as the homologous ancestor gene to all MHC (22, 84).

Primitive fish do not seem to have developed easily detectable MHC components, and very little or nothing is known about the genetic control of graft rejection, C' and immune response genes in these animals. They lack the capacity to reject grafts acutely and the reactivity of

their lymphocytes in MLR is at best on the border-line of detection
(12). Although at least some teleost fish show MLR, and all of them
exhibit acute graft rejection, no data on the genetic linkage of these
two phenomena are available. It is only known that allograft rejection
is under the control of multiple H-loci (48).

Urodele amphibians seem to be more related to primitive fish because
of their behavior in graft rejection. They reject graft chronically
and show poor MLR (11, 14). In phylogeny, the occurrence of an order
of vertebrate showing chronic graft reaction and minimal MLR after
another order of vertebrates which exhibits some MHC components, certainly suggests independent acquisition of MHC components (12). However we may miss the MHC organization in urodeles because of secondary
effects (e.g. the lack of an MHC independent gene permitting development of a proper T cell function). Unlike urodeles, anurans do possess
an MHC homologue, since there exists a single genetic region which controls acute graft rejection, MLR, and some red blood cell antigens (24).
Nothing is known about the state of duplication of this complex nor
about linkage with immune response genes and complement components.
One should not forget that the three functional markers used to characterize MHC in the South African frog may correspond to fewer than
three genetic loci.

Unfortunately we still lack information about an MHC in reptiles, except for the fact that they show only chronic graft rejection and acute
as well as chronic graft versus host reactions (2).

The next (phylogenetically) MHC homologue is found in birds, where Ir
genes, strong histocompatibility antigens, MLR, acute graft rejection
and some blood group antigens are under the control of a single genetic
region, which does not yet seem to represent a duplicated locus (for
references, 90). The maximal complexity of the MHC architecture seems
to be the province of mammals (for references, 24, 51) (duplicated
strong histocompatibility antigens, MLR (I) locus, Ir genes, complement
components (structural or regulatory genes)). It should be stressed,
however, that the simplicity of non-mammalian vertebrates' MHC may
simply reflect the lack of extensive investigation in the field. The
small amount of available data only allows us to say that the basic
architecture of the MHC is not restricted to warm-blooded vertebrates
and that it was probably constituted before the emergence of anuran
amphibians. Regardless of its mode of appearance, the very presence
of an MHC argues that it is necessary or useful to the success of the
vertebrates. One may hypothesize that further studies, especially in
those species showing MLR and acute graft rejection will show a genetic
linkage of these two functions and that these linked loci will be the
embryo of the MHC in all species. One also waits with interest for
studies dealing with MHC and genetic control of complement activity
in lower vertebrates.

The Use of Natural and Laboratory-Created Polyploid Individuals in
the Study of MHC Evolution

In order better to understand the evolution of the MHC, it appeared
worthwhile to study it in animals where polyploidization has occurred
spontaneously, because of the possible implication of tetraploidization
in gene duplication (64). Such species exist in the Amphibia, especially
in Xenopus. In this genus it is possible to create artificial polyploids in the laboratory and to study evolution in a more active way
than by comparing immune parameters in living species.

Several species of Xenopus showing polyploidization relationships
occur naturally (H.R. Kobel, M. Fischberg, manuscript in preparation):
X. tropicalis (2n = 20 chromosomes), X. laevis (2n = 36), X. vestitus
(2n = 72), X. ruwenzoriensis (2n = 108). The DNA content roughly correlates with chromosome numbers (C. Thibaut, personal communication).
This polyploidy is thought to be an allopolyploidy. Moreover, crosses
between X. laevis and other species like X. gilli (2n = 36) produce
female hybrids that are fertile, but which lay diploid eggs. When these
eggs are fertilized by normal haploid spermatozoa from a male of a
third species, they will develop into triploid trispecies hybrids (52).
The genetic control of several component of the MHC has been studied in
both natural and artificial polyploids (Du Pasquier, Kobel, Fischberg
and Miggiano, in prep.) in order to answer the following question: is
a single MHC detected in polyploid animals, and if so how did it come
about? X. laevis possesses about twice as many chromosomes as X. tropicalis and already represents a polyploid species. However, it expresses
only 1 MHC, as evidenced by segregation in F_1 sibs of patterns of acute
graft rejection, MLR, and some red cell antigens (24). Similarly, the
natural tetraploid species X. vestitus only expresses a single MLR
locus, as does the hexaploid X. ruwenzoriensis when it is assayed by
conventional MLR techniques. However, refinement in MLR methodology
which increased the discrimination capacity of the assay revealed that
even though one could not detect more than one locus in X. laevis and
X. vestitus, more than one was expressed in X. ruwenzoriensis. However,
the necessity of using an improved technique to detect the other MLR
loci indicates that one locus was preferentially expressed in this
species. In summary, in most of natural polyploid species studied only
one MHC is expressed.

The situation in laboratory-created triploids is different. Specifically,
in two types of hybrid, laevis-gilli-clivii (LGC) and laevis-gilli-
laevis (LGL), it was obvious from the following data that all the con-
stituting MLR and strong histocompatibility antigen haplotypes were
codominantly expressed. All the eggs of an LG diploid hybrid female
are genetically identical and can be either activated by irradiated
sperm to produce isogenetic animals, or fertilized by non-irradiated
sperm to produce triploids that share all the LG part of the genome
with the gynogens. Such diploid gynogens rapidly reject grafts from
LGC, whereas LGC permanently tolerate grafts from LG. The only explanation is that the C haplotype is expressed in LGC (otherwise LG would
not reject an LGC graft) and that all LG antigens are expressed in LGC
(otherwise C would not be tolerant of LG). The results of these grafting
experiments were confirmed by MLR.

Thus, the mechanism by which natural polyploid individuals express only
one MHC locus it not an instant one comparable to the lack of expression of the second X chromosome in mammals. This suggests that X.
ruwenzoriensis, which has not yet achieved diploidization of its MHC,
may be a relatively recent polyploid species.

One can imagine that reduction of the number of expressed loci (as detected by segregation patterns in F_1) could occur following unequal
crossing-over or a translocation which would place two MHC haplotypes
on the same chromosome. This mechanism could also explain the duplication of the locus. In this context it will be extremely interesting
to compare the polyploid species with X. tropicalis which has only 20
chromosomes.

This example gives an idea of what can be done with artificial and
natural polyploid individuals. Obviously their interest is not restricted to MHC, and one can think of numerous experiments concerning
the control of Ig synthesis in this system.

Summary and Conclusion

From this brief survey the following points are worth reamphasizing:

1. There may be molecules involved in the defense mechanisms of invertebrates such as Echinodermata that show homology with complement and immunoglobulins, as suggested by functional tests and by amino acid composition comparisons.

2. The lymphocyte of primitive animals has properties of both B and T mammalian lymphocytes.

3. A spectrum of antigens similar to that recognized by the mammalian immune system can be recognized by the immune system of small animals expressing no more than 10^4 antibody-forming clones.

4. The basic architecture of the mammalian major histocompatibility complex has been detected at the level of anuran amphibians. The expression of a single complex is not due to an instant regulation mechanism, as shown by studies on artificial and natural polyploid forms of anuran amphibians.

Acknowledgment. I thank Dr. N. Cohen for his discussions and help during the preparation of the manuscript.

References

1. Ambrosius, H., Fiebig, H.: In: L'Etude phylogénique et ontogénique de la réponse immunitaire et son apport a la théorie immunologique. Liacopoulos, P., Panijel, J. (eds.). Paris: Inserm 1973, pp.135-146.
2. Borysenko, M., Tulipan, P.: Transplantation 16, 496-504 (1973)
3. Carton, Y.: Ann. Biol. 12, 139-184 (1973)
4. Carton, Y.: Ann. Immunol. (Inst. Pasteur) 125C, 731-745 (1974)
5. Cathou, R.E., Holowka, D.A.: In: Advances in Experimental Medicine and Biology 64, Immunologic Phylogeny. Hildemann, W.H., Benedict, A.A. (eds.). New York, London: Plenum Press 1975, pp. 207-215
6. Charlemagne, J., Tournefier, A.: J. Immunogenetics 1, 125-129 (1974)
7. Charlemagne, J., Tournefier, A.: In: Advances in Experimental Medicine and Biology 64, Immunologic Phylogeny. Hildemann, W.H., Benedict, A.A. (eds.). New York, London: Plenum Press 1975, pp. 251-255
8. Clem, L.W., McLean, W.E., Shankey, V.: In: Advances in Experimental Medicine and Biology 64, Immunologic Phylogeny. Hildemann, H.H., Benedict, A.A. (eds.). New York, London: Plenum Press 1975, pp. 231-239
9. Cohen, N.: In: Biology of Amphibian Tumors. Mizell, M. (ed.). New York: Springer 1969, pp. 153-168
10. Cohen, N.: J. Am. Vet. Ass. 159, 1662-1671 (1971)
11. Cohen, N.: Am. Zool. 11, 193-205 (1971)
12. Cohen, N., Collins, N.: In: The Organization of the Major Histocompatibility Systems. Götze, D. (ed.). Springer, in press
13. Cohen, N.: In: The Lymphocyte: Structure and Function. Marchalonis, J.J. (ed.). New York: Marcel Dekker, in press
14. Collins, N.H., Manickavel, V., Cohen, N.: In: Advances in Experimental Medicine and Biology 64, Immunologic Phylogeny. Hildemann, H.H., Benedict, A.A. (eds.). New York, London: Plenum Press 1975, pp. 305-314

15. Cooper, E.L.: In: Cont. Topics Immunobiology. 2. Thymus Dependency. Davies, A.J.S., Carter, R.L. (eds.). New York, London: Plenum Press 1973, pp. 13-38
16. Cooper, E.L. (ed.): Cont. Topics Immunobiology. 4. Invertebrate Immunology. New York: Plenum Press 1974
17. Day, N.K.B., Gewurz, H., Johannsen, R., Finstad, J.: Good, R.A.: J. Exp. Med. 132, 941-950 (1970)
18. De Joannes, A.E., Hildemann, W.H.: In: Advances in Experimental Medicine in Biology 64, Immunologic Phylogeny. Hildemann, W.H., Benedict, A.A. (eds.). New York, London: Plenum Press 1975, pp. 151-160
19. Delanney, L.E., Collins, N.H., Cohen, N., Reid, R.: In: Advances in Experimental Medicine and Biology 64, Immunologic Phylogeny. Hildemann, W.H., Benedict, A.A. (eds.). New York, London: Plenum Press 1975, pp. 315-324
20. Du Pasquier, L.: Curr. Top. Microb. Immunol. 61, 37-88 (1973)
21. Du Pasquier, L., Weiss, N.: Europ. J. Immunol. 3, 773-777 (1973)
22. Du Pasquier, L.: Arch. Biol. (Brussels) 85, 91-103 (1974)
23. Du Pasquier, L., Haimovich, J.: Europ. J. Immunol. 4, 580-583 (1974)
24. Du Pasquier, L., Chardonnens, X., Miggiano, V.C.: Immunogenetics 1, 482-494 (1975)
25. Du Pasquier, L., Chardonnens, X.: Immunogenetics 2, 431-440 (1975)
26. Du Pasquier, L., Wabl, M.R.: In: The Generation of Antibody Diversity: A New Look. Cunningham, A.J. (ed.). New York: Academic Press 1976, pp. 151-164
27. Du Pasquier, L., Horton, J.D.: Immunogenetics 3, 105-112 (1976)
28. Du Pasquier, L., Haimovich, J.: Immunogenetics, in press (1976)
29. Ellis, A.E., Parkhouse, R.M.E.: Europ. J. Immunol. 5, 726-728 (1975)
30. Emmrich, F., Richter, R.F., Abrosios, H.: Europ. J. Immunol. 5, 76-78 (1975)
31. Etlinger, H.M.: Function and structure of rainbow trout leukocytes. Doctoral thesis, Univ. Washington, Seattle (1975)
32. Ferreira, A., Nussensweig, V.: J. Exp. Med. 141, 513-517 (1975)
33. Fu, S.M., Kunkel, H.G., Brusman, H.P., Allen, F.H.Jr., Fotino, M.: J. Exp. Med. 140, 1108-1111 (1974)
34. Geczy, C.L., Green, P.C., Steiner, L.A.: J. Immunol. 111, 1261-1267 (1973)
35. Gigli, I., Austen, K.T.: Ann. Rev. Microbiol. 25, 309-332 (1971)
36. Goldstein, L.: Transpl. Proc. 2, 191-193 (1970)
37. Goldstine, S.N., Collins, N.H., Cohen, N.: In: Advances in Experimental Medicine and Biology 64, Immunologic Phylogeny. Hildemann, W.H., Benedict, A.A. (eds.). New York, London: Plenum Press 1975, pp. 343-351
38. Hadji-Azimi, I.: Immunology 21, 463-474 (1971)
39. Hadji-Azimi, I., Michea-Hamzehpour, M.: Immunology 30, 587-592 (1976)
40. Haimovich, J., Jaton, J-C., Pink, J.R.L.: Europ. J. Immunol. 4, 290-295 (1974)
41. Hauenschild, C.: Roux Arch. Entwicklungsmechanik 147, 1-41 (1954)
42. Hildemann, W.H., Reddy, A.L.: Federation Proc. 32, 2188-2194 (1973)
43. Hildemann, W.H.: Nature (Lond.) 250, 116-120 (1974)
44. Horton, J.D., Rimmer, J.J., Horton, T.L.: J. Exp. Zool., in press
45. Houdayer, M., Fougereau, M.: Ann. Inst. Pasteur. 123, 3-28 (1972)
46. Ivker, F.B.: Biol. Bull. 143, 162-174 (1972)
47. Jurd, R.D., Stevenson, G.T.: Comp. Biochem. Physiol. 53A, 381-387 (1976)
48. Kallmann, K.D.: Transpl. Proc. 2, 263-271 (1970)
49. Karp, R.D., Hildemann, W.H.: In: Advances in Experimental Medicine and Biology 64, Immunologic Phylogeny. Hildemann, W.H., Benedict, A.A. (eds.). New York, London: Plenum Press 1975, pp. 137-147

50. Kehoe, J.M.: In: Advances in Experimental Medicine and Biology 64, Immunologic Phylogeny. Hildemann, W.H., Benedict, A.A. (eds.). New York, London: Plenum Press 1975, pp. 197-205
51. Klein, J.: In: Advances in Experimental Medicine and Biology 64, Immunologic Phylogeny. Hildemann, W.H., Benedict, A.A. (eds.). New York, London: Plenum Press 1975, pp. 467-478
52. Kobel, H.R., Du Pasquier, L.: Immunogenetics 2, 87-91 (1975)
53. Linthicum, D.S.: In: Advances in Experimental Medicine 64, Immunologic Phylogeny. Hildemann, W.H., Benedict, A.A. (eds.). New York, London: Plenum Press 1975, pp. 241-255
54. Litman, G.W.: In: Advances in Experimental Medicine and Biology 64, Immunologic Phylogeny. Hildemann, W.H., Benedict, A.A. (eds.). New York, London: Plenum Press 1975, pp. 217-228
55. Manickavel, V.: Studies on skin transplantation immunity in the lizard Calotes versicolor (Daud). Doctoral Thesis, Annamalai Univ., India (1972)
56. Manning, M.J., Collie, M.H.: In: Advances in Experimental Medicine and Biology 64, Immunologic Phylogeny. Hildemann, W.H., Benedict, A.A. (eds.). New York, London: Plenum Press 1975, pp. 353-362
57. Maramorosch, K., Shope, R.E. (eds.): Invertebrate Immunity. New York, London: Academic Press 1975
58. Marchalonis, J.J., Weltmann, J.K.: Comp. Biochem. Physiol. 38B, 609-625 (1971)
59. Marchalonis, J.J., Atwell, J.L.: In: Phylogenic and Ontogenic Study of the Immune Response and its Contribution to the Immunological Theory. Liacopoulos, P., Panijel, J. (eds.). Paris: Inserm 1973, pp. 153-174
60. Marchalonis, J.J., Cohen, N.: Immunology 24, 395-407 (1973)
61. Marchalonis, J.J., Cone, R.E.: Australian J. Exp. Biol. Med. Sci. 51, 461-488 (1973)
62. Oka, H., Watanabe, H.: Proc. Japan. Acad. 33, 657-659 (1957)
63. Ohno, S.: Evolution by Gene Duplication. Berlin, Heidelberg, New York: Springer 1970
64. Parkhouse, R.M.E., Askonas, B.A., Doumashkin, R.R.: Immunology 18, 575-584 (1970)
65. Pernis, B., Brouet, J.C., Seligman, M.: Europ. J. Immunol. 4, 776-778 (1974)
66. Pink, J.R.L., Askonas, B.A.: Europ. J. Immunol. 4, 426-430 (1974)
67. Pitchappan, R.M.: Studies on the development and immune functions of the thymus in the lizard Calotes versicolor, Doctoral thesis, Madurai Univ., India (1975)
68. Roch, P.G.: Thesis 3-Cycle E.S. 1077, Univ. Bordeaux (1973)
69. Ruben, L.N., Hoven, A., Dutton, D.W.: Cell. Immunol. 6, 300-314 (1973)
70. Ruben, L.N., Selker, E.V.: In: Advances in Experimental Medicine and Biology 64, Immunologic Phylogeny. Hildemann, W.H., Benedict, A.A. (eds.). New York, London: Plenum Press 1975, pp. 387-395
71. Sailendri, K.: Studies on the development of lymphoid organs and immune responses in teleost Tilapia mossambica (Peters). Doctoral thesis, Madurai Univ., India (1973)
72. Saluk, P.H., Drauss, J., Clem, L.W.: Proc. Soc. Exp. Biol. Med. 133, 365-369 (1970)
73. Shen, T.W., Boyse, E.A., Cantor, H.: Immunogenetics 2, 591-596 (1975)
74. Sigel, M.M., Ortiz-Muniz, G., Lee, L.C., Lopez, D.M.: In: L'Etude phylogénique et ontogénique de la réponse immunitaire et son apport à la théorie immunologique. Liacopoulos, P., Panijel, J. (eds.). Inserm. Paris: 1973, pp. 113-119
75. Silver, J., Hood, L.: Proc. Nat. Acad. Sci. 73, 599-603 (1976)
76. Siskind, G.W., Benacerraf, B.: Advan. Immunol. 10, 1-50 (1969)

77. Steiner, L.A., Mikoryak, C.A., Lopes, A.D., Green, C.: In: Advances in Experimental Medicine and Biology 64, Immunologic Phylogeny. Hildemann, W.H., Benedict, A.A. (eds.). New York, London: Plenum Press 1975, pp. 173-183
78. Terhost, C., Parham, P., Mann, D.L., Strominger, J.C.: Proc. Nat. Acad. Sci. 73, 910-914 (1976)
79. Tonegawa, S., Steinberg, C., Dube, S., Bernardini, A.: Proc. Nat. Acad. Sci. 71, 4027-4031 (1974)
80. Tournefier, A., Charlemagne, J.: In: Advances in Experimental Medicine and Biology 64, Immunologic Phylogeny. Hildemann, W.H., Benedict, A.A. (eds.). New York, London: Plenum Press 1975, pp. 161-171
81. Tournefier, A.: In: L'Etude phylogénique et ontogénique de la réponse immunitaire et son apport à la théorie immunologique. Liacopoulos, P., Panijel, J. (eds.). Paris: Inserm 1973, pp. 105-112
82. Turpen, J.B., Volpe, E.P., Cohen, N.: Science 182, 931-933 (1973)
83. Turpen, J.B., Cohen, N.: Cell. Immunol. 1976, in press
84. Urbain, J.: Arch. Biol. (Brussels) 85, 139-150 (1974)
85. Vaerman, J.P., Picard, J., Heremans, J.T.: In: Advances in Experimental Medicine and Biology 64, Immunologic Phylogeny. Hildemann, W.H., Benedict, A.A. (eds.). New York, London: Plenum Press 1975, pp. 185-195
86. Wolski, K.P., Schmid, F.R., Mittal, K.K.: Tissue Antigens 7, 35-38 (1976)
87. Wu, C.Y., Cinader, B.: Europ. J. Immunol. 2, 398-405 (1972)
88. Weiss, N., Horton, J.D., Du Pasquier, L.: In: L'Etude phylogénique et ontogénique de la réponse immunitaire et son apport à la théorie immunologique. Inserm. Liacopoulos, P., Panijel, J. (eds.). Paris: 1973, pp. 165-174
89. Yocum, D., Cucjens, M., Clem, L.W.: J. Immunol. 114, 925-927 (1975)
90. Ziegler, A., Pink, J.R.L.: J. Biochem. 251, 5391-5396 (1976)

Characteristics of B Lymphocytes and Their Mechanism of Activation

G. Möller

Introduction

The function of the lymphoid system is to defend the individual against otherwise lethal microbial infections. Although the important cells in the system — the small lymphocytes — are all morphologically similar, the immune system consists of two main cell types, which share few, if any, characteristics. One cell type — the thymus-derived (T) lympho-cyte — is responsible for the defence against viral, parasitic and other nonbacterial infections. The properties of this cell are not the topic of this paper. The other cell is generally referred to as the B (bone marrow derived) lymphocyte and is responsible for the defence against bacterial infections and is the only cell which can synthesize immunoglobulins. The B lymphocytes develop from stem cells in the bone marrow and — at least in birds — they have to pass the Bursa of Fabri-cius in oder to differentiate into immunocompetent cells. Presumably there is an analogous organ in mammals, although it is not morphologi-cally defined, but it is not unlikely that the entire gastrointestinal tract fulfils this function.

Characteristics of B Cells

The function of B cells is to recognize antigens and as a consequence of this develop into antibody secreting cells, which release antibodies at a high rate. These antibodies are thereafter distributed in the en-tire organism as an efficient defence system against bacterial infec-tions. The B lymphocytes do not reciroulate extensively, but rather export their product, in contrast to T cells, which continuously re-circulate from the lymphoid organ via the lymph into the blood stream. Consequently the lymph and the blood are predominantly made up of T cells, whereas approximately half of the cells in the spleen are B cells.

The B lymphocytes recognize antigens by membrane receptors which are identical in structure to the antibodies released from the cells after activation (see Transpl. Rev. 5 (1970) for review). These receptors are anchored at the membrane and have a very slow turnover rate. The immunoglobulin (Ig) receptor is the main characteristic of B cells and distinguishes them sharply from T cells, which completely lack surface Ig receptors (see Transpl. Rev. 14 for review). The obvious function of the Ig receptors is to bind antigen, but the subsequent steps lead-ing to activation are not equally obvious, as will be discussed later. Each B lymphocyte only expresses one type of antibody with regard to antigen binding capacity or, in other words, the variable regions of all receptors on one B cell are identical, but different from those present on other B lymphocytes. Since the B lymphocytes are competent to recognize a wide variety of antigens, it follows that one particular B cell recognizing an individual antigenic determinant must be rare.

Actually, the estimates of the frequency of precursor cells for different antigens have given figures of 1 cell per 10^6 to 10^5 lymphocytes.
Since each B cell has genes coding for immunoglobulin molecules on
both its chromosomes, but only one chromosome is expressed, it follows
that a mechanism exists to prevent the expression of the allelic genes
(allelic exclusion). The mechanism by which allelic exclusion operates
is not known as yet. Subpopulations of B lymphocytes possess other
characteristics that can be easily observed (see Transpl. Rev. 16, 17,
24, 25, 26 for review). Some cells have receptors for the Fc part of
the IgG molecules, as well as for various complement components. The
function of these receptors is not well understood. The B cells also
express various differentiation antigens, which can be used to enumerate
them or to distinguish B cell subclasses.

Response to Antigen
===================

Injection of an antigen into an animal promptly results in the secretion of specific antibodies, first of IgM class, later of IgG and other
immunoglobulin classes. It was realized some years ago that two types
of antigens exist. Thus certain antigens were capable of inducing an
immune response in animals (or in vitro) lacking T cells and were therefore termed thymus-independent (TI) antigens. Obviously, other types
of antigens failed to do so, and these antigens were therefore termed
thymus-dependent (TD) antigens. Such antigens can only induce a specific antibody response if both B cells, T cells and macrophages are present (see Transpl. Rev. 1 for review). This need for cellular interaction for the induction of a specific immune response immediately
suggests that the activation mechanism of B lymphocytes must involve
something other than the interaction between Ig receptors and the antigen. Actually the existence of TD antigens strongly argues that by itself the binding of antigens to B cells cannot deliver triggering signals to B cells.

Conceptual Problems of B Cell Activation
==

A major problem in immunobiology is the mechanism of B cell activation
(Transpl. Rev. 23). There are two conceptually important considerations
in this field. The first one is that the B cells are not activated by
binding TD antigens to their surface Ig receptors (in the absence of T
cells), whereas they are activated when TI antigens bind to the same
receptors. The second is that the organism is competent to react against
all foreign material, but normally does not react against antigens
within the organism. This self non-self discrimination is a fundamental
property of the immune system and it necessarily must be explained at
the cellular level. One explanation is that binding of antigen can
either activate or inactivate (tolerize) the same cell (see Transpl.
Rev. 23 for discussion). This is the explanation offerend in the two
signal theory of lymphocyte activation, which states that when antigen
interacts with the Ig receptors the cell receives a tolerogenic signal,
which irreversibly leads to a loss of ability to synthesize antibodies
to that antigen, presumably by killing the cell (Bretscher and Cohn,
1970). Activation requires a second signal, which is either delivered
by cooperating cells, such as T cells in the case of TD antigens, or
by properties intrinsic to TI antigens, e.g. TI antigens would be themselves provide the second signal. The hypothesis also states that the

second signal by itself cannot activate the cells, as a safeguard
against the risk of developing autoantibodies.

An alternative view on the self/non-self discrimination is that this
function is only carried out by T cells (Coutinho, 1975; Möller, 1975).
Thus, T cells would become tolerant to self antigens (by as yet unknown mechanisms), whereas the B cells do not have this capacity. Since
self antigens are thymus-dependent, this does not result in any major
danger for the formation of autoantibodies. Even if such antibodies
can be formed, which will be shown below, this is probably of minor
importance, since it is self reacting T cells with their killing capacity that constitute the major danger to the well-being of the individual. What was outlined above constitutes the problems in the area.
Now I will proceed to summarize the experimental evidence for one particular and general mechanism of B lymphocyte activation.

Polyclonal B Cell Activators

It was discovered some years ago that certain substances were competent
by themselves to induce antibody synthesis in resting B lymphocytes
(Andersson et al., 1972). However, these substances did not act like
antigens but had the capacity to induce antibody synthesis in B cells
of any immunological specificity. Therefore they were termed polyclonal
B cell activators (PBA) (Coutinho et al., 1974b). The list of PBA substances is now very long and includes various bacterial polysaccharides,
such as lipopolysaccharides from gram-negative bacteria (LPS), pneumococcal polysaccharides, purified protein derivative of tuberculin (PPD),
whole bacteria of a variety of origins, mycoplasma organisms and a
variety of other bacterial products, but also small molecular weight
synthetic substances, such as polyenes. Most PBA known are polysaccharides, but this is not a necessary characteristic. They are quite
often of high molecular weight, but this is also not prerequisite,
since very small MW substances (800-3000) may have PBA properties (for
review see Coutinho and Möller, 1975).

PBA will activate B lymphocytes to perform the function for which they
are genetically programmed and actually can perform according to their
state of differentiation. Thus, a PBA will reveal not only the genetic
repertoire of the lymphocytes, but also their state of differentiation.
The response of a particular cell may be only induction of antibody
synthesis with little or no DNA synthesis, whereas another cell will
only be activated into increased DNA synthesis. Most cells express both
parameters after activation. The main conceptually important conclusion
from these findings is that substances that do not bind to the variable
region of the Ig receptors have the ability to initiate antibody synthesis in these cells. Apparently the Ig receptors are not necessary
for triggering, a conclusion that could also have been reached by the
known inability of TD antigens to activate B cells after binding to
the Ig receptors. However, the latter finding was previously interpreted as causing a tolerogenic signal to the B cells, whereas a second
signal was needed for activation to occur. Since activation can occur
without any recognition by the Ig receptors, it is obvious that at
least the variable regions of the Ig receptors are not necessary for
triggering to occur.

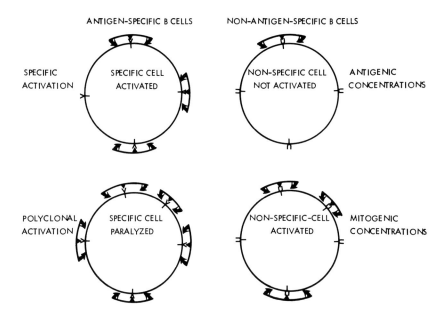

Fig. 1. Schematic outline of one nonspecific signal hypothesis showing focusing function of specific Ig receptors. Thymus-independent antigens display both antigenic determinants (<u>triangles</u>) and mitogenic properties (<u>arrows</u>)

TI Antigens are PBA

Since PBA could activate B cells without any participation of the anti-gen combining sites of the Ig receptors it seemed possible that the role of the Ig receptors was only to bind antigen to the correct cells, whereas the triggering signal would be delivered by other structures on the antigen molecule. Since TD antigens cannot activate lymphocytes after binding to the Ig receptors, these antigens would not possess triggering properties. In contrast, TI antigens are competent to di-rectly activate specific lymphocytes into antibody synthesis. Conse-quently, it seemed likely that all TI antigens were at the same time polyclonal B cell activators. Direct experiments confirmed this notion, and today all TI antigens have been found to be PBA (Coutinho and Möller, 1973a). Thus, PBA and TI antigens share the capacity of being directly competent to activate B cells, without involvement of various accessory cells, the difference being that TI antigens induce a spe-cific immune response, whereas PBA activate B cells of all Ig specifi-cities. This difference was ascribed to specific function of the Ig receptors as outlined in Figure 1 (Coutinho et al., 1974a). Thus, at low TI antigen concentrations the specific B cells would selectively concentrate the antigen to the correct cells, which would be activated by the nonspecific signal intrinsic to the TI antigens. At these low TI antigen concentrations other B cells would not bind sufficient sub-stance to their membranes to become activated. However, with increasing TI antigen concentrations the nonspecific cells also would accumulate sufficient material to be activated. However, the activation is now polyclonal, whereas the specific cells would be turned off by super-optimal concentrations of the inducer. This one nonspecific signal hypothesis (Coutinho and Möller, 1974) thus ascribes a totally passive

120

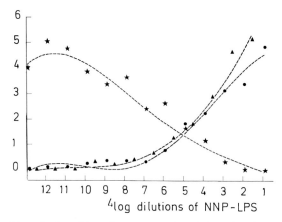

Fig. 2. Results of one experiment testing hypothesis outlined in Fig. 1. Normal
mouse spleen cells were cultured for 2 days in presence of different concentrations
of NNP-LPS in serum-free medium. Activation of DNA synthesis, as well as induction
of antibody production against an irrelevant antigen (SRC), was determined. In addition, high affinity antibody synthesis against the specific hapten NNP was measured. ▲———▲ = cpm/culture (Y:0 to 84,000); ●———● = anti-SRC PFC/culture (Y:0
to 300); ✱———✱ = high-avidity anti-NNP PFC/culture (Y:0 to 1,500)

role to the Ig receptors, triggering being achieved nonspecifically
by properties intrinsic to TI antigens which are all PBA.

In this case of TD antigens it is postulated that T cells and macrophages will deliver the triggering signals to the responding B cells,
the Ig receptors again only serving to bridge the B cells to the T
cells.

This concept was tested experimentally by adding the hapten-PBA complex NNP-LPS to spleen cells in culture. As seen in Figure 2, low concentrations resulted in specific anti-NNP responses, whereas high concentrations caused polyclonal antibody synthesis in the total absence
of specific anti-NNP responses (Coutinho et al., 1974a).

It was also found that the addition of free hapten did not prevent induction of specific anti-NNP responses, but the dose response curve
was now shifted so that specific antibody synthesis occurred at polyclonal concentrations of the NNP-LPS complex, in agreement with the
passive focusing role of Ig receptors postulated in the hypothesis
(Coutinho et al., 1974a).

Evidence for the One Non-Specific Signal Concept

1. Polyclonal B cell activators can induce immune reactivity (DNA synthesis, antibody synthesis) in resting B cells, although there is no
complementarity between the Ig receptors and the PBA. Thus, induction
occurs in the total absence of any participation of the Ig receptors.
Free hapten or hapten TD conjugates do not prevent PBA induction of
polyclonal antibody synthesis, nor did they increase anti-hapten antibody synthesis (Coutinho et al., 1974a; Möller et al., 1975a).

2. Mouse strains that for genetic reasons lack the ability to respond to LPS by induction of polyclonal antibody synthesis also fail to mount a specific immune response to any antigenic determinant present on or conjugated to LPS (Coutinho et al., 1975b; Coutinho and Gronowicz, 1975; Skidmore et al., 1975; Watson and Riblet, 1974). However, the mice respond normally to the same determinants coupled to other PBA or on to DT antigens. Thus, the V gene repertoire is normal, but the ability to become activated by the PBA has been lost.

3. All TI antigens are PBA (Coutinho and Möller, 1973a, b). This intrinsic activating property of TI antigens is responsible for their capacity of directly activating B cells in the absence of helper cells, and is the only common property of TI antigens.

4. Haptens coupled to TD carriers or to Sepharose beads fail to induce antibody synthesis in vitro even though 10^9 fold variations in hapten concentrations were employed (Möller et al., 1975a). However, the same haptens coupled to LPS consistently induced antibody responses. Thus, the nature of the carrier (PBA or not) determined whether a response occurred, not the epitope density. The role of epitope density in facilitating binding to specific B cells and therefore cause focusing of hapten-PBA conjugates has been discussed elsewhere (Coutinho and Möller, 1975).

5. Since specific anti-hapten responses to hapten PBA conjugates are induced by signals given by the PBA property of the carrier, it is to be expected that the addition of unconjugated PBA would markedly influence the anti-hapten response. It was shown that suboptimal concentrations of LPS synergized with low concentrations of NNP-LPS in the induction of specific anti-NNP responses (Coutinho et al., 1975a, b). In contrast, hapten TD conjugates never caused a synergistic effect with LPS (Möller et al., 1975a).

6. Hapten-LPS conjugates are immunogenic only if LPS possesses PBA properties, but not if lipid A has been removed, which abolshes PBA activity, even though the epitope density was the same in both cases (Jacobs and Morrison, 1975). Similarly, the addition of polymyxins to LPS markedly decreases the PBA property of LPS and also suppresses the immunogenicity of hapten-LPS conjugates (Smith et al., 1975).

7. It should only be possible to interfere with specific TI and polyclonal responses by blocking the PBA receptor structure on the responding B cells, but not by blocking the Ig receptors. It was shown that haptenic PBA molecules, such as low MW dextran, suppressed specific anti-hapten responses as well as polyclonal antibody synthesis (Coutinho et al, 1974b). These substances also suppressed secondary anti-sheep red cell responses.

8. Hapten-LPS conjugates stimulate the synthesis of high affinity anti-hapten antibodies at low conjugate concentrations and polyclonal antibody synthesis (with the exception of anti-hapten antibodies) at high concentrations. The addition of free hapten to the system entirely changed the dose response curve and high affinity anti-hapten antibodies were now formed only by high conjugate concentrations (Couthino et al., 1974a) to the same extent as when similar concentrations of nonconjugated LPS are added. This demonstrates the passive focusing role of Ig receptors and the exclusive triggering role of the LPS carrier and is incompatible with two signal models postulating that both hapten and LPS (signal 1 + 2) are needed for induction of anti-hapten antibodies.

9. When cells responsive to a certain PBA are removed by a hot pulse of thymidine, no response to a hapten coupled to that PBA can be induced (Coutinho, unpublished). Thus, hapten specific cells are included in the PBA responsive subpopulation. Similarly, when hapten specific cells responding to immunogenic concentrations of hapten-PBA conjugates are removed by a hot pulse, specific anti-hapten responses cannot be induced by polyclonal doses of that PBA (Möller et al., 1976). However, these cells can respond with specific antibody synthesis to a different hapten coupled to the same PBA.

Subpopulation of B Cells

It has been found that each PBA acts on a partly overlapping subpopulation of B lymphocytes (Gronowicz and Coutinho, 1974; Gronowicz et al., 1974; Gronowicz and Coutinho, 1975a). Some act on a comparatively large population, such as LPS, and others on a much smaller population. Each target B cell subpopulation differs with regard to its stage of differentiation. Thus, dextran sulfate acts on immature B cells, LPS on more mature and PPD on the most differentiated cells, even though overlap is the rule and is sometimes considerable. In addition, each subpopulation is characterized by its own stereotype response after activation by the corresponding PBA, e.g. the response is determined by the differentiation stage of the target B cell. To date no PBA has been found to activate all B cells. Therefore, it is expected — and has actually been found — that only a proportion of the available B cells are activated by each PBA. This in no way reflects on the generality of the concept, which deals with the nonspecific and single signal responsible for activation. When comparisons are made between the strength of a certain TI antigen as regards immunogenicity and the intensity of the PBA property (excluding form the discussion most of these comparisons, which are made between induction of antibody synthesis on one hand and of DNA synthesis on the other) it is often found that immunogenicity is strong, but the PBA property weak.

This is to be expected form the concept. A particular subpopulation that is the target of a certain PBA has a random set of Ig variable regions of different B cells, or in other words, the distribution of antibody specificities within each subset is the same. However, the size of the subpopulation in terms of number of cells varies greatly between different subsets. Therefore, the strength of the PBA property will mainly depend on the number of cells within each subset, whether induction of polyclonal antibody synthesis or DNA synthesis is measured. However, immunogenicity will not vary to the same extent, because the same V gene repertoire is present in all subsets and the ultimate outcome of the immune response is the result of strong amplification forces, usually due to clonal proliferation. Since the PBA property is measured at day 2 or 3 and the immune response at later times; since the immune response has been amplified by clonal proliferation (but not the polyclonal response to the same extent); since the subpopulation for the polyclonal response may be small or large, whereas the targets for specific antigen will exist in any subset (small or large) and the quantitative differences in number will not be expressed because of the long and pronounced clonal amplification, it is expected that there should be no correlation between strength of immunicity and PBA property. But the triggering signal is still the same in both situations (for discussion see Möller et al., 1976).

Thus, even these simple considerations are sufficient for the conclusion that so many factors determine the strength of expression of PBA

property in specific versus polyclonal responses that a simple relationship is not to be expected.

Role of Passive Focusing Receptors

The Ig receptors are passive receptors that have an auxiliary role in
the immune response. They are efficient, however, because of the high
affinity of the interaction between the antigenic determinants and the
combining site of the Ig receptor. This is illustrated by the fact that
optimal polyclonal induction with e.g. LPS occurs at about 100 microgram/
ml, whereas optimal specific TI immune responses to a determinant present on LPS occurs at 10^3 to 10^9 lower concentrations depending on the
number of epitopes coupled to each LPS molecule. Still, it is the same
PBA property of LPS that induces the cell in both cases. Obviously,
binding of the PBA (TI antigen) to the PBA receptors triggers the cells,
whereas any binding to the cell of PBA does not trigger, as illustrated
by lymphocytes from the nonresponder strain C3H/HeJ, which can bind
LPS by the Ig receptors, but are not activated because the PBA receptors
do not function or are missing. It follows that any property of the PBA
that facilitates binding to a B cell with functional PBA receptors for
that particular PBA will increase the efficiency of activation of that
cell. The great efficiency of the Ig receptors in this passive binding
event has already been illustrated. However, other receptors can have
an analogous role. This is illustrated by the fact that PBA that has
fixed complement triggers B cells at lower concentrations that PBA without complement (Möller and Coutinho, 1975). Still, it has been clearly
shown that complement receptors by themselves cannot activate B cells
after interaction with complement (Möller and Coutinho, 1975). Therefore, it is the increased binding of the PBA-complement complex effectuated by the complement receptors on B cells that is responsible for
activation of lower PBA concentrations. The efficiency of the complement receptor binding is much smaller than that of the Ig receptors,
and can only decrease the optimal stimulating concentrations 10- to
100-fold.

Obviously any property of the target cells or of the PBA itself that
influences binding will have an effect on triggering. This effect will
be most pronounced for polyclonal induction, since the affinity of a
particular PBA for its corresponding triggering receptor is low. Thus
properties such as charge, lipid solubility, possibility for multiple
bond formation (high MW), presence of certain chemical groups increasing
binding to B cells, and other physical properties will influence polyclonal induction (Coutinho et al., 1974c; Coutinho and Möller, 1975).
However, these factors will not markedly affect the strength of the
specific TI immune responses, because the affinity of the Ig binding
for the antigenic determinants of the PBA is so large that any minor
factors that influence binding to the non-specific PBA receptors will
not show up. This illustrates again that a relationship should not exist
between the strength of polyclonal induction and immunogenicity.

Thymus-Independence is the Ultimate Test for PBA Properties of a Substance

Since the binding between the antigenic determinants and the Ig receptors is of such remarkable affinity, it follows that the Ig receptors
will always be competent to focus the antigen (PBA) to the specific cell

much more efficiently than any cell lacking these receptors. If the PBA for any number of reasons bind with very low affinity to the non-specific cells (lacking specific Ig receptors), it can become practically impossible to raise the PBA concentrations to such levels that triggering concentrations are reached at nonspecific cells. The specific cells will always have bound a sufficient number to become triggered. This concept can be easily illustrated. Suppose that a TI antigen is optimally effective at concentrations of 0.01 mg/ml when inducing specific TI immune responses. In order to demonstrate optimal polyclonal induction it may be necessary to raise the PBA concentrations 10^6 to 10^7-fold, or in other words to add 10,000 to 100,000 mg/ml. This cannot be done in vitro or in vivo. Other possibilities would be to utilize different auxiliary binding receptors such as the complement receptors, but because of the low affinity of this interaction it would still require the addition of about 100 to 1,000 mg/ml. It is difficult indeed to beat the Ig receptors when it comes to high affinity binding. This may be the reason for the reluctance of immunologists to accept that the Ig receptors only have a passive, non-triggering role.

In view of the above excercise it is to be expected that several TI antigens will be found that have no demonstrable PBA property. This is in accordance with postulates of the one nonspecific signal concept, and cannot be used as an argument against it. However, the reports published to date on this issue can be traced to poor methodology and with more refined techniques PBA activity has been demonstrated with all TI antigens so far tested, although sometimes the PBA properties have been weak.

Tolerance and the One Nonspecific Signal Concept

It follows from this concept that the immunoglobulin receptors cannot be responsible for the signal delivery in induction or paralysis. Since thymus-dependent (TD) antigens lack the inbuilt triggering property of thymus-independent antigens, TD antigens cannot directly activate B cells. Therefore, tolerance to a thymus-dependent antigen cannot exist at the B cell level, if tolerance is defined as a change from the resting state to a tolerant state as a consequence of a signal delivery. A formal proof of the existence of an induced state of tolerant B cells (or the absence of immunocompetent B cells) would be to activate B cells from tolerant animals by PBA and fail to detect induction of antibodies against the tolerogen. Only one such study has been performed as yet (Gronowicz and Coutinho, 1975b) and it was indeed found that cells from animals tolerized by injections of free hapten were activated by PBA to anti-hapten antibody synthesis.

We induced immunological tolerance in adult mice by the injection of 5 mg of deaggregated hapten-protein conjugate (Fig. 3). The tolerant state was confirmed 5 to 19 days later by the failure of such animals to mount an immune response against an aggregated form of the same thymus-dependent hapten-protein conjugates as well as by the inability of spleen cells from tolerant animals to respond to a thymus-independent hapten charrier conjugate. Even though the animals were fully tolerant, their spleen cells were activated by LPS in vitro to produce normal numbers of plaque-forming cells against the hapten (Fig. 4) (Möller et al., 1976).

The finding that spleen cells from tolerant animals could be activated by LPS into synthesis of antibodies against the tolerogen indicates

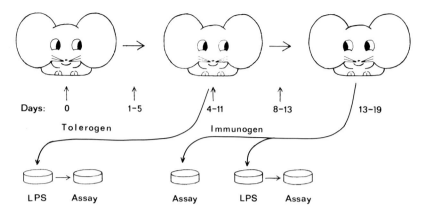

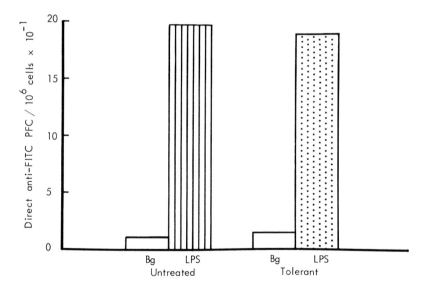

Fig. 3. Schematic outline of experimental procedure to reveal presence of immuno-competent cells in tolerant animals. (A x B10.5M)F_1 mice were injected twice intra-venously with deaggregated $FITC_{10}$-HGG (2.5 mg/mouse per injection) and at days in-dicated spleens were removed and put in serum-free cultures with 100 µg/ml LPS. 2 days later cultures were assayed for number of anti-FITC PFC. Remaining mice plus previously untreated mice were immunized twice with aggregated $FITC_{10}$-HGG together with partussis bacteria at days indicated and thereafter spleen were removed and directly tested for number of anti-FITC PFC and put in serum-free cultures with or without 100 µg of LPS. Anti-FITC PFC response determined 2 days later. Cells from completely untreated animals also put in culture as a control

Fig. 4. Effect of LPS on spleen cells from untreated animals or animals tolerized by 2 injections of deaggregated FITC-HGG (day 0 and 5) and put in serum-free cultures at day 11 with or without LPS. Anti-FITC response determined 2 days later. Open bars represent background response in untreated serum-free cultures, striped and dotted bars represent response to 100 µg LPS

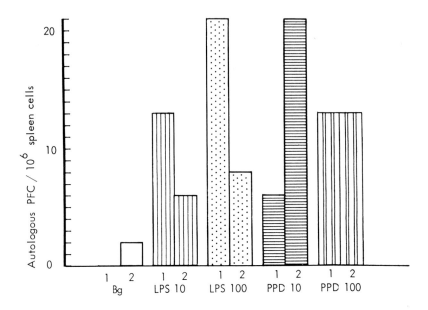

Fig. 5. Induction of autologous PFC in serum-free cultures with bovine lymphocytes stimulated with different concentrations of LPS and PPD. Numbers below bars indicate cow number. Values expressed as PFC/10^6 lymphocytes

that tolerance to thymus-dependent antigens does not affect B cells, but presumably only T cells. It appears that the only stringent test for the existence of B cell tolerance is the inability of polyclonal B cell activators to activate antibody synthesis against the tolerogen. The findings make it unlikely that B cell tolerance to autologous thymus-dependent antigens exists and further indicate that such antigens cannot deliver activating or tolerogenic signals to B cells, although they are competent to combine with and block the Ig receptors.

The critical test for the one nonspecific signal hypothesis is to show that tolerant B cells to autologous antigens do not exist. Since auto-antigens must be in prolonged contact with possible self reactive B cells, the two signal concept necessarily predicts that such B cells must be irreversible tolerized. Therefore, we activated bovine spleen cells in culture with the two PBA, namely LPS and PPD. This resulted in an increased thymidine uptake and the development of plaque-forming cells against untreated autologous red cells (Fig. 5) as well as against hapten-coated sheep cells (Hammarström et al., 1976). The induction of self-reacting antibodies by PBA shows that B cells are not tolerant to their own thymus-dependent antigens, and, therefore, that tolerogenic signals are not generated by the interaction between auto-antigen and the Ig receptors. These findings are only compatible with the one nonspecific signal concept.

Summary

The function and characteristics of B lymphocytes have been summarized. B cells defend the organism against bacterial infections and are the

only cells in the immune system competent to secret antibodies. The
mechanism by which interaction between antigen and B cells leads to
activation and antibody synthesis has been analyzed and it was shown
that B lymphocytes are triggered by nonspecific signals not delivered
to the B cells by their immunoglobulin receptors.

The mechanism of immunological unresponsiveness (tolerance) has also
been anaylzed and it was shown that B lymphocytes cannot be rendered
tolerant even to antigens within the organism. Presumably T lymphocytes
alone are responsible for the self/nonself discrimination.

Acknowledgments. This work was supported by grants from the Swedish Cancer Society
and the Swedish Medical Research Council. The technical assistance by Miss Susanne
Bergstedt is gratefully acknowledged. The work and thoughts summarized in this article
represent a joint effort by many collaborators, as can be seen from the reference
list.

References

Andersson, J., Sjöberg, O., Möller, G.: J. Immunol. 2, 349 (1972)
Bretscher, P.A., Cohn, M.: Science 169, 1052 (1970)
Coutinho, A.: Transpl. Rev. 23, 49 (1975)
Coutinho, A., Gronowicz, E.: J. Exp. Med. 141, 753 (1975)
Coutinho, A., Gronowicz, E., Bullock, W., Möller, G.: J. Exp. Med. 139,
    74 (1974a)
Coutinho, A., Gronowicz, E., Möller, G.: Progr. Immunol. 11, 2, 167
    (1974b)
Coutinho, A., Gronowicz, E., Möller, G.: Scand. J. Immunol. 4, 89
    (1975a)
Coutinho, A., Gronowicz, E., Sultzer, B.: Scand. J. Immunol. 4, 139
    (1975b)
Coutinho, A., Möller, G.: Nature (New Biol.) 245, 12 (1973a)
Coutinho, A., Möller, G.: Europ. J. Immunol. 3, 60 (1973b)
Coutinho, A., Möller, G.: Scand. J. Immunol. 3, 133 (1974)
Coutinho, A., Möller, G.: Advan. Immunol. 21, 113 (1975)
Coutinho, A., Möller, G., Richter, W.: Scand. J. Immunol. 3, 321 (1974c)
Gronowicz, E., Coutinho, A.: Europ. J. Immunol. 4, 771 (1974)
Gronowicz, E., Coutinho, A.: Transpl. Rev. 24 (1975a)
Gronowicz, E., Coutinho, A.: Europ. J. Immunol. 5, 413 (1975b)
Gronowicz, E., Coutinho, A., Möller, G.: Scand. J. Immunol. 3, 413
    (1974)
Hammarström, L., Smith, E., Primi, D., Möller, G.: Induction of anti-
    bodies to autologous red blood cells in bovine spleen cells by poly-
    clonal B cell activators. Submitted for publication, 1976
Jacobs, D.M., Morrison, D.C.: J. Immunol. 114, 360 (1975)
Möller, G.: Transpl. Rev. 23, 126 (1975)
Möller, G., Coutinho, A.: J. Exp. Med. 141, 647 (1975)
Möller, G., Coutinho, A., Gronowicz, E., Hammarström, L., Smith, E.:
    In: Mitogens in Immunobiology. In press, 1976
Möller, G., Coutinho, A., Persson, U.: Scand. J. Immunol. 4, 37 (1975a)
Möller, G., Gronowicz, E., Persson, U., Coutinho, A., Möller, E.,
    Hammarström, L., Smith, E.: J. Exp. Med. In press, 1975b
Skidmore, B.J., Chiller, J.M., Morrison, D.C., Weigle, W.O.: J. Immunol.
    114, 770 (1975)
Smith, E., Hammarström, L., Coutinho, A.: J. Exp. Med. In press
Transplantation Reviews. Möller, G. (ed.): Vol. I (1969), Vol. 14 (1973),
    Vol. 16 (1973), Vol. 17 (1973), Vol. 23 (1975), Vol. 24 (1975),
    Vol. 25 (1975), Vol. 26 (1975)
Watson, J., Riblet, R.: J. Exp. Med. 140, 1147 (1974)

128

Discussion

Dr. Pernis: If membrane immunoglobulins are only playing an antigen-concentrating role, but by binding antigen, themselves do not signal anything to the B cells, how then can you explain the fact that treatment of B cells with an anti-immunoglobulin antibody together with the mitogen LPS blocks polyclonal stimulation? If the immunoglobulin were just antigen-concentrating molecules the process of activation by mitogen should not be disturbed by antibodies against immunoglobulin receptors.

Dr. Möller: I would say that the experimental evidence on this point is very contradictory. Taylor found that anti-immunoglobulin had no effect whatsoever on LPS induction. On the other hand we have heard from various people, not the least from the Basel Institute, that anti-immunoglobulin antibodies can suppress. I can see a variety of suppressive mechanisms resulting from loading up the surface of a B cell when you get cap formation. The cell is simply not very happy in all this and doesn't want to be activated. I think that this is controversial and also that nearly all work on the action of anti-idiotype antibodies on idiotype-bearing cells has shown that this neither stimulates nor suppresses if you work with pure B cells. Another point is that if you work in a mixed population like T cells, B cells and macrophages, then anti-immunoglobulin on the B cell would, I guess, recruit some Fc-binding T cell, which may even kill the B cell as in allotype suppression or in other ways influence them.

Dr. Melchers: Unfortunately, I have a different view. It has been observed by many laboratories that in vivo as well as in vitro B cells exposed to anti-immunoglobulin antibodies binding to the immunoglobulin receptors on the surface of B cells, will render these B cells unreactive to further stimulation, either by antigen or by polyclonal activators. The view which I take is that the immunoglobulin molecule in the surface membrane must possess the property of changing its conformation, in a functional sense, and may be able to associate itself with reactions in the cell leading to growth, in which case it is not being occupied by the anti-immunoglobulin molecule. It may also associate itself, in another functional conformation, which may be reflected in a structural conformation, with reactions in the cell leading to inhibition of growth. Our experiments would argue for a functional allosteric property of the immunoglobulin molecule, modulating growth signals, the basis of which could be structural changes of conformations in the immunoglobulin molecule, as suggested by the work of Dr. Huber and Dr. Pecht.

Dr. Möller: May I add that there is a phenomenon called receptor cell blockade which indicates that if you load up the cell surface with antigenic materials as antibody to immunoglobulin receptor, you do not change the resting state, or whatever, of the cells but in the case of antibody-secreting cells you prevent secretion of antibodies. The mechanism of it is not known but since the doses involved are very high it may very well be rather trivial, and some of the experiments with anti-immunoglobulin could be due to this.

Dr. Melchers: At the stage when you load the resting small B cells with anti-immunoglobulin antibodies and then subsequently expose them to a polyclonal activator, you can wash out excess antibody. These small B cells, then, will not develop into plasma cells, which would be the cells on which effector cell blockade should occur.

Dr. Henning: May I just add a short question to that point? What happens if you develop the cells with Fab fragments: would they inhibit?

Dr. Melchers: No, they don't inhibit at all, or very much less efficiently.

Dr. Möller: Which shows that the Ig receptor isn't really involved then!

Dr. Diamantstein: May I ask you only one question? You thought the T-independent antigens do not react with T cells? As far as I know, if you inject a polyclonal mitogen like pneumococcal polysaccharide, then they react with T cells and you get very good T-suppressor cells. Second, you can get very good cell-mediated immunity with T-independent antigens.

Dr. Möller: The last statement surprised me because, in general experience, it has been difficult to get delayed hypersensitivity to polysaccharide and LPS. I agree that if you inject into an animal certain of these antigens you will get different effects. But all these substances can interact with macrophages. Some of them, no doubt, activate macrophages and we also know that macrophages can in turn activate T cells. So it is very doubtful if we are talking about a direct interaction with T cells rather than one via macrophages.

Dr. Hoffmann: In your discussion you haven't distinguished between a signal that tells a B cell to proliferate and possible signals that might tell a B cell to differentiate into an antibody-forming cell.

Dr. Möller: Same thing - different cells!

Dr. Hoffmann: I was just going to comment on the accumulating data that suggest that the nonspecific signal could be one which induces the cell to differentiate into an antibody-forming cell, and that there is also a gradual increase in the amount of data suggesting that the anti-immunoglobulin stimulation of B cells in a sufficiently polyvalent form will tell a B cell to proliferate.

Dr. Möller: I don't really want to go into this too deeply. The alternatives now vary from 1 to 5 signals in order to activate B cells, I think one is likely to be correct because it is simplest. But what I haven't gone into at all is that all the PBAs do not act on all B cells; there are sub-populations and these sub-populations go through differentiation events. The least mature cells divide, the more mature produce antibodies after activation. Not only do they go through this differentiation, but in doing so they also change their susceptibility to different PBAs. My point is that adding something twice doesn't mean two signals, which is a mistake quite often made. Two different substances can give the same signal.

Dr. Pecht: One thing that remains quite in the dark is the mechanism of how all these PBAs are carrying out a function. I mean, you tended to indicate there was no specificity involved in this whatsoever. Would you care to comment on this?

Dr. Möller: There is a specificity since there is a mouse strain that on the basis of a single genetic defect lacks responsiveness to LPS. This indicates the existence of LOS receptors. There is probably a family of receptors to different PBA and they are genetically determined and they have a low affinity for the PBA. We ourselves and several others are working to try to characterize these receptors. All the B-cell surface structures we know such as Ig, C'3 etc. have no function yet, and all the functional receptors have an unknown structure.

Dr. Rajewsky: I would like to say one thing which you will certainly agree with, I think, namely that the question of whether tolerance can be induced in B cells to thymus-dependent antigens or not, is certainly a very controversial issue. There are experiments, I think, which one can certainly interpret along the idea that tolerance can be induced, and maybe tolerance could be induced say in the neonatal phase at the B-cell level to thymus-dependent antigens. For example the recent experiments of Nossal's group would argue in that sense or one could cite experiments in which high doses of antiidiotypic antibody, without the induction of suppressor cells, seem to suppress idiotype - specific precursor cells, and so forth. I think a general problem in experiments concerning this matter ist the question of affinities like in the experiments which you showed us. Is it so that the cells which you still see after an attempt to induce tolerance really have the full spectrum of affinities which they exhibit if no tolerance is induced? We really cannot go here into a detailed discussion of all this. All I want to say is that this is certainly a controversial field still.

Dr. Möller: Yes, it all depends on your interpretation. If you interpret tolerance as the inability to show that the cells are activated when you give antigen, I agree. Then it exists. But my definition of tolerance is that the cells have received a signal which irreversibly inactivates them and that they can't be rescued by giving activating signals. This has only been done in two studies, those which I showed you and previously by Coutinho and Gronowicz, and in both cases it was found that the cells were there. But certainly you couldn't show that they were there by straightforward immunological methods (by giving antigen), because the Ig receptors are blocked by antigen.

T Lymphocytes

H. Wigzell

Introduction

The small lymphocytes are the cells in the body of an immunocompetent
individual, which are able to initiate a specific immune response
against foreign material due to their own, predetermined immune reactivity. Although these cells look strikingly similar when observed
under conventional light microscopy unsurpassed heterogeneity resides
in this cellpopulation as to surface markers and functional activities.
Two major groups of lymphocytes comprise this diverse population, the
B and the T lymphocytes (17). These two types of cells differentiate
according to their own distinct pathways and no evidence exists that
a T lymphocyte can convert itself into a B lymphocyte or vice versa.
The functions of the B lymphocytes are seemingly not manifold, but
appear to reside in their ability to produce humoral antibodies of
conventional immunoglobulin type. The T lymphocytes, on the other hand,
are known to express a variety of functions including killer cell activity against foreign cells, helper activity for B cells to allow the
latter to make antibodies, suppression of the specific helper activity
etc. (17, 19, 20).

It has only been possible during the very last few years to analyze
the functional heterogeneity of the T lymphocytes in relation to the
actual subgroup of T cells that perform the relevant functions, and a
major part of the present lecture will deal with a brief description
of these matters. Two major parameters will be dealt with: the surface
structures and functional makers that design two distinct groups of
immunocompetent T lymphocytes and a lengthy presentation on how T
lymphocytes cognize and recognize antigens.

The Life History of a T Lymphocyte

Both B and T lymphocytes have their stem cells in the bone marrow and
production of new, immunocompetent cells of the respective types takes
place throughout life at an only slightly decreasing rate until very
advanced age. The pro-T cell or pro-thymocyte can be shown to be able
to migrate to the thymus, were under hormonal influence differentiation
may take place, allowing the now immunocompetent T lymphocyte to leave
the thymus. During this maturation, which most likely involves many
steps, surface markers as well as functional characteristics of the
cells change in measurable way. Through the use of antisera on these
differentiation T cell antigens it is possible to describe this process
and actually isolate the functionally intact subgroups of T cells for
further analysis (7, 9). Figure 1-3 depict in a schematic way the distribution patterns of antigens and functions on peripheral T lymphocytes in the mouse where three groups of cells can be described. Two
of these have immune function to our present knowledge.

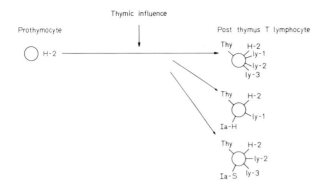

Fig. 1. Development of surface antigens on thymic prothymocytes when developing into peripheral T lymphocytes under thymic influences. Scheme denotes mouse lymphocytes

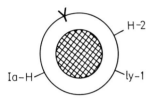

Helper cell
MLC reactive cell
Will never be cytotoxic

Fig. 2. Surface markers of mouse helper T cells. Symbols denote antigen-binding receptor, H-2 serologically defined antigens, the Ly1 antigenic marker significant for this kind of immunocompetent T cell. Ia-H represents the suggestive evidence from some workers that mouse helper T cells may express their own unique Ia antigens

The first group with specific immune function is, according to its best-analyzed function, called helper T cells; that is they are endowed with the pharmacological ability, via an antigen-specific reaction to produce a catalytic signal to either B or T (17, 19) lymphocytes to respond to a given immunogen. It is still unclear whether all B lymphocytes are able to respond to this helper T cell signal, or whether there exist so-called T-dependent and T-independent B lymphocytes. The helper T cell, however, only performs the function of a catalyst; it will not in any direct manner change the specificity of the B cell product, the humoral antibodies (Fig. 2).

The capacity of helper T lymphocytes to allow other T cells to respond to an immunogen is most clearly shown in reactions against transplantation antigens of the major histocompatibility complex (MHC) locus antigens (19, 23). Here, T cells with the differentiation antigens signifying them to belong to the helper type can be shown to respond by proliferation, preferentially towards the Ia-antigens of the MHC locus (7, 8, 19). A remarkably high figure of responding T cells can be noted in these reactions even when using responding T cells from unimmunized donors, suggesting that these anti-Ia reactions in a sense constitute highly potent T cell-triggering substances (23). However, the induced proliferation is dependent upon specific immunocompetent T cells reacting against the Ia-antigens, and will not occur in stages of immune tolerance. During this proliferation helper T lymphocytes can be shown to release soluble factors that will now allow T lymphocytes with sur-

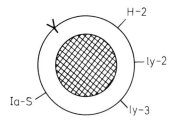

Fig. 3. Surface markers on mouse killer T cells.
Symbols denote the antigen-binding receptor,
the serologically defined H-2 antigens and the
Ly-2 and Ly-3 antigens that are characteristic
for this cell type. Also included are Ia-S as
denoting the possibility that killer or maybe
in particular suppressor cells have certain
special Ia antigens expressed on their outer
surface

Killer cell
Suppressor cell

face markers of the second group (shortly to be described) to enter
into proliferation and development of function. It is also quite likely
that helper T lymphocytes may help to recruit other helper T lymphocytes by the same mechanism.

No formal proof exist that the same helper T cells can help both B and
T lymphocytes to respond to an antigen, but all available evidence is
in agreement with such an assumption. Final evidence for this would
come when the specific catalyzing reagents are available in a form of
reasonable purity as to substitute for the actual helper T cells in
these processes. Already at present, sizeable although preliminary
data exist on the occurrence of such helper T lymphocyte factors, some
being reported to be nonspecific as to the antigen, whereas other are
claimed to have antigen-binding specificity (15, 21). The antigen-
specific factors released from helper T cells will be discussed later
in this lecture.

The second group of the immunocompetent T lymphocytes house cells with
two distinct functions; the killer T cells and the suppressor T lymphocytes with the surface markers as depicted in Figure 3. Again, it is
not known for sure whether the very same cell can be a killer T or/and
a suppressor cell and further work is necessary to sort this out. The
killer T lymphocytes are in the priming event seemingly dependent upon
helper T lymphocytes to get started, but it is unclear whether the
same requirements hold during a secondary encounter with the immunogen.
The potential T killer cells express the depicted surface antigens already at time zero, that is before contact with the immunogen and will
retain the differentiation antigens of e.g. Ly type during the immunization procedures. Cell division is required for the virgin T killer
cell to become a killer, but a memory killer cell does not require
this (22). Killing is effectuated via contact with the target cell and
the lytic stage is reached after a rapid (within minutes) delivery of
a "lethal kiss" (13, 19). A killer T cell can go on killing many target
cells and does so in a unidirectional manner, killer cells themselves
being susceptible to attack by other killer T cells with receptor for
their surface antigens. No evidence exists that the T killer cell releases any soluble factor necessary for lysis of the target cells to
take place.

Killer T cells can be shown to use their own, actively synthesized
receptors for antigen when making the proper combination with the target
cells (13). A finding with far-reaching implications, however, is the
fact that killer T lymphocytes may perhaps be entirely designed to kill
other cells via the serological defined antigens of the major histo-

compatibility complex genes (18). Accordingly, virus-infected cells
(25) or hapten-coupled cells used as a immunogen will only be lysed
if the later target cells carry both e.g. virus antigens plus some
major histocompatibility antigens which are identical to the cells
used for immunization. As virus infections normally occur within an
individual this would mean that the individuals own virus-infected
cells are recognized as the major target for the killer cells. Virus-
infected cells from another individual, using the same virus but now
present in cells differing at these major antigens will not be lysed.
It is thus very likely that when testing for e.g. tumor-immune reactions using killer cells in human tumor-immune research, great care
must be taken to ensure that the tumor cells to be tested have "normal"
histocompatibility antigens in common with the lymphocyte donor. Similar reasoning may have to be applied in tests involving in vitro
delayed hypersensitivity. Two major concepts to explain this need for
similarity in MHC antigens between target and killer exist: according
to one, all antigens recognized as foreign by the killer T cells function via modification of MHC structures, the "altered self" hypothesis.
According to another model a dual recognition exists, which may include
two steps, a specific binding of conventional serological nature between the antigen-binding receptors on the T cells and e.g. the virus-
induced antigen, but also including another essential requirement,
namely a minimum necessary degree of identity between ments to exclude
any of these hypothesis are lacking but the results strongly indicate
one point: the T lymphocytes are particularly apt to react against MHC
antigens, the helper T cells especially against the MLC-inducing structures within the Ia regions in the mouse H-2 region, whereas the killer
T cells are particularly prone to kill via reactions towards the serologically defined H-2K or H-2D ends of the H-2 locus (18, 19, 25).
When analyzing the actual receptor sites of immunocompetent T lymphocytes, as will be discussed later, it also becomes apparent that this
high reactivity against MHC antigens also is reflected in the high
frequency of T cells bearing receptors for these antigens.

The suppressor T cells have been reported by some workers to function
via release of specific factors with ability to inactivate selectively
and irreversible relevant helper T lymphocytes (20). So far no evidence
exists that suppressor T cells exert any direct inhibitory activity on
specific B lymphocytes.

Both groups of immunocompetent T lymphocytes (realizing that within
each group additional sub-groups may certainly hide) express immunological memory; that is they can be triggered into exerting their
function via blast transformation and proliferation when brought in
proper contact with the immunogen but can then revert back to a resting,
long-lived memory cell (1). Whereas helper cells seem to be induced
with low concentrations of antigen, predominant production of suppressor T cells frequently occurs using high doses of antigen; a fact
which appears practical from the point of view of regulation, as to
avoiding damaging effects of hyperimmunization. Whereas helper cells
and suppressor cells normally function in well-regulated harmony,
several diseases of auto-immune character both in mouse and men are
blamed on deficiency with regard to suppressor T cell functions. It
is to be expected that the analysis of the factors generating helper
versus suppressor T cells will be a major field in cellular immunology
in the near future. It is already known, however, that a helper T
lymphocyte normally never converts into a suppressor T cell or vice
versa. Examples of experiments performed as to prove this point are
shown in Figure 4 and 5, where purified T lymphocytes of either type
were inoculated into syngeneic (= histocompatible) T deficient mice.
The reconstituted mice could then be shown to function as "helper" and

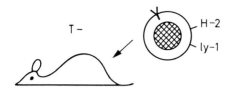

Fig. 4. Demonstration that transfer into a thymus-deficient mouse of T lymphocytes of "helper" type as assessed by surface markers results in the production of a stable "helper" mouse for several months. No switch of these normel "helper" T cells into killer or suppressor T cells could be demonstrated

Wait a few months, give antigen
Helper T cells ┼┼┼
MLC   ┼┼┼
Killer   ──
Suppressor   ──

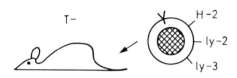

Wait a few months, give antigen
Helper T cells ──
MLC   ──
Killer   ┼┼┼
Suppressor   ┼┼┼

Fig. 5. Demonstration as in Fig. 4, but this time that transferred killer or pre-killer T cells will not change into helper cells after prolonged time in vivo

"killer" mice respectively when tested for immune competence several months later (Cantor, H., personal communication).

Specific Antigen-Binding Receptors Produced by T Lymphocytes

The B lymphocytes as a population within an individual are able to express a remarkable diversity in recognizing various chemical groups belonging to an immunogenic molecule. The exact number of possible antigenic sites generated by the combinations of heavy and light chain variable polypeptide sequences in the serum immunoglobulin molecules of a given individual are unknown, but minimum estimates start at the level of 10^4 and are frequently much higher. The group of T lymphocytes are able to express a discriminatory ability for antigen which is in the same order of magnitude as that encountered for the B cells (20), and with largely overlapping, although not necessarily identical spectra of antigen cognition. It would thus seem both economical and logical that the same system for generation of diversity should be responsible for the creation of the antigen-binding receptors on both types of lymphocytes.

The antigen-binding receptors for antigen on T lymphocytes have, however, until very recently been highly elusive (10). Most people, including the laboratories of the talker, have been unable to detect immunoglobulins of conventional type as a product of T lymphocytes, but certain other workers have reported T lymphocytes to produce Ig

136

Table 1. Gene loci involved or implicated as participating in the creation of antigen-binding receptors on lymphocytes[a]

1. Gene locus coding for heavy chains of immunoglobulins

2. Gene locus coding for kappa light chains of immunoglobulins

3. Gene locus coding for lambda light chains of immunoglobulins

4. Major histocompatibility complex locus genes

[a] No indications exist that any of these loci are genetically linked to each other.

molecules of IgG-like nature and with conventional light chains of kappa type (14). During the last four years a new family of genes (besides the conventional Ig determining genes) has emerged as a possible candidate for the creation of the antigen-binding receptors on T lymphocytes. The genes, called IR genes (= immune response) of the type linked to the MHC locus of the species can be shown to play a decisive role in determining whether a T lymphocyte population would be allowed or not to react to certain antigens (2). Minor changes in the chemical structure of an antigen are enough to allow a responder T cell population of the IR-controlled type to become a non-responder or vice versa, indicating a highly discriminatory ability of these IR-gene-controlled reactions. The four "competing" gene families participating in the creation of the T cell receptors for antigen are depicted in Table 1. It should be noted that no indications exist that any of the four gene clusters are genetically linked to each other.

Evidence that IR gene products constitute the antigen-binding T cell receptors did originally only stem from experiments of genetic linkages (2). Recently several groups have reported on the isolation of antigen-binding receptors from helper or suppressor T cells, with capacity to exert their various immune functions and displaying antigenic determinants of Ia type (15, 21). As these Ia antigens are coded for by genes present amongst the IR genes (IR and Ia genes may be identical in many circumstances) such results constitute strong support for the IR genes as vital participants in the build-up of the T cell receptors for antigen. Actual proof that these molecules are produced by T lymphocytes is still lacking as e.g. analyzed by internal labeling experiments but the data already look very impressive. The biochemistry of these T cell-derived molecules indicates a molecular weight in the range of 25,000 to 60,000 daltons. No antigenic determinants of immunoglobulin type have been found on these molecules. A complicated interaction between T cell factor of helper type and acceptor molecules on B lymphocytes has been suggested by absorption experiments. It should be noted that the B lymphocyte acceptor molecule is also seemingly coded for by IR genes, although genetically and serologically distinct from those of the T helper factor (15). As these interactions most likely will be presented in detail elsewhere during the meeting I will not discuss these processes any further.

Using another approch other workers, including our group, have accumulated data suggesting certain striking similarities between B and T lymphocyte receptors for antigen, thus arguing for the T receptor to be at least partly coded for by genes linked to known immunoglobulin gene clusters. The present approach has been based on the assumption that the striking unique characteristics of the immunoglobulin molecules reside in their variable polypeptide parts. As these variable

regions are known to be directly involved in the creation of the antigen-binding sites, preservation of genes governing such variable
regions would have great survival value and data exist suggesting such
a preservation of variable genes during phylogeny (24). As for constant
Ig regions a similar demand for retention would not previal and thus
different genes may here contribute in the creation of the constant
regions of antigen-binding receptors in the T and B lymphocyte groups.
Using such assumptions, antibodies have been induced against the antigen-binding regions of immunoglobulins; such anti-antibodies are
called anti-idiotypic antibodies. We have thus attempted to make anti-
idiotypic antibodies against B or T lymphocyte receptors with specificity for the same antigens using as a target antigens MHC antigens.

Shared idiotypic determinants between molecules have previously been
demonstrated to have a similar counterpart in sharing of amino acid
sequences. If T and B lymphocyte receptors for antigen express shared
or identical idotypic antigens this would thus suggest a corresponding
similarity at level of DNA. Using a genetic system allowing direct
immunization against T cell receptors (16) via the use of inbred animals differing at the major histocompatibility locus, we have succeded
in making anti-idiotypic antibodies directed against T cell receptors
or IgG molecules directed against the same transplantation antigens
(3-6). Here we could show that such anti-idiotypic antibodies in the
presence of complement would wipe out the relevant T cells, leaving
other T cells reacting against unrelated antigens functionally intact.
Most important was the finding that highly purified IgG molecules of
the relevant antigen-binding specificity remove all anti-idiotypic
reactivity of anti-T cell receptor antibodies when testing against the
T lymphocytes (3). Conversely, pure T cells with the relevant idiotypic,
antigen-binding receptors would remove all anti-idiotypic antibodies
when testing the latter against idiotype-positive IgG molecules. As
such complete overlap was found in three different anti-T cell idiotypic sera, this strongly suggested the involvement of the same variable
gene subsets in the B and T receptors for antigen. Similar conclusions
were reached using anti-idiotypic antibodies to induce idiotype-positive suppressor or helper T lymphocytes to become switched into immune
function via the combination between the idiotype-positive receptors
and the added anti-idiotypic antibodies (11). Again, the results indicated sharing of idiotypes or even it not identity between the B
and T lymphocyte receptors for antigen. Further experiments in this
stimulation system have also yielded results suggesting that the inheritance of T cell idiotypic receptors is linked to the heavy chain
Ig gene cluster (12). Our own data on the linkage between T cell idiotype and IR or Ig gene clusters have so far only revealed an absence
of linkage to the histocompatibility-linked IR genes. These data are
thus in striking contradiction to those reporting that Ig-related genes
may play no role in the creation of the T cell receptors but that only
IR-gene products of Ia-linked type constitute such receptors.

It was thus essential to try to purify the T cell receptor molecules
using anti-idiotypic immunosorbants for further characterization. This
has been found quite successful, as the frequency of idiotype-positive
lymphocytes in the presence of an antigenic system is very high even
in the normal individual (4). This, presumably via membrane turn over,
would lead to release of detectable amounts of idiotypic, antigen-
binding receptor molecules that can be extracted in "pure" form from
large volumes of normal serum or supernatant from normal lymphocytes.
These idiotypic-positive molecules have the following characteristics
(6): In the normal serum four differently sized groups of molecules
can be found. They all display antigen-binding activity and relevant
idiotypic markers. Peak I has a molecular size of around 160,000,.

peak II around 150,000 daltons, peak III around 70,000 and peak IV 30-40,000 daltons. Using supernatants from either pure T or B lymphocytes and external or internal radiolabeling procedures it became clear that peak I is of B cell origin whereas peaks II-IV are synthesized by T lymphocytes. Biochemical and serological analysis of the different peaks revealed that peak I consisted of "conventional" 7 - 8S IgM molecules. Peak II could be shown to consist of two polypeptide chains, each somewhat larger than 70,000 daltons in size = peak III molecules. The peak III molecules could be shown to be sensitive to proteolysis thereby producing peak IV molecules (6). No evidence was found that any light chains of Ig type were linked to the T cell-derived molecules. Serological analysis of the T cell molecules (peaks II-IV) only demonstrated that the molecules could be removed by the proper anti-idiotype immunosorbant, but no antigenic determinants of conventional constant Ig type could be disclosed on these molecules. Attempts were also made to analyze if such molecules contain Ia antigenic determinants but no such structures were found.

We interpret these data accordingly: the relevant normal immunocompetent T cells express on their surface and are able to release molecules with antigen-binding capacity. Each T cell expresses only receptors with one kind of antigen-binding sites (5). The released molecules, in comparison to those of Ig molecules with the same (similar) antigen-binding specificity released from B cells, have one thing in common, a striking similarity as to tertiary structures in their antigen-binding sites. Genetic linkage data of other groups (12) and our own preliminary biochemical analysis data (3, 6) suggest that the T and B lymphocyte receptors for antigen most likely use identical genes coding for, in case of the B cell molecules, the variable regions of the heavy chains. The rest of the T cell chain(s) belong to a distinct class of proteins, not belonging to the known immunoglobulin classes. From this point of view one might call the structure an immunoglobulin or, as used before, IgT. However, contrariwise to other workers demonstrating such "IgM" molecules we find no cross-reactions to IgM, nor do we find any evidence in the soluble products for light chains. In fact, one striking fact about the T cell receptor is, in the present system, the ability to express highly significant antigen-binding ability even when present as a single chain polypeptide. Whether this is due to the target antigens (the MHC antigens) in the experimental system or a general feature of the T receptor remains to be analyzed.

Is it then possible to bring the reports on Ia-containing antigen-specific molecules and the present, Ig-like molecules together in a useful synthesis? If one accepts the different results at face value this is not an easy task, and might substantiate the concept that T lymphocytes are indeed able to use two genetically distinct systems in the generation of diversity of their receptors. Alternatively, one would argue that at the membrane level it is possible that the a two chain unit may be around with one chain of "Ia" and one of "Ig" type. No evidence of such molecules exists as yet, however, and as the isolated molecules of both types are reported to express antigen-binding specificity, this will not help to solve the question.

My personal view at present, biased as it is, is that the T lymphocytes make use of variable genes of Ig type to make up receptors of the type described as peak II molecules. These receptors would thus constitute the receptors, through which the T cells can recognize a highly diversified antigenic spectrum, perhaps frequently with low avidity but yet strongly enough to trigger the relevant T lymphocytes. Enhancing or suppressing factors in the interactions between cells leading to triggering would then be the described Ia factors, but according to this reasoning these factors would not be truly antigen specific but of a lower degree of diversity than the Ig-type receptors. Only further experiments will sort this out. The eventual explanation why the T cell receptor has been so elusive may well prove to be that there is no such ting as a singular T cell receptor for antigen, but rather a multicomplex structure with several contributing units.

References

1. Andersson, L.C., Häyry, P.: Transpl. Rev. <u>25</u>, 121 (1975)
2. Benacerraf, B., McDevitt, H.O.: Science <u>175</u>, 273 (1972)
3. Binz, H., Wigzell, H.: J. Exp. Med. <u>142</u>, 197 (1975a)
4. Binz, H., Wigzell, H.: J. Exp. Med. <u>142</u>, 1218 (1975b)
5. Binz, H., Wigzell, H.: J. Exp. Med. <u>142</u>, 1231 (1975c)
6. Binz, H., Wigzell, H.: Scand. J. Immunol., in press (1976)
7. Cantor, H., Boyse, E.A.: J. Exp. Med. <u>141</u>, 1376 (1975a)
8. Cantor, H., Boyse, E.A.: J. Exp. Med. <u>141</u>, 1390 (1975b)
9. Cantor, H.: personal communication
10. Crone, M., Koch, C., Simonsen, M.: Transpl. Rev. <u>10</u>, 36 (1972)
11. Eichmann, K., Rajewsky, K.: Europ. J. Immunol. <u>5</u>, 661 (1975)
12. Hämmerling, G.J., Black, S.J., Berek, C., Eichmann, K., Rajewsky, K.: J. Exp. Med. <u>143</u>, 861 (1976)
13. Kimura, A.K., Wigzell, H.: In: Contemporary Topics in Molecular Immunology. New York: Plenum Press 1976, Vol. VI in press
14. Marchalonis, J.J., Cone, R.E., Atwell, J.L.: J. Exp. Med. <u>135</u>, 956 (1972)
15. Munro, A.J., Taussig, M.: Nature <u>256</u>, 103 (1975)
16. Ramseier, H., Lindenmann, J.: Transplant. Rev. <u>10</u>, 57 (1972)
17. Separation of T and B lymphocyte subpopulations, Transplant. Rev. Möller, G. (ed.). 1975a, vol. 25
18. Shearer, G.M., Lozner, E.C., Rehn, T.G., Schmitt-Verhulst, A.M.: J. Exp. Med. <u>141</u>, 930 (1975)
19. Specificity of effector T lymphocytes. Transplan. Rev. Möller, G. (ed.). 1975b, vol. 26
20. Herzenberg, L.A., Okumura, K., Cantor, H., Sato, V.L., Chen, F.W., Boyse, E.A., Herzenberg, L.A.: J. Exp. Med. <u>144</u>, 330 (1976)
21. Takemori, T., Tada, T.: J. Exp. Med. <u>142</u>, 1241 (1975)
22. Wagner, H., Röllinghof, M.: Europ. J. Immunol. <u>6</u>, 15 (1976)
23. Wilson, D.B., Blyth, L.L., Nowell, P.C.: J. Exp. Med. <u>128</u>, 1157 (1968)
24. Zinkernagel, R.M., Doherty, P.C.: Nature (Lond.) <u>248</u>, 701 (1974)

Discussion

Dr. Porter: Jensenius, Williams and Mole are trying to do the same kind of experiments but are using anti 'a' locus allotype in the rabbit to estimate the quantity of variable region of the heavy chain on T cells. They cannot find more 'a' locus allotype than would be associated with the estimated amount of contaminating immunoglobulin heavy chains.

Dr. Wigzell: The A allotype in rabbits is not a simple allotype in the sense that it is a marker of *all* V regions, that is my point. We have tried to do exactly the same experiments as you, looking for whether the T cells of rabbits will actually release molecules carrying the 'a'locus allotype, and we have also failed, so far.

Dr. Henning: I am not going to ask you how pure your reagents are, I want to ask you: Have you done the following experiments? You say your receptors shown on the gels have been shed. Have you compared that to cell-bound receptors?

Dr. Wigzell: That is exactly what we are doing now. As yet, we really don't know.

Dr. Albert: Have you been able to measure the difference in affinity between antigen and idiotype, and anti-idiotype and idiotype? Is there a difference?

Dr. Wigzell: We have not done this because of the complexity of our antigens. We have the transplantation antigens here. There are some suggestive data from Klaus Rajewsky's group suggesting they might have a molecule of a kind similar to that I was describing here, the affinity of which for hapten was about a 1000 times lower than a normal antibody; thus, the anti-idiotypic antibody in such a system would be supposed to be about 1000 times better. I wouldn't like to say more.

Dr. Krawinkel: We have indeed isolated receptor molecules, presumably from the T-cell surface, with properties very similar to those of the molecules described by Hans Wigzell.

Dr. Wigzell: That molecule of yours also lacks immunoglobulin markers, of the constant part, and lacks Ia markers?

Dr. Krawinkel: It seems not to carry any known class-specific immunoglobulin determinants, and we haven't been able to detect Ia determinants so far.

Dr. Hämmerling: Hans, the frequency of your idiotypic T cells with specificity for a given haplotype is very high, around 5%, and so far you did not find any overlap among T cells which are specific for other haplotypes. If this is true also for T cells with specificity for all other transplantation antigens, how can the animal deal with the rest of the antigenic universe? It is obvious that these high frequencies tell us that the T cell recognition system somehow is uniquely tuned up to react against major histocompatibility antigens. Do you consider the possibility that within this 5% population recognizing the transplantation antigens there exists additional heterogeneity allowing some receptors to react also with conventional antigens?

Dr. Droege: Dr. Wigzell, you said that the properties of your T-cell product are similar to the properties of the product which has been investigated by Drs. Krawinkel and Rajewsky, but that it is difficult to reconcile the results with observations in other experimental systems. My question is: Do you think that this product is possibly produced by a T-cell subset, which may be different, for example, from helper or killer T cells?

Dr. Wigzell: Well, in this case you will get into even more extreme problems with the high frequency of 5%. I think our idiotype positive T cells belong to the conventional subsets.

Tumors of Immunoglobulin-Producing Cells and Thymus-Derived Lymphocytes

M. Potter

Introduction

The immunoglobulin-producing cells differentiate so that one specialized cell and its progeny by mitosis are limited to produce immunoglobulin molecules with a single binding specificity. The binding specificity of an immunoglobulin molecule is determined by the interaction
of the V_L and V_H regions of the light (L) and heavy (H) chains. There
is now compelling evidence to support the concept of Dreyer and Bennett
(1965), that single L or H chains are each controlled by two separate
genes V_L or V_H and C_L or C_H genes, and the differentiation of the im-
munoglobulin-forming clones begins with the process that activates a
single V_L and V_H gene in a cell. This type of differentiation might
have very little special significance, were it not for the fact that
a single mammalian (vertebrate?) haplotype can probably generate 10^3
to 10^4 different immunoglobulin producing clones. The population size
of each specialized clone in the organism is usually relatively small,
e.g. 10^3 to 10^4 cells, making it difficult to isolate in quantity individual clones and their products. Further the immunoglobulin producing clone does not develop in a single step but develops sequen-
tially in time and space in the organism.

Tumors in immunoglobulin producing cells may themselves arise clonally
at different developmental stages, as "growth mutants", and can be
regarded in relation to the immunoglobulin producing cell clone as a
clone within a clone. To the biochemist the transplantable or in vitro
cultured tumor arising in an immunoglobulin producing cell is a convenient means for amplifying a specific stage of development and a
specific cell product. While this method has been very succesful with
immunoglobulin secreting cells, many workers are approaching tumors
of other cell types in the immune system for a source of special products; in particular, the tumors of B lymphocytes, i.e. the precursors
of immunoglobulin secreting cells and the thymus derived lymphocytes
(T lymphocytes). In both of these tumor types, the immunoglobulin in
B lymphocytes and the specific antigen-binding receptors of T lymphocytes, are chiefly associated with the plasma membrane and three are
relatively very few molecules per cell.

The purpose of this paper is to review the biological properties of
tumors of Ig producing and thymus derived cells and relate them to
both normal development and the abnormal condition in which they were
induced.

Commonly studied tumor inducing systems, e.g. radiation, chemically or
virally induced thymic lymphocytic tumors; mineral oil induced plasmacytomas; Abelson virus induced B lymphocytic tumors, or spontaneous
tumors in inbred strains of mice (e.g. AKR, C58) or rats (Lou W1) are
the usual source of transplantable tumors. Though these are currently
the best studied models they may provide only a limited sample of
available tumor types.

Tumors of B Lymphocytes
========================

B Lymphocyte Development

The activation of a single V_L and a single V_H gene in a lymphoid cell
from an uncommitted precursor marks the beginning of the immunoglobulin
producing clone.

Very little is known about the mechanism of this differentiation but
some facts are deduced from the outcome.

First a mammalian haplotype genome e.g. as in the inbred BALB/c mouse
contains multiple V_L and V_H genes (Hood et al., 1973; Potter et al.,
1976). Second, usually only a single V_L or V_H on only one of the respective autosomal chromosomes is activated in a cell (phenomenon of
allelic exclusion; Mage, 1975). Once V genes are activated, the cell
is committed to use these as templates for immunoglobulin synthesis
throughout the further development. Following the postulate of Dreyer
and Bennett (1965) it is thought by many that the V gene and its corresponding linked C gene, which are physically separated on the same
chromosome, are brought together to form a continuous covalent template
by cleavages and ligations of V and C chromosomal genes. The mRNA for
immunoglobulin light chains has been shown to be a continuous covalent
structure and is assumed to have been transcribed from a covalent DNA
(Swan et al., 1972). The V gene - C gene template in the plasma cell
is a highly stable if not irreversible product of differentiation that
despite the remarkable selection process involved in its formation (i.e.
the exclusion of all the other very closely linked and related V genes)
is maintained through many normal cell divisions and more dramatically
remains stable throughout the extended cell divisions that occur in
transplantable plasma cell tumors.

Third, C gene activation is just becoming better understood. For L
chain genes the process is less complex. During the V differentiation
a V_{Kappa} or V_{lambda} gene with its corresponding C_{kappa} or C_{lambda} is
activated. V_{kappa} genes never associate with C_{lambda} and vice versa.
Either a kappa or a lambda locus is activated, not both. Since kappa
and lambda may have nonoverlapping functions this just reflects the
selection of an appropriate V_L gene.

For H chain genes, the differentiation process is more involved. Em-
pirically one V_H may be associated with more than one C_H, and the expression depends usually on the stage of development. In early stages
V_H-C_μ or -C_δ are produced. Later there is a switch to $V_H C_\gamma$, α or ϵ
chains. Some tumors in fact produce $V_H C_\mu$ plus $V_H C_\gamma$ (Wang et al., 1970)
or even $V_H C_\alpha$ plus $V_H C_\gamma$ (see Morse et al., 1976). One but by no means
the only explanation is that during V_H differentiation the V_H gene
fuses with DNA of several C genes to make templates that are subsequently differentiated by specific signals.

The immunoglobulin clone is depicted in Figure 1. Two major stages
are: the generative stage, and the antigen-reactive stage. In the generative stage the uncommitted cell differentiates into a V determined
cell. This step occurs in different tissue sites depending upon the
stage of ontogeny and the species. In Avians, primitive lymphoid cells
develop in the bursa of Fabricius a small lymphoid organ in the terminal part of the gut. After sexual maturity the bursa atrophies and
the bone marrow takes over this function. In mammals, the fetal liver
is an early site of V differentiation (Owen et al., 1975); postnatally
the bone marrow and possibly the spleen take over this function. While

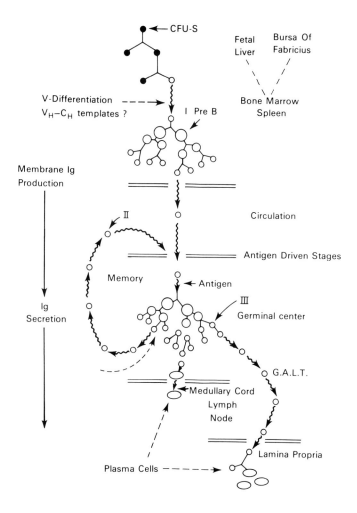

Fig. 1. B lymphocyte development. For description see text. CFU-S: colony forming unit-stem cell; V: variable region; I, II, III: probable counterparts in scheme for type I, II, III B lymphocytes as described by Melchers et al (1975a); G.A.L.T.: gut associated lymphoid tissue

there is no agreement on how many different types of V_L and V_H genes can be generated in an organism, a plausible number might be in the order of 100 V_L and 100 V_H. If only one of these different V_L and V_H genes independently was activated in a cell, then there could be 10,000 different combinations based on random pairing. Some form of active proliferation occurs before (as arbitrarily shown in the figure) or after V differentiation to expand the population of potential immunoglobulin producing cells and also to make available all of the possible types of V_L - V_H combinations. The mammalian organism and probably all vertebrates as well begin to generate these clones in fetal life (Owen et al., 1975) and probably continue to generate them throughout life. Trentin et al (1967) have shown the lymphoid cell precursor in an adult can regenerate the entire complement of V antibody forming cells.

The candidate for the ultimate precursor cell is the colony forming
unit stem cell CFU-S that can give rise to erythroid-granulocytic development as well. The conversion of the CFU-S to a V differentiated
cell has been difficult to study. Cells in the Ig-producing series are
large lymphocytes, and a closely related cell called pre B (PB) cells
(Lafleur et al., 1972). Some of these early cells divide rapidly in
the bone marrow of adults and produce large numbers of small lymphocytes or resting virgin B cells (Miller and Phillips, 1975; Osmond and
Nossal, 1974; Osmond et al., 1975). Marrow resting virgin B cells have
a variable (usually low) amount of cell surface Ig which increases
somewhat on maturation. Most small lymphocytes do not divide in the
marrow, but are released into the circulation, where they now recircu-
late and populate peripheral tissue. The marrow also contains lymphocytes with high density cell surface Ig. These cells increase with
age and are thought to be memory cells (Osmond and Nossal, 1974). Pre
B in the marrow apparently do not divide to produce or make small
lymphocytes with surface Ig, but can give rise to Ig-producing lymphocytes in lethally irradiated recipients. Miller and Phillips (1975)
estimate the ratio of CFU-S/PB:B_2 per 10^3 nucleated cells in the marrow is: 0.5/13/140. Melchers et al. (1975) have studied membrane IgM
synthesis on B lymphocytes and described three types in nude mice.
Type I cells produce 7-8S IgM rapidly with a turnover time of 1 to 3 h,
are found in the bone marrow predominantly and are probably the counterpart of the virgin B lymphocyte (B_2 in Miller and Phillips' scheme).
Melchers et al., 1975 estimate that the amount of immunoglobulin produced could be synthesized from a single messenger RNA. The type II
lymphocytes are found in the lymph nodes and circulation, these cells
produce IgM slowly and are probably memory cells. Type III cells actively secrete IgM, and are most commonly found in the spleen. Early
Ig-synthesis involves the IgM and IgD classes; both of these forms
have additional domains in the heavy chains (C_H4) which have been pre-
liminarily associated with membrane binding. IgM appears first in
ontogeny (Vitetta and Uhr, 1975). Both IgM and IgD may appear on the
same cell (Abney et al., 1976). During this period little or no cell
proliferation is thought to occur and the cells are believed to be in
a resting stage.

Tumors of B lymphocytes have several special advantages, three of which
are mentioned here because of relevance to current problems in immunology and cancer research. First, tumors of B lymphocytes may be useful
in studying the molecular biology of V differentiation, specifically
as a source for enzymes, and altered forms of DNA that may be involved
in the process of differentiating the specific V genes. Second, B cell
tumors may permit the isolation and characterization of cell surface
structures that appear during the cell differentiation process and
which are involved in regulating proliferative activities. Third, the
recent successes with temporary cultures of B lymphocytes in semisolid
agar containing 2-mercaptoethanol (2-ME) by Metcalf et.al. (1975) and
Kincade et al. (1976), and in suspension cultures grown at proper cell
densities and supplemented with 2-ME and mitogens (Melchers et al.,
1975b; Zauderer and Askonas, 1976; Askonas et al., 1976; Kearney and
Lawton, 1975a, b) provide a source of normal B lymphocytes. These cells
can be compared with neoplastic counterparts and it can be asked in
this differentiated system: what is the biochemical basis of the neoplastic change, and can it be regulated?

While tumors of B lymphocytes have been known for many years, very few
inductive systems are available. Most emphasis in the following discussion will be placed on a system that has recently provided a source
of tumors of B lymphocytes in mice, the Abelson virus induced lymphosarcomas. Two other systems, the chemically induced lymphocytic neo-

plasms in SJL/J mice (Haran-Ghera and Peled, 1973) and the bursal
lymphomas in fowls will be briefly discussed.

Bursal Lymphomas in Fowls

The bursa of Fabricius develops in vivo between the 11th and 12th day
as a bud from the cloacal endoderm. This bud progressively extends and
invaginates and begins to form lymphoid follicules. Lymphocyte proliferation and development begins about the 12th day in follicles and
almost immediately IgM synthesis can be detected (Cooper et al., 1972).
Using rosettes as a means of detection Lydyard and Cooper (1976) have
demonstrated the sequential appearance of specific antigen binding
clones within follicles. Specific types of clones develop simultaneously in different follicles, and the clones expand in the embryo in
the absence of antigen. The importance of the bursa to experimental
immunology is that in embryonic and newly hatched chicks the bursa is
virtually the sole anatomic source of immunoglobulin producing cells
(see Cooper et al., 1969 for ref.). Bursal cells are released where
they migrate to germinal centers and bone marrow. Toivanen et al.
(1972a, b) have studied the precursor-Ig-committed cell relationship
in the bursa. For this purpose they treated chicks with cyclophospha-
mide which destroyed bursal lymphocytopoiesis, but left behind a rem-
nant which could be regenerated with viable precursor cells. They
found that bursal lymphocytopoiesis and immunoglobulin production could
only be restored with embryonic or chick bursal cells but not by yolk
sac, embryonic or post embryonic liver, spleen, thymus or bone marrow.
Adult bone marrow was able to restore immunoglobulin production but
not bursal lymphocytopoiesis. These findings indicate that V differentiation probably take place in the bursa itself, and may depend upon
factors supplied by the epithelium.

Tumor of bursal lymphocytes can be induced by injecting strains of
type-C Avian leukosis virus (e.g. RPL-12, RPL-39, RAV-1 and F42) (Cooper
et al., 1974) into newly hatched white leghorn chicks. In a sequential
study of tumor development Cooper et al. (1968) found that the earliest
tumor developed in bursal follicles at 8 weeks of age. In "transformed
follicles" immature, primitive lymphoblastic cells were found in increased numbers. The entire follicle was greatly enlarged. Between 8
and 20 weeks only single follicles in a bursa were involved, thereafter
multiple follicles, and finally evidence of visceral lymphomatosis
appeared. Bursectomy several months after viral infection prevents the
development of visceral lymphomatosis (Peterson et al., 1966).

Cooper et al. (1974) have studied immunoglobulin synthesis by bursal
lymphomas. White Leghorn Hybrids of line 15 and 17 were injected with
Avian leukosis viruses between 1 and 14 days of age. The tumors were
obtained from 17- to 21-week-old chickens that were selected for pal-
pable bursal tumors or for evidence of visceral lymphomatosis. The
cells from the tumors were examined for immunoglobulin production by
immunofluorescence staining with anti μ, γ, and α antisera. Of 27 tumors
studied, all were histologically large lymphocytic tumors with cell
surface IgM that varied in intensity. Also there were varying degrees
of cytoplasmic staining indicating the tumors were possibly also producing some IgM for secretion. Further, Copper et al. (1974) found
that there were elevations in the serum IgM levels in these chickens,
suggesting that the tumors were in fact capable of secreting IgM and
that the tumors were possibly polyclonal. Electronmicroscopy revealed
the cells were large lymphocytes containing many polyribosomes and
some rough endoplasmic reticulum. No evidence of IgA or IgG synthesis
was observed. From these findings it would appear that the bulk of the
tumors were of cells in 'generative' stages of differentiation.

Abelson Virus-Induced Lymphosarcomas and Plasmacytomas in Mice

Origin of Abelson Virus. The Moloney leukemia virus is a highly infectious C type RNA virus that induces predominantly thymic lymphocytic neoplasms in mice of many strains. In a study designed to determine the leukemogenic properties of Moloney leukemia virus in mice whose thymuses were maintained in an atrophic state by chronic corticosteroid administration, Abelson and Rabstein (1970a, b) found a mouse with an unusual lymphosarcomatous process. This mouse had been treated with prednisolone since birth and inoculated with Moloney leukemia virus at one month of age. When the mouse was 93 days of age enlarged cervical and inguinal lymph nodes were observed; thereafter the mouse developed tumors that appeared over the thorax and hip bones. Histologically the tumor process consisted of a lymphocytic neoplasm that extensively involved the marrow cavities of the cranial, vertebral and hip bones and peripheral lymph nodes. The thymus was not involved. The remarkable aspect of this tumor process was its rapid development within 65 days after the injection of virus.

Cell free extracts prepared from the original tumor and from transplants yielded an agent that reproduced the same type of bone marrow and lymph node tumors when injected into newborn or adult BALB/c mice. In these mice also the thymus was never involved and the tumors appeared with extraordinary rapidity. The explosive proliferation of tumor cells in the bone marrow cavities was responsible for the appearance of paraskeletal tumors in the meninges along the vertebra and long bones. Many mice develop paraplegia from compression of the spine.

Virological Studies. Virological studies of the Abelson virus indicated that the virus was a new strain of type C RNA Murine leukemia virus. Scher and Siegler (1975) established that the Abelson virus contained two type C RNA viral genomes; one as expected was the NB tropic Moloney leukemia virus and the other was a new component that was found to be replication defective. With the same types of methods that had been used with other complex C type RNA viruses in mice (the Murine sarcoma viruses and the Friend erythroblastic leukemia virus) Scher and Siegler (1975) were able to dissociate the two components. It was first shown that Abelson virus transformed flattened BALB/c and NIH Swiss 3T3 fibroblasts into foci of rounded highly refractile elongated cells in 3 days. The morphologically transformed cells produced typical Abelson virus and further upon transplantation to syngeneic recipients grew into fibrosarcomatous tumors. The formation of foci then provided an assay for transformation by Abelson virus. The Moloney virus component can be readily assayed and quantitated by the XC plaque test.

By infecting 3T3 cells at limiting dilution, cells producing only Moloney virus were obtained. While this finding strongly suggested dissociation of the two components, convincing proof came from the isolation of non-Moloney virus producing transformed cells. From these cells complete Abelson virus could be recovered or rescued by infection with Moloney virus or Gross MuLV. These findings then indicated the Abelson virus contained a complete replicating C type virus which in the case of the original isolate was the Moloney MuLV and a second element which was replication defective and transforms fibroblasts. The Abelson defective component then uses the genes of the Moloney virus to replicate (helper effect); these are integrated in the viral particles that are formed so that the released virus transmits both elements. Detailed virological studies of this remarkable and awesome virus have not been completed. It is not known for example whether a complete replicating xenotropic mouse MuLV could be coupled with the Abelson defective component. Such a virus could hypothetically replicate and possibly transform cells from species other than the mouse. If a xenotropic Abelson virus could induce lymphocytic neoplasms in vitro in other species, it could be a hazardous laboratory agent.

Transformation of Lymphocytes in vitro. In addition to its ability to transform fibroblasts in vitro, Abelson virus (MLV-A) can transform lymphoid cells from fetal mouse liver (Rosenberg et al., 1975). While a morphological manifestation of lymphoid transformation was not noted in vitro, transformed cells were detected by the appearance of rapid proliferation when supplemented with mercaptoethanol, an agent that is required by normal B lymphocytes (Metcalf et al., 1975) in vitro.

In liquid cultures grown in plastic petri plates, fetal liver cells infected with MLV-A, adherent cells and granulocytic cells were first observed in the fluid phase (Rosenberg et al., 1975). After 10 to 15 days in transformed cultures a new population of rapidly proliferating cells appeared which 5 to 7 days later attained a population density of 3 to 5 x 10^6 cells/ml. These cells could be propagated for up to 2 months by the addition of 50 μM mercaptoethanol to the medium. The transformed lymphocytes were shown to have surface Ig by immunofluorescence when grown in the absence of mercaptoethanol, but with cells grown in the presence of mercaptoethanol less surface Ig was demonstrable.

When the rapid growing lymphoid cells were injected into mice, they produced lymphosarcomas. The rapidity with which these lymphosarcomas appeared suggested to Rosenberg et al. (1975) that the lymphosarcomatous process was due to transferred cells and not to induction by virus propagated in vitro. This however is not proof that the cells induced in vitro were in fact the cells that formed tumors in the recipient, since a large dose of virus can mimic this effect. Sklar et al. (1974) have transformed spleen cells in vitro with Abelson virus by exposing the cultured cells for 4 to 6 days in vitro and then transferring them to F_1 hybrids, compatible recipients which have an unusual chromosome marker. In this study transformation was demonstrated to have occurred during the culture period.

The development of a rapidly proliferating lymphoid cell population after 10 to 14 days in vitro is compelling evidence that the cells were transformed in the cultures (Rosenberg, 1975), and that the whole process of lymphoid neoplastic development can be studied in vitro. Rosenberg and Baltimore (1976) have also developed a quantitative assay for transformation of lymphocytes by Abelson virus. Using semi-solid cultures of adult bone marrow cells in agarose medium they obtained large colonies of lymphocytes in 12 to 15 days that could be readily subcultured.

Spectrum of Tumor Induced in vivo by Abelson Virus. In vivo MLV-A induces lymphosarcomas in newborn mice of many strains including BALB/c NIH/Swiss, C3H, C57BL/6 and DBA/2 (Abelson and Rabstein, 1970b). In adult mice the most susceptible strain is BALB/c and most other strains tested, with the possible exception of DBA/2, are resistant. We have found C57BL for example to be completely resistant, while adult F_1 hybrids of BALB/c and C57BL/6 are susceptible (R. Risser and M. Potter, unpublished observations).

We have studied the tumor-inducing properties of MLV-A in adult BALB/c mice that have been previously treated with an intraperitoneal injection of pristane (Potter et al., 1973). We had previously observed

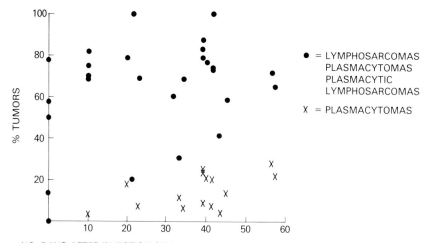

Fig. 2. Incidence of lymphosarcomas and plasmacytomas wihtin 125 days in pristane conditioned BALB/c mice infected with Abelson virus. Summary of 26 experiments involving BALB/c 571 mice infected intraperitoneally with Abelson virus. Percentage of mice developing both tumor types indicated by ●; percentage of mice with plasmacytomas indicated by X. Plasmacytomas were not observed in mice given only virus. However, progressively after injection of 0.5 ml pristane i.p. a higher incidence of plasmacytomas was obtained

that MLV-A induced lymphosarcomas in newborn BALB/c mice appeared to grow more rapidly in mice that had been conditioned with an intraperitoneal injection of pristane (Sklar et al., 1974). Some of the pristane treated mice bearing IP transplants of MLV-A induced lymphosarcomas developed plasmacytomas coincidentally. This finding led to determining whether MLV-A might induce plasmacytomas in pristane conditioned mice. It was found that MLV-A induced lymphosarcomas in pristane conditioned mice in high yield and in addition a modest number of plasmacytomas. Remarkably these plasmacytomas appeared very rapidly after the injection of virus, as did the lymphosarcomas. Latent periods after virus could be as short as 25 to 30 days. The true latent period however must be calculated from the time of injection of the pristane and in the initial study (Potter et al., 1973) pristane was given 40 days before virus injection. The earliest appearing plasmacytoma following a single injection of pristane was in 138 days. The short latent periods of 70 to 80 days indicated that the MLV-A was accelerating the induction process. We have for the past three years been attempting to determine the effect of the length of pristane conditioning on the yield of plasmacytomas in hopes of finding a rapid method that produces a high yield of tumors. A study involving 571 mice in 26 different experimental groups is summarized in Figure 2. With virus preparations containing 10^{5-6} PFU/0.1ml by the XC test of Moloney virus and 10^{4-5} FFU (focus forming units, i.e. transformed foci on NZB/Q cells) the yield of tumors usually ranges from 55 to 100 %. The largest group (83 %) of these are lymphosarcomas; 17 % are plasmacytomas. Plasmacytomas appear to develop more frequently in mice that have been injected for over 20 days with pristane. As may be seen, in many experimental groups no plasmacytomas were obtained. The frequency of plasmacytomas increased as the time of conditioning was lengthened.

Table 1. Heavy chain classes expressed in plasmacytomas

	Man multiple myeloma	Lou/Ws rat spont. Ileocecal	BALB/c Mouse mineral oil pristane	BALB/c Mouse pristane Abelson virus	NZB Mouse mineral oil pristane
No.	270	250	558	93	230
	%	%	%	%	%
IgG	55.0	48.5	19.7	16.6	52.5
IgA	25.0	0.6	45.5	50.5	20.9
IgE	0.3	34.2	0.0	0.0	0.0
IgM	0.0	3.3	1.1	0.0	1.7
IgD	1.5	0.0	0.0	0.0	0.0
L chain	16.0	0.0	2.2	0.0	1.7
None	2.2	0.0	24.9	30.1	3.9
Polysecret.	0.0	0.0	6.6	3.2	9.6
Untypable	0.0	0.0	0.0	0.0	10.0
	Acute Leukemia Group B (1973)	Bazin et al. (1973b)	Morse et al. (1976)	Morse et al. (1976)	Morse and Weigert (1976)

The heavy chain classes represented in the myeloma proteins produced
in Abelson virus induces plasmacytomas are shown in Table 1 (Morse et
al., 1976). As may be seen, the distribution of heavy chain classes
is essentially the same as that seen in tumors induced by mineral oils
or pristane alone. This is evidence in favor of the hypothesis that
MLV-A is superimposing its effects on a process in progress in BALB/c
mice. If Abelson virus were inducing plasmacytomas by an independent
means then one should expect to see a different distribution of heavy
chain classes and possibly also induce plasmacytomas in strains other
than BALB/c that are susceptible to this virus but not to mineral oil
induced plasmacytomas. We have only preliminary evidence on this point.
The (BALB/c x DBA/2)F$_1$ hybrid is a good test animal for this purpose
as it is highly susceptible as an adult to Abelson virus lymphosarcoma
induction but very resistant to mineral oil or pristane plasmacytoma
induction (see Table 2). These mice were pre-treated with pristane,
and Abelson virus was injected at various intervals after conditioning.
Thus far we have only obtained lymphosarcomas in these hybrids with
Abelson virus.

A principle question now is: what relationship do the Abelson virus
induced lymphosarcomas have to the Ig clonal development and are they
related in any way to the plasmacytomas? Studies on these questions
have not yet been resolved but some information is available for eva-
luation.

We (N. Wivel and M. Potter, 1976) have recently recognized two morphological forms of lymphosarcomas, (1) a cytoplasmically undifferentiated
form whose cytoplasm contains only a few ribosomes and virtually no
rough endoplasmic reticulum (all the cells in a tumor of this type may

Table 2. Susceptibility to plasmacytoma development in different strains of mice given 3 intraperitoneal injections of mineral oild or pristane

Strain	No. PCT Total	%	Strain	No. PCT Total	%
BALB/c	219 373	58	DBA/2 SWR	0/36 0/30	0 0
C57BL/6,Ka	3 103	3	A/He	0/26	0
(B6 × BALB/c)F$_1$	16 227	8	AL/N	1/59	0
F$_1$ × BALB/c	15 71	20	[a]NZB	18/51	35
Bailey RI					
CXBD	15/53	28	(BALB/c × DBA/2)F$_1$	0/82	0
CXBE	11/63	17	[a](BALB/c × NZB)F$_1$	79/163	49
CXBG	51/85	36			
CXBH	0/58	0			
CXBI	2/63	3	[a](BALB/c × SJL/J)F$_1$	2/53	4
CXBJ	23/69	47	[a](BALB/c × NZC)F$_1$	0/56	0
CXBK	2/46	5	[a](BALB/c × C3H)F$_1$	1/28	3

[a]Data of N. Warner.
Data summarized from Potter (1975); Potter et al. (1975); Warner (1975).

have this common morphology); (2) a plasmacytic lymphosarcoma which is similar in size and general morphology to the undifferentiated type but in which 80 % of the cells have a moderate amount of rough endoplasma reticulum, many polysomes and a Golgi apparatus (Fig. 3). These cells in light microscopy with Wright's stained preparations have a plasmacytic appearance. Some tumors are composed entirely of undifferentiated lymphocytes, others entirely of plasmacytic lymphocytes, and many are mixed (Wivel and Potter, 1976).

Immunoglobulin Synthesis. Immunoglobulin synthesis by Abelson lymphosarcomas has been studied by Premkumar et al. (1975) in three tumors, ABLS-1, ABLS-5 and ABLS-8, which were originally induced by Sklar in newborn BALB/c mice (Sklar et al, 1974) and were established in transplant in pristane conditioned mice and then adapted to tissue culture. These three tumors in vitro produce small quantities of immunoglobulin, most of which is 8S IgM that is inserted into the plasma membrane. A small amount of 8S IgM immunoglobulin can be measured in the extracellular pool. The rate of incorporation of S^{35} methionine into intracellular and extracellular IgM molecules was determined for MOPC 104E

Fig. 3. Electronenmicrographs of typical cell types from the Abelson virus induced lymphosarcoma. (1) ABLS 140 and the plasmacytic lymphosarcoma 17,000x. (2) ABPL-2 = 11,000x. Insert bar = 1μ. Electronenmicrographs made by Dr. Nelson Wivel, National Cancer Institute

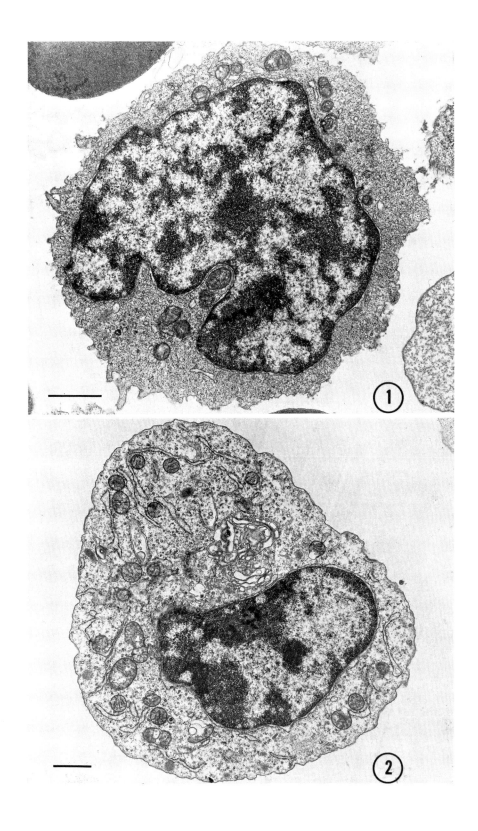

plasmacytoma and ABLS-5. The time at which the intracellular and extracellular pools reached an equilibrium was determined for both cell
tpyes, and it was found the MOPC 104E culture reached an equilibrium
in 4 hours while ABLS-5 cultures did not reach equilibrium until 24 h.
The rate of immunoglobulin production was very much slower in the
ABLS-5 lymphosarcoma. The amount of IgM actually produced was estimated
to be 100 times less on a per cell basis in ABLS-5 than with MOPC 104E.
The ABLS-8 lymphosarcoma produces a small amount of IgG simultaneously
with the IgM. A comparable study of immunoglobulin synthesis in plasmacytic lymphosarcomas is currently underway.

Several studies have been made on the cell surface antigens and markers
on Abelson virus induced lymphosarcomas. Sklar et al. (1974) found
that these tumors lacked theta antigen. More recently Warner et al.
(1975) have found that about 20 % of ABLS ascites cells were weakly
stained with fluorescent antibodies to IgM or mouse Ig. Warner et al.
(1975) also found weak Fc receptor activity. In incomplete studies
(Sklar et al., 1974) found no C3' receptors on Abelson lymphosarcomas,
but quite recently we have observed that the plasmacytic Abelson
lymphosarcomas do have C3' receptors. These were detected by formation
of rosettes with Zymosam particles and mouse complement.

Abelson lymphosarcomas possess a large amount of membrane associated
protein called MAID (Membrane associated immunoglobulin detaining protein) which has a molecular weight of about 35,000 and in SDS polyacrylamide gel electrophoresis of membrane extracts migrates in the
region between gamma H and L chains. To resolve Ig peaks it is necessary to remove MAID by a series of extractions with antigen-antibody
complexes.

The only Ly (lymphocyte alloantigen) thus far found on Abelson lymphosarcomas is Ly-4.1 (quoted by Warner, and determined by Dr. I.McKenzie).

ABLS appear to be proliferating virgin B lymphocytes with small amounts
of surface Ig. The plasmacytic lymphosarcoma may be derived from a more
advanced step of early generative B-lymphocyte development i.e. as the
controls regulating the formation of rough endoplasmic reticulum and
the Golgi apparatus have been activated.

It is of special interest to those interested in the genetic basis of
tumor susceptibility that the adult BALB/c mouse is susceptible to the
induction of Abelson virus lymphosarcomas as well as plasmacytomas,
this suggests that there may be a fundamental genetic link between the
two processes.

Chemically Induced Lymphocytic Neoplasms in SJL/J Mice

Much of the research on chemically induced leukemia in mice has been
concerned with thymic lymphocytic tumors which in young mice are the
most frequently observed inducible tumor type. In recent years the
interest in chemical leukemogenesis has shifted to working with newborn mice, where doses of carcinogens can be better regulated (Ball
and Dawson, 1969). In addition, much work was done in the area of
chemical leukemogenesis before the functional differences between T
and B lymphocytes were known. Haran-Ghera and Peled (1973) found that
thymectomized SJL/J mice were highly susceptible to the induction of
lymphocytic neoplasms by weekly feeding of 7,12 dimethylbenzanthracene
(DMBA). In thymectomized mice alone the incidence of lymphocytic tumors
was 52 %; this could be moderately increased to 70 % by exposing the
thymectomized mice to total body X-irradiation prior to DMBA adminis-

tration. Tumor cells from these lymphocytic tumors were studied by direct immunofluorescence with anti-Thy-1 and anti-Ig sera, and it was found that most of the cells had surface Ig and only a very few were Thy-1 positive. Linker-Isreaeli and Haran-Ghera (1975) in a more detailed study of the surface markers in these tumors found evidence for the presence of Fc and C3 receptors. Further surface immunoglobulin was found on 15 to 75 % of cells within a tumor. When the cells were incubated in vitro for two days and trypsinized it was found that surface Ig reappeared after 24 h, indicating the tumors were able to regenerate Ig. This potentially promising system offers a new source of B lymphocytic neoplasm in mice, and the general methodology might be applicable to other strains.

Comment

It is difficult with the present data accurately to align tumors of B-lymphocytes with normal counterparts. The Abelson virus-induced lymphocytic tumors appear to express low amounts of surface Ig, and have variable amounts of rough endoplasmic reticulum. These tumors thus appear to lack the ability to synthesize and secrete the large amounts of Ig produced in later stages of development. The maturation of this step (increased Ig synthesis and secretion) appears to be blocked or repressed by the neoplastic state. Abelson lymphosarcomas are most likely tumors of PreB, or virgin lymphocytes, some may be even tumors of memory cells. The Avian Bursal tumors appear to more heterogeneous in Ig production, as they are often associated with elevated levels of serum IgM.

Tumors of Immunoglobulin Secreting Cells

Immunoglobulin-Secreting Tumor Systems

Activation of the B lymphocyte is initiated by encounter with antigen participating T lymphocytes, macrophages and mitogenic substances. The mechanism of the process is not agreed upon and there may be several different pathways for activating B lymphocytes to proliferate and undergo further differentiation. Regardless, the outcome of activation is vigorous cell division, morphologic change into a large lympho-blastic type cell, development of the organelles for immunoglobulin secretion (endoplasmic reticulum and Golgi apparatus) and a change in the character of immunoglobulin production from membrane incorporation to a secretory program. The IgA, G (and subclasses) E classes appear. Further some cells actually secrete IgM or even IgD immunoglobulins. The heavy chain class differentiations of immunoglobulin producing cells might be termed physiological classes, e.g. the IgA immunoglo-bulins are involved with production of secretory immunoglobulin in the lamina propria of the gut and respiratory tracts (see Lamm, 1976, for review). IgE which adheres to mast cells via a specific IgE receptor is involved in allergic reactions (Ishizaka, 1972). IgE in man is produced in gut and respiratory tract associated lymphoid tissue, e.g. tonsils, adenoids, bronchial and mesenteric lymph nodes, Peyer's patches, and lamina propria of the gut (Ishizaka, 1972). IgG is produced predominantly in lymph nodes and spleen.

These heavy chain class differentiations appear to have separate regulators that govern tissue distribution and cell proliferation. It is not surprising then that segments of the immunoglobulin secreting cell population appear to be selectively involved in plasmacytoma formation. The H chain classes of myeloma proteins associated with five different clinical and experimental plasmacytoma systems are shown in Table 1. As may be seen there are marked differences in the different systems. BALB/c has a remarkable predilection to develop IgA producing plasmacytomas, while NZB produces predominantly IgG class proteins (H. Morse and M.G. Weigert, unpublished observations). Rat immunoglobulin producing tumors have an extraordinarily high incidence of IgE producers. In human myeloma the prevailing class in IgG (Acute Leukemia Group B, 1973). Bazin et al. (1973a) discovered that spontaneous ileocecal tumors in strain Lou/Wsl that arise in the ileocecal lymph node are immunoglobulin secreting tumors. Tumors like these have been observed for many years in different strains of laboratory rats. The inbred Lou/Wsl, however, developed an incidence between 10 and 20 % from 12 to 16 months of age. The histogenesis of these tumors has not been studied but it would appear because of their origin in a specific lymph node that drains the cecal region in the rat, that the precursors of these cells are involved with some specific aspect of a gut associated immunity. Further support of this notion derives from the fact that an extraordinarily high percentage of these tumors produce IgE type immunoglobulins (Table 1). Investigation of this model system offers great promise for learning more about cellular aspects of the IgE response and how this might be involved in a neoplastic transformation of a segment of the immunoglobulin secreting population.

One of the most useful by-products of Ig secreting tumors is to find one that has antigen binding activity. These proteins are essentially homogeneous antibodies and can be used for studying the three-dimensional structure of antigen binding sites by X-ray crystallography (when the protein can be crystallized), the physical chemistry of hapten-protein interaction, and the antigenic structure of the V regions (idiotypes).

Antigen binding activity has been detected in 1 to 5 % of BALB/c myeloma proteins to a selected series of antigens (Potter, 1970, 1971). A current summary of common activities found both at NIH and at the Salk Institute by Dr. Melvin Cohn is listed in Table 3. These activities were determined by screening myeloma sera with available test antigens and others which were prepared from the normal bacterial flora of the gut (Potter, 1970, 1971). As may be seen in Table 3, for some antigens, e.g. those containing phosphorylcholine, bacterial levan, and various B1-6-linked galactans, there are myeloma proteins from 8 or more independently derived tumors that bind each of them.

An intriguing aspect of both BALB/c plasmacytomas is the high incidence of polysecreting tumors (Morse et al., 1976) that can be identified in primary sera from primary hosts bearing these tumors. There are several reasons for this: first, many tumors may be biclonal, composed of more than one V differentiated cell type; second, one tumor actually secretes two immunoglobulins, that differ in C_H class but share the same V_H.

A fundamental question concerns the switch from membrane Ig synthesis (IgM, IgD) to a secretory program (IgA, IgG, IgE or even IgM and IgD). In a given Ig clone it is assumed the same V_H genes are ligated on to different C_H genes during differentiation and that in the switching process separate signals activate specific types of synthesis. The mechanism of this process is not established. The first convincing evidence to support it came from the study of the Til biclonal myeloma in which cells were found that produced alternatively IgM or IgG with the same V regions on each (Wang et al., 1970). Morse et al. (1976)

Table 3. Antigens and haptens bound by BALB/c myeloma proteins

Antigens	Haptens	No. Proteins
B512 dextran	1,6 glucan	3
B1355 dextran	1,3 glucan	3
Aerobacter levan, inulin	β2,6 β2,1 fructan	2,11
Gum ghatti, wheat, Hardwood galactan	β1,6 galactan	8
Pneumococcus C polysacch. Antigens from Lactobacillus Aspergillus, Trichoderma, and Ascaris Proteus morganii l.p.s.	Phosphorylcholine	8
Salm. tranoroa, Tel Aviv, Proteus mirabilis l.p.s.	α-CH$_3$ D-galactoside α-CH$_3$ D-mannoside	1 1
Salmonella flagellin	?	2
Streptococcal A. poly sacch.	N-acetyl D-glucosamine	1
Salm. weslaco l.p.s.	N-acetyl D-mannosamine	1
Mima polymorpha l.p.s.	?	1
E. coli	?	1
Nitrophenylated proteins	Nitrophenyl	3
Dansylated proteins	Dansyl	1
Polyvinyl pyrrolidone		1
Dextran sulfate		1 ?
Strain A L/N rbc antigen		1

have recently described a transplantable mouse plasmacytoma TEPC 609 that produces two secretory immunoglobulins, IgA and IgG2b.

By immunofluorescence staining single cells in TEPC 609 can be found that have both IgA and IgG2b myeloma proteins in their cytoplasm (Morse et al., 1976). These tumors should provide an unusual source of RNA and DNA components that reflect the mechanism of the IgC$_H$ switching process.

The following discussion will concern some details on the peritoneal plasmacytoma system in BALB/c mice.

Peritoneal Plasmacytomas in Mice

The induction of plasmacytomas in mice has been reviewed several times recently (Potter, 1972, 1975; Warner, 1975) and a brief summary of basic facts will be presented followed by some new information on pathogentic mechanisms (Cancro and Potter, 1976).

A method for inducing plasmacytomas in mice was fortuitously discovered in 1958 (Merwin and Algire, 1959) when it was found that BALB/c mice implanted with Millipore diffusion chambers containing allogeneic tumor tissue evoked the formation of peritoneal plasmacytomas and fibrosar-

Table 4. Substances injected or implanted intraperitoneally into BALB/C mice that evoke plasmacytomas

---

Solids

Millipore diffusion chambers + tissue

Millipore diffusion chambers

Millipore membranes

Plastic discs

Plexiglas borings

Liquids (0.5 ml x 1 or x3 q30 to 60 d)

Mineral oils: Bayol F, Primol D, Drakeol 6VR

Pure Alkanes: 2,6,10,14 tetramethylpentadecane (pristane)

              7N-hexyloctadecane, phytane

Adjuvants: Complete Freund's Adjuvants

---

comas. This finding immediately focused attention on the BALB/c mouse and very soon other agents such as mineral oil containing adjuvants, mineral oils and pure alkanes which could be injected intraperitoneally were also found to induce plasmacytomas in BALB/c mice. A list of active components is given in Table 4 (for references see Potter, 1975). While these substances are very different in chemical composition, they appear to have in common the inability to be removed from the peritoneum once implanted or injected. In addition they all stimulate the formation of a granulomatous process which is fibroblastic in character with the solid materials, and an oil granuloma with the liquid materials. The oil granulomas which have been most extensively studied arise on the messenteries, abdominal walls and diaphram. The plasmacytomas appear to develop in this abnormal tissue and usually remain there.

Susceptibility to Plasmacytoma Induction is Strain Dependent. The ability of mineral oils and plastics to induce plasmacytomas in mice is restricted to only two inbred strains thus far, BALB/c and NZB (Merwin and Redmon, 1963; Potter et al., 1975; Potter, 1967; Warner, 1975) or to recombinant inbred strains derived from BALB/c (Table 2). Two possibilities are that susceptible strains have a specific type of C type RNA virus incorporated into their genomes which because of the integration sites in the genomes or by virtue of some special intrinsic property of the viral genome itself, determine susceptibility. An alternative view is that specific alleles of germ line genes present in these two strains influence susceptibility. For example there might be differences in the immune responses that are determined by immunoglobulin structural genes or regulators of the immune response. There is virtually no hard evidence to support either of these factors.

Early Plasmacytomas are Conditioned Neoplasms. The plasmacytoma cells are shed into the peritoneal fluid where they further, rather regularly, cause erosion of peritoneal blood vessels with bleeding. Free tumor cells for diagnosis and propagation can be obtained by paracentesis. Very early in our studies on transplanting plasmacytomas we found that free ascitic plasmacytoma cells were ineffective in initiating transplant lines. For this reason the tumors were transplanted subcutaneously

Table 5. Transplantation of 10^5 primary plasmacytoma cells to conditioned mice

Conditioning		Effects			No. mice with progressively growing tumors	
Day O	Day 3 transplant +	Total APC x 10^6	Granuloma formation	No. tumors tested	Total	(%)
Pristane 0.5		3.86	+	23	227/228	99
Pristane 0.5 Cortisone		0.72	–	11	32/150	21
Pristane 0.5	Cortisone	3.63	+	11	150/150	100
None		0.43	–	23	5 /212	3
Thioglycollate Medium (TGM) Continuous		5.22	–	6	0/60	0
TGM-APC Cortisone Pristane 0.5		N.D.	+	3	30/30	100

Summarized from Potter et al. (1972) and Cancro and Potter (1976)

by implanting small pieces of involved oil granulomatous tissue from the mesenteries by a trochar. The reason for the inefficiency of the free plasmacytoma cells to establish transplants was not explained until 1972, when it was found that primary plasmacytoma cells grew very well when they were transplanted back into the same type of microenvironment from which they were taken (Potter et al., 1972). This environment could be simply produced by injecting mineral oil or the pure alkane pristane into the recipient 1 to 60 days before transplantation. We have now had experience with many tumors and found this to be a very general characteristic of early (primary) plasmacytomas; in fact when 10^5 viable plasmacytoma cells are injected into normal recipients the tumors almost universally fail to grow, but when the same numbers or less are implanted into conditioned mice virtually 100 % grow progressively (Table 5).

Takakura et al. (1966) found that when BALB/c mice were injected with mineral oil and then treated continuously with cortisone they developed a very reduced incidence of plasmacytomas. Oil granulomatous tissue did not form in these treated mice. Cortisone has many effects on lymphoreticular cells but very specifically and effectively it appears to shut off the influx of blood monocytes into the peritoneal cavity (Thompson and Van Furth, 1970, 1973) by blocking the release of precursors of blood monocytes from the bone marrow. Corticosteroids apparently do not affect the proliferation or function of macrophages in the inflamed peritoneal cavity (Thompson and Van Furth, 1970). These findings prompted us to determine if corticosteroids would influence the development and physiology of the conditioned state. Accordingly hydrocortisone acetate was administered to mice immediately after the injection of pristane and then 3 days later 10^5 primary plasmacytoma cells were injected intraperitoneally into the mice. In all of the 23 primary plasmacytomas tested in 212 mice, only 1 grew in 100 days in the normal recipients while all 23 grew in the pristane conditioned mice. There was a remarkable reduction in the number of progressively growing plasmacytomas in the hydrocortisone treated mice (Table 5): they grew in

158

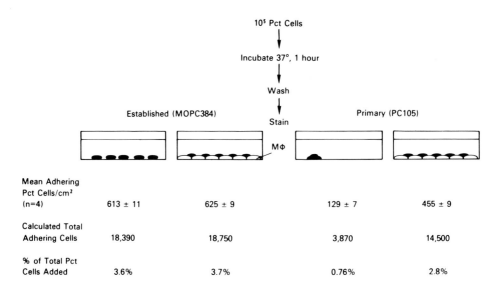

Fig. 4. Effect of attached macrophages on adherence of established and primary plasmacytoma cells (PCT) in vitro. 10^5 PCT from an established transplant line (MOPC 384)
and from a primary tumor (Pc 105) were added to 60 mm Falcon dishes with and without
a layer of attached macrophages. After 1 h incubation at $37^{\circ}C$ plates were washed and
number of attached plasmacytoma cells enumerated. MOPC 384 cells attached equally
well to both types of plates while PC 105 attached about 4x more efficiently to plates
containing attached macrophages. This property is characteristic of all pristane induced primary plasmacytomas so far tested. Established transplants acquire the ability
to attach to both types of plates (Cancro and Potter, 1976)

only 20 % of the mice. To determine if the effect of the hydrocortisone
was toxic to the primary plasmacytomas, pristane was administered three
days before transplantation and hydrocortisone was given simultaneously
with the transplant. In this situation 100 % of the mice developed progressively growing transplants. Thus hydrocortisone was not toxic to
the plasmacytoma cells.

The effects of hydrocortisone on the formation of oil granuloma and the
appearance of free adherent peritoneal exudate cells were then studied.
We found a dramatic increase in the total number of free adherent peritoneal exudate cells (APC) from an average of 430,000 in the normal
mouse peritoneum to 3.86×10^6 in the pristane conditioned mouse peritoneum. These cells remained at this high level for at least 60 days
after a single injection in pristane. The simultaneous administration
of hydrocortisone with the pristane reduced the numbers to near normal
levels. However, if hydrocortisone was administered 3 days after the
pristane then high levels of free APC were again obtained. To determine
whether the free APCs were responsible for the conditioned environment
essential to primary plasmacytoma growth, mice were injected with
thioglycollate medium (TGM), a material that evokes a brisk production
of APC. Chronic TGM administration elevated the free APC levels but did
not result in the formation of an oil granuloma or the development of
a conditioned environment. When TGM induced APC were passively transferred into a mouse that had been injected with pristane and simultaneously treated with hydrocortisone, the conditioned environment and

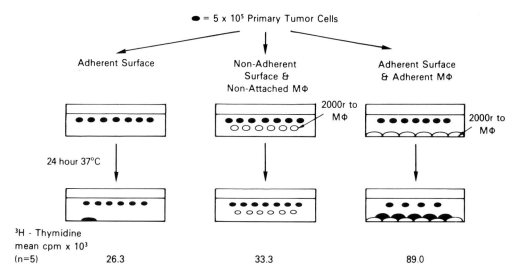

Fig. 5. Effect of adherent cell layers on DNA synthesis by primary plasmacytoma cells. Primary plasmacytoma cells (PCT) which inefficiently adhere to 60 mm Falcon plates incorporate ^{3}H-thymidine during 4-h pulse after 24 h at 37°C (plates on left). About the same degree of thymidine incorporation is obtained when cells were added to dishes with nonadherent surfaces and 10^4 irradiated macrophages (middle). Over three times (3.4 x) as much incorporation was obtained when the PCTs adhered to attached irradiated macrophages (right) (Cancro and Potter, 1976)

oil granuloma were produced (Table 5). These findings indicated that the granulomatous tissue itself was the essential requirement for the primary plasmacytoma and not just the elevation in free APC, and further that this could be produced by macrophages that were adherent to the mesenteric and peritoneal tissues.

An in vitro system to assay primary plasmacytoma growth was then developed. Primary plasmacytoma cells were placed in Falcon plastic dishes containing medium, or in dishes that had been seeded with a layer of adherent peritoneal macrophages or 3T3 fibroblasts. After 1 hour the dishes were washed and the number of adhering plasmacytoma cells were enumerated. With primary plasmacytoma cells 4 to 5 times as many cells were found to be adhered in the dishes with an underlying adherent layer. For controls we used established plasmacytomas and found that there was no difference between the number of adhering cells in the two types of culture dishes (Fig. 4).

We studied the ability of primary plasmacytoma cells under varying conditions to incorporate tritiated thymidine into DNA (Cancro and Potter, in prep. 1976) (Fig. 5). Primary plasmacytoma cells overlayed on an adherent layer actively incorporated 2- to 4-fold more tritiated thymidine into DNA, while primary plasmacytoma cells in plastic dishes without underlying adherent cell layers or primary plasmacytoma cells added had a very much reduced rate of incorporation. To control the influence of the macrophages and to rule out the possibility that a soluble factor was being released from the macrophages, primary plasmacytoma cells and macrophages were placed in Petri dishes (i.e. untreated plastic dishes) in which adherence does not take place, and

160

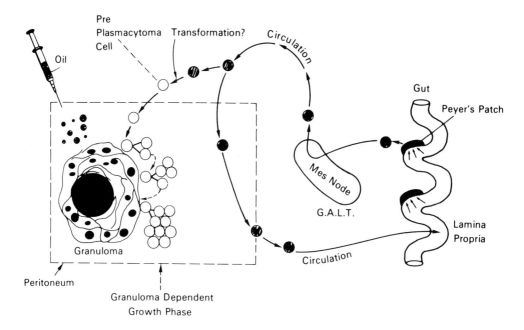

<u>Fig. 6.</u> Hypothetical scheme of oil-induced peritoneal plasmacytoma development in
BALB/c mice. Plasmacytoma depicted as arising from IgA secretory system of cells,
since about 60 % of all oil-induced plasmacytomas express IgA. Other sources of
precursors are not excluded. Basic hypothetical events in scheme: transformation of
a precursor into a preneoplastic cell, and selection growth and development of this
cell on oil-granuloma

the DNA synthesis was similar to that obtained with primary plasmacytoma cells added alone to regular plastic dishes. These findings indicated that the adherence of primary plasmacytoma cells to underlying
layers of macrophages or other adherent cell layers enhanced subsequent
DNA synthesis.

All of these findings suggest that the primary plasmacytoma in the mouse
is a conditioned or dependent tumor.

<u>Comment.</u> The immunoglobulin secreting tumors of plasma cells appear to
develop from abnormalities affecting the late stages of development and
differentiation. In this phase the cells, after a direct extension as
in the migration from cortex to medullary cords in the lymph nodes, or
by entering the circulation and then homing into the submucosal tissues
of the gut and respiratory tracts, mature to plasma cells gradually,
cease to divide and are finally eliminated. The neoplastic change in
effect abrogates this process and blocks the capacity of the cells to
complete these final steps. The formation of an abnormal tissue microenvironment such as the one induced by the injection of oily non-metab-
olizable substances into the peritoneum plays a selective role in neoplastic development by permitting abnormal cell populations to expand.
This expansion appears not to affect the plasma cell population as a
whole, because the oil granumola is not populated by large numbers of
hyperplastic plasma cells in all stages of differentiation and development, such as might be seen in the medullary cords of lymph nodes or
lamina propria. Thus one is led to the hypothesis that the oil granu-

loma is selecting out specific cell types from the circulating population at large. These cells upon attachment to the subtratum recycle,
and if they can reattach they are able to continue in cycle. During
this period it is conjectured that the continuous divisions provoke
the cells adaptively to develop the machinery for this mode of life.
Thus we were able to show the gradual acquisition during transplantation of a cell type that can efficiently attach to plastic dishes
(Cancro and Potter, manuscript in preparation).

The initial defect probably occurs before the first cell enters the
abnormal peritoneal environment, and could be caused by a variety of
several different mechanisms. A hypothetical scheme of plasmacytoma
development, in an IgA precursor, is shown in Figure 6.

Tumors of T Lymphocytes

Development and Differentiation of Lymphocytes in the Tumors

It is now well established that those lymphocytes which develop in the
thymus (T lymphocytes) and are released into the circulation have different physiological roles in immune responses from the B lymphocytes
that are triggered to become immunoglobulin secreting cells. The functions of T lymphocytes in immune responses are: (1) to act as helper
cells by specifically triggering B lymphocytes that bind a mutual antigen; (2) to recognize non-self antigen determinants on cellular
structures and via cell - cell interactions kill the foreign cell (i.
e. cytotoxic cell); this function is usually assayed in graft versus
host reaction or allogeneic graft rejections; and (3) to liberate
lymphokines that: (a) attract macrophages to inflammatory sites, (b)
activate macrophages to increase their phagocytic activities, and (c)
non-specifically stimulate lymphocyte proliferation during immune responses.

The early development and differentiation of the T lymphocytes in
vertebrates takes place in the thymus (Fig. 7). These lymphoid organs
have a network of epithelial cells that extends throughout the cortex
and medulla and presumably provides an essential microenvironment for
thymocytopoiesis (Goldstein, 1975). In vitro the polypeptide hormone,
thymopoietin, induces percursor cells (pre-T cells) to produce theta
antigen (Thy-1) and other thymus specific cell surface components such
as the TL antigen (Basch and Goldstein, 1975). The precursors of thymocytes (thymic lymphocytes) are generated in the bone marrow and
migrate to the thymus in embryonic or adult life. In the absence of
thymic injury this process may occur very slowly in the normal thymus.
Thymic lymphocytopoiesis begins with proliferation of subcortical
lymphocytes (see Cantor and Weissmann, 1976, for review and references)
which are cells capable of self-renewal. The cells derived from division of subcortical lymphocytes undergo rapid mitosis, a process which
pushes cells towards the thymic medulla. Bryant (1972) has estimated
that the mouse thymus produces 50×10^6 new thymocytes each day, enough
to replenish the circulating pool 5 times over. More thymocytes are
produced than are used each day. Many are thought to be eliminated in
the thymus, many are released into the circulation. In the periphery
the T lymphocytes carry out the essential functions in immune responses.
The thymus in the normal state is not an immune organ (in fact antigens
are partially excluded from the thymus by a blood thymus barrier (Clark,
1973), but rather a generative lymphocytopoietic tissue.

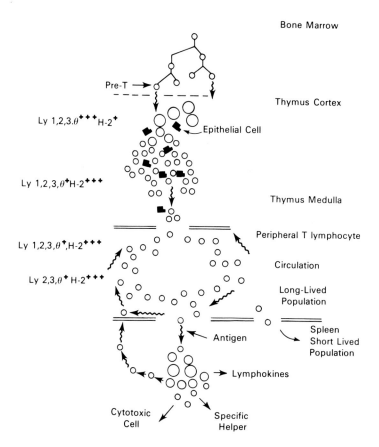

Fig. 7. T lymphocyte development (see text)

The localization of T lymphocyte generation in a single tissue site
has provoked many studies on the role of the thymus in immunity. For
many years it was known that thymectomy in adults did not impair immune functions or survival; however, Miller (1961) demonstrated that
neonatal thymectomy resulted in severe immunological impairment.

The chief function of the thymus is to generate diverse types of T
lymphocytes and release them to the periphery. This process occurs
most actively in the mouse in the first few days of life; neonatal
thymectomy (Miller, 1961) results in severe immuno-incompetence. The
thymus in adult life continues to generate new T lymphocytes but the
organism is apparently less dependent on this source. Adult thymectomy
(see Cantor and Weissmann, 1976) does not produce immediate incompe-
tence, and indeed the normal levels of recirculating, long-lived (in-
frequently cycling) T lymphocytes are maintained for many months (see
Cantor and Weismann, 1976). Adult thymectomy in a matter of several
weeks reduces a splenic population of T lymphocytes. It was postulated
that these (rapidly dividing) cells were recently released from thymus,
and were still dependent on the thymic epithelium.

Adult thymectomy followed by total body X-irradiation does produce immunological imcompetence (Miller et al., 1963), and this can be cor-

rected by a thymic graft. Thus the thymus can throughout life provide a reserve of competent T lymphocytes.

Tumors of T Lymphocytes and Thymocytes

Thymic lymphocytic tumors in mice occur spontaneously in 90 % of the mice of so-called high leukemic strains such as AKR and C58 (see Rowe, 1973) or can be induced by a variety of methods such as total body X-irradiation (Kaplan, 1967) chemical carcinogens (Kirshbaum, 1955; Joshi and Frei, 1970; Ball, 1968), leukemogenic transmissable type C RNA viruses such as Moloney leukemia virus (Moloney, 1962) or Gross passage A (Gross, 1959; Axelrad and Van der Gaag, 1962). A most important feature in some of these systems is that the lymphocytic neoplasms can be prevented by thymectomy (Furth, 1946; Law and Miller, 1950; Kaplan, 1967), indicating the leukemogenic process takes place in the thymus. Thymectomy is not effective in all situations, as for example when it is delayed 30 days after injecting newborn mice with Gross passage A virus (Gross, 1959) or when an overwhelming amount of a chemical carcinogen is administered (Kirschbaum, 1955; Nishizuka and Shisa, 1972).

There have been many studies on the preneoplastic changes in the AKR thymus. Cortical atrophy and medullary hyperplasia have been described in great detail by Metcalf (1966) as preceding the actual appearance of leukemia in the AKR thymus. When the thymic cortex atrophies spontaneously as it does in 4- 5-month-old AKR mice there is a collapse of the epithelial cells around the medulla. Siegler and Rich (1963) and Siegler et al. (1966) have shown this can occur unilaterally. Thymic lymphocytopoiesis in the gland with the atrophic cortex is apparently nonfunctional. It is in this kind of abnormal microenvironment that the thymic lymphocytic tumors appear. The late atrophy of the cortex is to be distinguished from the acute atrophy (lymphocytolysis) induced by corticosteroids, irradiation, starvation and toxic chemicals. The thymus cortex can regenerate following limited acute challenges or even after prolonged stress induced by corticosteroids. Most leukemogens induce a temporary cortical depletion followed by recovery (Joshi and Frei, 1970; Ball, 1968) and then 2 to 4 months after recovery a second more irreversible type of cortical atrophy ensues. It is in this second phase that the thymic tumors appear. Chazan and Haran-Ghera (1976) have demonstrated that: (1) after exposure to leukemogenic doses of fractional total body X-irradiation, (2) after single doses of total body X-irradiation and injection of radiation leukemia virus or (3) at the time of maximal spontaneous leukemia development in strain AKR mice; the thymus glands were regenerated or contained thymocytes with relatively high surface H-2 alloantigen (characteristic of peripheralized T lymphocytes). These cells appeared in irradiated mice in glands that had apparently regenerated a cortex and attained a normal size.

Most T lymphocytic tumors have been isolated from the thymic tumors. Tumors of peripheralized T cells probably exist but have not been as extensively studied or sought. In fact, many lymphocytic tumors that arise in thymectomized mice (Kirshbaum, 1955; Hiai et al., 1972; Haran-Ghera and Peled, 1972) may lack T lymphocyte markers and be tumors of B lymphocytes (Haran-Ghera and Peled, 1972). The fact that a large segment of the peripheral T lymphocyte population is long-lived and (slowly dividing) may make them relatively refractory to neoplastic-induction, a mitogenic stimulus may be required. Dexter et al. (1974) have provided suggestive evidence that methylnitrosourea can induce T-cell leukemias in thymectomized mice. In their experiments lympho-

cytic tumors with long latent periods were obtained in 68 % of mice,
while none were seen in thymectomized bone marrow reconstituted mice.

Biological Properties of Tumors of T Lymphocytes

Cytotoxic and Immunosuppressive Activities in T Lymphocytic Neoplasms.
Several different suppressive or cytotoxic activities in T cell lymphocytic tumors have been described. Roman and Golub (1976) have shown
that AKR leukemic cells suppress an in vitro Mishell-Dutton immune response to SRBC. In these experiments equal numbers of normal spleen
cells and leukemic cells were added to the cultures. Large numbers of
leukemic cells were required to induce the effect. Irradiation of the
leukemic cells with 8000R did not alter their ability to suppress the
response. Direct contact between the leukemic cells and responding
cells was required as suppression was not observed when the leukemic
cells were separated from the responding cells by an 0.2 μm nucleopore
filter. Suppression by H-2^k AKR leukemic cells could be induced in
cultures of spleen cells from stains of mice carrying H-2^k or parts
of the H-2^k locus, but was not obtained with allogeneic cells. AKR
leukemic cells were less suppressive for an AKR anti-TNP-LPS immune
response in vitro.

Stocker et al. (1974) demonstrated a different type of immunosuppres-
sion with cells from an X-ray induced thymic, lymphocytic neoplasm,
WEHI-22 of BALB/c origin. WEHI-22 is an unusual tumor that has Thy-1-
antigen and Ig on its surface (Harris et al., 1973). Stocker et al.
(1974) found that culture fluid concentrate from WEHI-22.1 cultures
inhibited Donkey RBC responses in vitro but not anti-DNP responses.
Four other thymic lymphocytic tumors, WEHI-7.1, WEHI-112.1, S49.1, and
EL-4.1 were not suppressive. This system acted across allogeneic H-2
barriers and thus was not like the syngeneic suppression observed by
Roman and Golub. Feldmann et al. (1975) studied the effects of the Ig
isolated from the WEHI-22 and El-4 lymphomas grown in vitro on antibody responses to DNP in the Mishell-Dutton system. The mice were
primed with TNP-KLH, and when TNP-KLH was used to stimulate the cultures, the IgM response was depressed and unexpectedly the IgG response
was stimulated. It was postulated here that the T lymphoma cells were
releasing factors that competed with those produced by the normal reacting T cells in the culture, possibly on a macrophage (Feldman et
al., 1975). The enhancement of the IgG response was not explained.

Proffitt et al. (1973, 1975a, b) have shown that Moloney virus (MuLV-M)
induced thymic lymphocytic tumor cells are autoreactive with normal
C_3H/HeJ embryo cells. In this system C_3H/HeJ mice were infected with
Moloney leukemia virus, and a strain of mice was developed that congenitally passaged the virus. C_3H/HeJ mice so infected develop a high
incidence of thymic lymphocytic neoplasms (90 to 100 % by 6 months of
age). In these mice the thymus at 3 to 4 weeks of age gradually became
depleted of cortical cells and replaced by large blast-like cells. At
16 weeks, the thymus became involuted. The thymic cells from the preleukemic thymuses and thymic lymphocytic tumor cells from these mice
were cytotoxic for uninfected C_3H/HeJ embryo cells, but not MuLV-M infected embryo cells. The ratio of tumor cells to target cells at which
maximal cytotoxicity was observed was from 100/1 to 1000/1. These cells
were found to be nonadherent, Thy-1-positive cortisone-resistant cells
(Proffitt et al., 1975b). These workers postulate that the thymus becomes replaced with these autoagressive cells and that this process
may be involved in the pathogenesis of the neoplastic population.

Table 6. Distribution of Ly antigens on T cell lymphocytic tumors in mice[a]

Thymic	Strain	Ly1	Ly2
BAL-EN-TL-3	BALB/c	−	+
BAL-EN-TL-4	BALB/c	−	+
BAL-EN-TL-5	BALB/c	−	+
P 1798	BALB/c	+	+
CD-EN-TL-101	CDF$_1$[a]	+	−
AKR-TL-12	AKR	+	−
L 4946	AKR	+	−
Generalized			
BAL-EN-LM-13	BALB/c	+	−
BAL-EN-LM-14	BALB/c	−	+
CD-EN-LM-101	CDF$_1$[a]	+	−
× (BALB/C × DBA/2)F$_1$			

[a]Data of B. Mathieson, M. Potter, R. Asofsky.
EN, tumor induced by ethylnitrosourea.

T Lymphocyte Alloantigens

a) Ly1, Ly2, Ly3. Cantor and Boyse (1975a, b) demonstrated that T cell specific differentiation alloantigens Ly1, Ly2, and Ly3 are expressed only on thymocytes and peripheralized T lymphocytes, and then went on to show that two specific subpopulations of T lymphocytes emerge, one that expresses only Ly1 and the other Ly2+3. With the use of cytotoxic antisera they were able to show that when pools of T lymphocytes were depleted of Ly1 cells by the addition of the antiserum and complement they were no longer able to help syngeneic B cells restore an immune response to SRBC in lethally irradiated recipients. By contrast, depletion of Ly2,3 cells from T cell pools did not affect the ability of the T cells to cooperate in the anti-SRBC response.

The Ly2,3-bearing cells were found to be killer cells. These were assayed by incubating H-2^b T cells with anti-Ly1 or anti-Ly2,3 cytotoxic sera and complement to kill the respective population, and then injecting the pool into lethally irradiated H-2^b/H-2^d F$_1$ hybrid mice. After 4 days the cells were harvested and texted for their ability to kill LSTRA(H-2^d) cells in vitro by a chromium release assay. T cell pools depleted of Ly1 cells were cytotoxic, while those depleted of Ly2 or Ly3 were not. These findings demonstrated for the first time a correlation of an alloantigen with a specialized subclass of T cells.

In studying the distribution of the Ly1, Ly2 and Ly3 antigens on lymphocytes it was also demonstrated by Cantor and Boyse (1975a, b) that thymocytes had all three antigens on their surface at one time.

Recently we have examined a series of lymphocytic neoplasms for Ly1 and Ly2 alloantigens (B. Mathieson, M. Potter, R. Asofsky, 1976). Seven of these transplantable tumors were initiated from thymic lymphocytic tumors, two were from peripheralized lymphomatous growths (Table 6).

166

Two of the tumors have been previously described; P1798 is a cortisone sensitive thymic lymphocytic tumor that arose in an estrogenized BALB/c mouse (Lampkin and Potter, 1958). L4946 arose spontaneously in an AKR mouse with a thymic lymphocytic neoplasm in the laboratory of Dr. L.W. Law. The other tumors were induced by intraperitoneal injection of ethylnitrosourea into adult BALB/c or (BALB/c x DBA/2)F$_1$ hybrid mice.

As may be seen from the table there were three types of tumors: Ly1$^+$, Ly2$^-$; Ly1$^-$, Ly2$^+$, and Ly1$^+$, Ly2$^+$. This indicated that cells arising within the thymus were differentiated for Ly antigens as well as peripheralized T lymphocytes.

b) Thy.1. Theta antigen or Thy.1 is an alloantigen found on thymocytes and peripheralized T lymphocytes. The biochemical nature of the antigenic molecule is not yet completely resolved, i.e. whether it is a glycolipid (Vitetta et al., 1973; Miller and Esselman, 1975) or glycoprotein (Atwell et al., 1973; Trowbridge and Hyman, 1975); there is agreement that the antigenic determinant is a carbohydrate. Thy.1 is found on brain as well as thymus (Miller and Esselman, 1975) but in lymphocytes it is a differentiation antigen that distinguishes T from B lymphocytes. The demonstration that thymopoietin can cause the appearance of Thy.1 on bone marrow prethymic cells raises the possibility that Thy.1 may develop in cells in the absence of thymic lymphocytopoiesis. However, during thymic lymphocytopoiesis, thymocytes appear to produce large amounts of Thy.1, but as they migrate towards the medulla and mature, less Thy.1 is produced and the H-2 antigen appears more abundantly on the surface. Medullary and peripheralized T lymphocytes have low Thy.1 and high H-2 on the cell membrane. Cells from congenitally athymic, nude mice may have a small amount of Thy.1 antigen (Roelants et al., 1975). These may represent pre-T cells. Trowbridge and Hyman (1975) have described mutant sublines of Thy.1 + tumors that are Ty.1 negative.

A Thy.1-positive IgM-positive lymphoma originating in a 14-month-old thymectomized AKR mouse has been described by Greenberg and Zatz (1975). In a study of 19 AKR thymomas Krammer et al. (1976) found that tumor line 2590 had cell membrane 10S Ig and also reacted with anti-Thy.1 antisera. Biosynthesis of the Ig was demonstrated by capping the surface Ig and then culturing the cells for 6 to 18 h, during which time the Ig reappeared, and by radiolabeling cell in culture and demonstrating freshly synthesized Ig. This tumor also possessed an Fc receptor. The WEHI-22 radiation induced thymic lymphoma also has Thy.1 and 7S surface Ig which has been shown to be biosynthesized in cultures (Hautstein et al., 1975).

Normal and neoplastic lymphocytes may simultaneously possess surface Ig and Thy.1 (Haustein et al., 1975; Roelants et al., 1975; Marchalonis, 1974). Some of these tumors have arisen in the thymus while others have occurred in thymectomized mice. It is clear that additional criteria for classifying the tumors as T cell tumors should be obtained, since Thy.1-positive cells may arise extrathymically and these tumors might be derived from prethymic bone marrow cells and spuriously express both surface Ig and Thy.1. Looking at the problem from another point of view, it may well be that only a few T lymphocytic tumors will produce enough Ig for detection, and that the observations of Marchalonis (1974), Haustein and Goding (1975), Greenberg and Zatz (1976) and Krammer et al. (1976) provided the best evidence that T lymphocytes can in fact produce some forms of Ig. The presence of the Ly1, 2, 3 antigens on these tumors will be a most convincing finding.

The Antigen-Binding Receptor on T Lymphocytes. It is well accepted that a major function of the activated thymus derived lymphocytes is

cooperation with B lymphocytes in specific immune responses. Specific
T cell cooperation or help that leads to triggering of determined B
lymphocytes depends upon antigen binding by the T cell or T cell product. The chemical identity of the antigen binding receptor on T lymphocytes has not been agreed upon. Candidates are: (1) immunoglobulin (Ig),
i.e. the same molecules produced in B lymphocytes and plasma cells;
(2) a special C_H class of Ig (IgT), but nonetheless molecules that
share the same V regions with Ig; (3) a molecule with a two-chain structure that if vastly different in sequence from immunoglobulins but
which binds antigens.

The physical relationships of antigen, T cell and B cell in B cell
triggering are also not established; two possibilities currently under
discussion are: (1) the antigen is mutually bound between T and B
cells (Mitchison et al., 1970); (2) the antigen binding receptor antigen complex is shed from the T cell and then bound to the B cell (Munro
et al., 1974; Munro and Taussig, 1975; see also Sachs and Dickler,
1975; Barton and Diener, 1975 for discussions).

There is conflicting information on the nature of the specific T cell
antigen binding receptor. One school has evidence that the receptor
is shed from T cells, is not precipitable with anti-Ig sera, has a
molecular weight of about 50,000 and appears to be controlled by genes
in the K end of the H-2 locus in the mouse (i.e. the Ir region) (Munro
and Taussig, 1975). Alternatively evidence has been presented that
the antigen binding receptor is in fact composed of Ig. Some of the
most compelling evidence for this has been the demonstration that
specific helper T cells in immune responses to streptococcal-A polysaccharide have on their surface an idiotype that is antigenically
the same as the idiotype on antibodies to A-CHO (Black et al., 1976;
Hammerling et al., 1976; see Wigzell, this volume). This idiotype is
controlled by genes that are non-H-2 linked and are linked to the IgC_H
(allotype) locus in the mouse.

The search for Ig on thymocytes has only provided suggestive support.
Haustein and Goding (1975) have radioiodinated the surface proteins of
T lymphocytes by a lactoperoxidase catalyzed reaction and found a small
peak that migrates between the μ and γ chain marker peaks. Marchalonis
(1974) has reviewed the evidence in support of immunoglobulin as the
specific antigen binding receptor. The small amount of immunoglobulin
on the surface of T lymphocytes has been difficult to demonstrate directly. There is indirect evidence that anti-immunoglobulin sera inhibit binding of antigen to T lymphocytes (Marchalonis, 1974). Marchalonis considers that much of the difficulty in demonstrating Ig on the
surface of T lymphocytes is due to intrinsic properties of the T
lymphocyte membrane itself, which has a thick coat of glycoprotein.
There are a few studies on Ig synthesis by tumors of T lymphocytes.
Haustein et al. (1975) have studied four Thy.1 + tumors established
in tissue culture and demonstrated an anti-IgM precipitable protein
on the surface of all four (WEHI-22, WEHI-7, S49 and EL-4). Premkumar
et al. (1975) have found an anti-Ig, a precipitable peak on P1798 cells
(an Ly.1,2,3 +, Thy.1 + thymic lymphcytic tumor). An underlying membrane protein (MAID) that is made in larger quantities than immunoglobulin and migrates in the same region complicates the resolution
in acrylamide gel electrophoresis.

Comment. Based on present knowledge the natural history of the thymic
lymphocytic neoplasm is still not completely clarified. In the usual
thymic lymphocytic tumors that develop between 6 and 12 months of age
the probable history is as follows. The leukemogen (virus, X-irradiation or chemical) causes an acute injury to the thymus and to the bone

marrow; while this is initially repaired in the first 4 to 6 weeks following the acute injury, the thymus at about 4 to 6 months of age undergoes a progressive cortical atrophy which can develop unilaterally. During and even before this period the character of the thymus population changes, in that there are progressively more thymocytes with a high content of H-2 and a relatively low content of Thy.1 (these resemble the medullary or mature thymocytes and are like the peripheral T lymphocytes). The medulla of the thymus may actually be considered a true peripheralized tissue and indeed in some thymuses germinal center formation within plasma cell development has been observed (Metcalf, 1966). The cortical changes might be explained by the hypothesis that the self-renewing lymphocytopoietic cells are depleted and exhausted. If this line of reasoning is correct, then why is leukemic development dependent upon the thymic environment? The evidence points to the influence of the thymic epithelial cell. The preleukemic cell must still depend on the influence of the lymphocytopoietic factors from the thymic epithelial cell for its early development. The evidence also suggests that leukemic cells in the thymus are able to complete several differentiations, as reflected in the change from high Thy.1, low H-2 to low Thy.1, high H-2, and the segregation of the Ly.1 and Ly.2,3 antigens.

By contrast, there may be an overwhelming development of thymic lymphocytic neoplasms particularly when the mouse has been given the leukemogen as a newborn animal. The histogenesis of the disease is less clear here and the role of cortical atrophy is less defined. Axelrad and Van der Gaag (1962) attribute the susceptibility of the mouse to viral leukemogenesis to a function of age and the presence of large undifferentiated lymphoblasts in the subcapsular zone of the thymus. They postulated that the virus in fact interfered with the ability of these cells to differentiate.

The nonthymic lymphocytic neoplasms with thymus markers such as Thy.1 may represent yet another group of T cell tumors. These may arise in peripheralized T cells that have developed in the thymus and pass into the periphery, where they continue to divide. Several workers have not found Thy.1-positive leukemia in situations where a large amount of carcinogen was administered to a young thymectomized mouse, as with the induction of tumors by the administration of N-butyl nitrosourea in the drinking water (Hiai et al., 1973).

Finally, there is the possibility that in the thymectomized mouse bone marrow precursor cells (pre-T) can be induced to produce Thy.1 antigen in the absence of the thymus, and that these cells become leukemic. Cells such as these may differ biologically from those passaged through the thymic lymphocytopoietic pathway.

In future studies an effort should be made to develop tumors of peripheralized T lymphocytes, since peripheralized T cells represent a more mature form in the T lymphocytes. Such tumors, and possibly even those of thymic origin, should be a potential source of antigen binding receptors.

The data on Ig production in T cell tumors is difficult to interpret. In cases where the cells are established in culture, the evidence is suggestive that a component reacting with anti-Ig sera is present. The specific binding properties of this Ig and its chemical structure should bring further insight into this problem.

Acknowledgments. The author wishes to thank his colleagues Michael Cancro, Dr. Nelson Wivel NCI; Drs. Richard Asofsky, Bonnie Mathieson, Herbert Morse and Rex

Risser NIAID for kindly supplying data prior to publication, and last but not least Ms. Linda Brunson for the preparation of the manuscript.

References

Abelson, H.T., Rabstein, L.S.: Cancer Res. <u>30</u>, 2213-2222 (1970a)

Abelson, H.T., Rabstein, L.S.: Cancer Res. <u>30</u>, 2208-2212 (1970b)

Abney, E.R., Hunter, I.R., Parkhouse, R.M.E.: Nature (Lond.) <u>259</u>, 404-406 (1976)

Acute Leukemia Group B: Arch. Intern. Med. <u>135</u>, 46-52 (1976)

Andersson, J., Lafleur, L., Melchers, M.: Europ. J. Immunol. <u>4</u>, 170-180 (1974)

Askonas, B.A., Roelants, G.E., Mayor-Whitney, K.S., Welstead, J.L.: Europ. J. Immunol. <u>6</u>, 250-256 (1976)

Atwell, J.L., Cone, R.E., Marchalonis, J.J.: Nature (New Biol.) <u>241</u>, 251-252 (1973)

Axelrad, A.A., Van der Gaag, H.C.: J. Nat. Cancer Inst. <u>28</u>, 1065-1093 (1962)

Ball, J.K.: J. Nat. Cancer Inst. <u>41</u>, 553-558 (1968)

Ball, J.K., Dawson, D.A.: J. Nat. Cancer Inst. <u>42</u>, 579-591 (1969)

Barton, M.A., Diener, E.: Transpl. Rev. <u>23</u>, 5-22 (1975)

Basch, R.S., Goldstein, G.: Ann. N.Y. Acad. Sci. <u>249</u>, 290-298 (1975)

Bazin, H., Deckers, C., Beckers, A., Heremans, J.F.: Intern. J. Cancer <u>10</u>, 568-580 (1973a)

Bazin, H., Deckers, C., Moriame, M., Beckers, A.: J. Nat. Cancer Inst. <u>51</u>, 1359-1362 (1973b)

Black, S.J., Hammerling, G.J., Berek, C., Rajewsky, K., Eichmann, K.: J. Exp. Med. <u>143</u>, 846-860 (1976)

Boylston, A.W., Mowbray, J.F.: Immunology <u>27</u>, 855-861 (1974)

Bryant, B.J.: Europ. J. Immunol. <u>2</u>, 38-45 (1972)

Cancro, M., Potter, M.: submitted for publ. (1976)

Cantor, H., Boyse, E.A.: J. Exp. Med. <u>141</u>, 1376-1389 (1975a)

Cantor, H., Boyse, E.A.: J. Exp. Med. <u>141</u>, 1390-1399 (1975b)

Cantor, H., Weissmann, I.: Progr. Allergy <u>20</u>, 1-64 (1976)

Chazan, R., Haran-Ghera, N.: Cell. Immunol. <u>23</u>, 356-375 (1976)

Clark, S.L., Jr.: In: Contemporary Topics in Immunobiology. New York: Plenum Press 1973, Vol. II, pp. 77-99

Cooper, M.D., Chain, W.A., Van Alten, P.J., Good, R.A.: Intern. Arch. Allergy <u>35</u>, 242-252 (1969)

Cooper, M.D., Lawton, A.R., Kincade, P.W.: In: Contemporary Topics in Immunobiology. New York: Plenum Press 1972, Vol. I, pp. 33-47

Cooper, M.D., Payne, L.N., Dent, P.B., Burmester, B.R., Good, R.A.: J. Nat. Cancer Inst. <u>41</u>, 373-389 (1968)

Cooper, M.D., Purchase, H.G., Bockmann, D.E., Gathings, W.E.: J. Immunol. <u>113</u>, 1210-1222 (1974)

Dexter, T.M., Schofield, R., Lajtha, L.G., Moore, M.: Brit. J. Cancer <u>30</u>, 325-331 (1974)

Dreyer, W.J., Bennett, C.J.: Proc. Nat. Acad. Sci. <u>54</u>, 864-868 (1965)

Feldmann, M., Kontiainen, S., Greaves, M.F., Hogg, N., Boylston, A.: Ann. N.Y. Acad. Sci. <u>249</u>, 424-437 (1975)

Furth, J.: J. Gerontol. <u>1</u>, 46-54 (1946)

Goldstein, G.: Ann. N.Y. Acad. Sci. <u>249</u>, 177-183 (1975)

Greenberg, R.S., Zatz, M.M.: Nature (Lond.) <u>257</u>, 314-316 (1975)

Gross, L.: Proc. Soc. Exp. Biol. Med. <u>100</u>, 325-328 (1959)

Hammerling, G.J., Black, S.J., Berek, C., Eichmann, K., Rajewsky, K.: J. Exp. Med. <u>143</u>, 861-869 (1976)

Haran-Ghera, N., Peled, A.: Nature (Lond.) <u>241</u>, 396-398 (1973)

Harris, A.W., Bankhurst, A.D., Mason, S., Warner, N.L.: J. Immunol. <u>110</u>, 431-438 (1973)

170

Haustein, D., Goding, J.W.: Biochem. Biophys. Res. Commun. 65, 483-489 (1975)
Haustein, D., Marchalonis, J.J., Harris, A.W.: Biochemistry 14, 1826-1834 (1975)
Hiai, H., Shisa, H., Matsudaira, Y., Nishizuka, Y.: Gann. 64, 197-201 (1973)
Hood, L., McKean, D., Farnsworth, V., Potter, M.: Biochemistry 12, 741-749 (1973)
Ishizaka, K.: In: The Biological Role of the Immunoglobulin E System. Washington, D.C.: U.S. Govt. Printing Office 1972, pp. 3-16
Joshi, V.V., Frei, J.V.: J. Nat. Cancer Inst. 44, 379-394 (1970)
Kaplan, H.S.: Cancer Res. 27, 1325-1340 (1967)
Kearney, J.F., Lawton, A.R.: J. Immunol. 115, 671-676 (1975a)
Kearney, J.F., Lawton, A.R.: J. Immunol. 115, 677-681 (1975b)
Kincade, P.W., Ralph, P., Moore, M.A.S.: J. Exp. Med. 143, 1265-1270 (1976)
Kirschbaum, A., Liebelt, A.G.: Cancer Res. 10, 689-692 (1955)
Krammer, P.H., Citronbaum, R., Read, S.E., Forni, L., Lang, R.: Cellular Immunol. 21, 97-11 (1976)
Lafleur, L., Miller, R.G., Phillips, R.A.: J. Exp. Med. 135, 1363-1374 (1972)
Lamm, M.E.: Advan. Immunol. 22, 223-290 (1976)
Lampkin, J.M., Potter, M.: J. Nat. Cancer Inst. 20, 1091-1111 (1958)
Law, L.W., Miller, J.H.: J. Nat. Cancer Inst. 11, 253-262 (1950)
Linker-Israeli, M., Haran-Ghera, N.: Immunochemistry 12, 585-588 (1975)
Lydyard, P.M., Grossi, C.E., Copper, M.D.: J. Exp. Med. 144, 79-97 (1976)
Mage, R.G.: Federation Proc. 34, 40-46 (1975)
Marchalonis, J.J.: J. Med. 5, 329-367 (1974)
Mathieson, B., Campbell, P., Potter, M., Asofsky, R.A.: In preparation
Melchers, F., Cone, R.E., Von Boehmer, H., Sprent, J.: Europ. J. Immunol. 5, 382-388 (1975a)
Melchers, F., Coutinho, A., Heinrich, G., Andersson, J.: Scand. J. Immunol. 4, 853-858 (1975b)
Melchers, F., Boehmer, H. von, Phillips, R.A.: Tansplantation Reviews 25, 26-58 (1975c)
Merwin, R.M., Algire, G.H.: Proc. Soc. Exp. Biol. Med. 101, 437-439 (1959)
Merwin, R.M., Redmon, L.W.: J. Nat. Cancer Inst. 31, 998-1017 (1963)
Metcalf, D.: In: The Thymus. Recent Results in Cancer Research. New York: Springer 1966, Vol. V, pp. 1-144
Metcalf, D., Nossal, G.J.V., Warner, N.L., Miller, J.F.A.P., Mandel, T.W., Layton, J.E., Gutman, G.A.: J. Exp. Med. 142, 1534-1549 (1975)
Miller, H.C., Esselman, W.J.: Ann. N.Y. Acad. Sci. 249, 54-60 (1975)
Miller, J.F.A.P.: Lancet 2, 748-749 (1961)
Miller, J.F.A.P.: Proc. Soc. Exp. Biol. Med. 112, 785-792 (1963)
Miller, R.G., Phillips, R.A.: Federation Proc. 34, 145-150 (1975)
Mitchison, N.A., Rajewsky, K., Taylor, R.B.: In: Developmental Aspects of Antibody Formation and Structure. Sterzl, J. (ed.). Prague: Czech. Acad. Sci. 1970, pp. 547-561
Moloney, J.R.: Federation Proc. 21, 19-31 (1962)
Morse, H.C. III., Pumphrey, J.G., Potter, M., Asofsky, R.: J. Immunol. in press (1976)
Morse, H.C. III., Weigert, M.: Personal communication
Munro, A.J., Taussig, M.J.: Nature (Lond.) 256, 103-106 (1975)
Munro, A.J., Taussig, M.J., Campbell, R., Williams, H., Lawson, Y.: J. Exp. Med. 140, 1579-1587 (1974)
Nishizuka, Y., Shisa, H.: In: Topics in Chemical Carcinogenesis. Nakahara, W. et al. (eds.). Baltimore: Univ. Park Press 1972, pp. 493-499
Osmond, D.G., Miller, R.G., Boehmer, H.V.: J. Immunol. 114, 1230-1236 (1975)

Osmond, D.G., Nossal, G.J.V.: Cell. Immunol. 13, 132-145 (1974)
Owen, J.J.T., Raff, M.C., Cooper, M.D.: Europ. J. Immunol. 5, 468-473 (1975)
Peterson, R.D.A., Purchase, H.G., Burmester, B.R., Cooper, M.D., Good, R.A.: J. Nat. Cancer Inst. 36, 585-598 (1966)
Potter, M.: Federation Proc. 29, 85-91 (1970)
Potter, M.: Ann. N.Y. Acad. Sci. 190, 306-321 (1971)
Potter, M.: Physiol. Rev. 52, 631-719 (1972)
Potter, M.: In: Cancer. A Comprehensive Treatise VI. Becker, F. (ed.). New York: Plenum Press 1975, pp. 161-182
Potter, M.: In: Methods in Cancer Research. Busch, H. (ed.). New York: Academic Press 1967, Vol. II, pp. 105-157
Potter, M., Padlan, E., Rudikoff, S.: J. Immunol. in press (1976)
Potter, M., Pumphrey, J.G., Bailey, D.W.: J. Nat. Cancer Inst. 54, 1413-1417 (1975)
Potter, M., Pumphrey, J.G., Walters, J.L.: J. Nat. Cancer Inst. 49, 305-308 (1972)
Potter, M., Sklar, M.D., Rowe, W.P.: Science 182, 592-594 (1973)
Premkumar, E., Potter, M., Singer, P.A., Sklar, M.D.: Cell 6, 149-159 (1975)
Proffitt, M.R., Hirsch, M.S., Gheridian, B., McKenzie, I.F.C., Black, P.H.: Intern. J. Cancer 15, 221-229 (1975a)
Proffitt, M.R., Hirsch, M.S., McKenzie, I.F.C., Gheridian, B., Black, P.H.: Intern. J. Cancer 15, 230-240 (1975b)
Roelants, G.E., Loor, F., Von Boehmer, H., Sprent, J., Hagg, L-B., Mayor, K.S., Ryden, A.: Europ. J. Immunol. 5, 127-131 (1975)
Roman, J.M., Golub, E.S.: J. Exp. Med. 143, 482-496 (1976)
Rosenberg, N., Baltimore, D.: J. Exp. Med. 143, 1453-1463 (1976)
Rosenberg, N., Baltimore, D., Scher, C.D.: Proc. Nat. Acad. Sci. 72, 1932-1936 (1975)
Rowe, W.P.: Cancer Res. 33, 3061-3068 (1973)
Sachs, D.H., Dickler, H.B.: Transpl. Rev. 23, 159-175 (1975)
Scher, C.D., Siegler, R.: Nature (Lond.) 253, 729-731 (1975)
Shen, F-W., Boyse, E.A., Cantor, H.: Immunogenetics 2, 591-595 (1975)
Siegler, R., Harrell, W., Rich, M.A.: J. Nat. Cancer Inst. 37, 105-121 (1966)
Siegler, R., Rich, M.A.: Cancer Res. 23, 1669-1678 (1963)
Sklar, M.D., Shevach, E.M., Green, I., Potter, M.: Nature (Lond.) 253, 550-552 (1975)
Sklar, M.D., White, B.J., Rowe, W.P.: Proc. Nat. Acad. Sci. 71, 4077-4081 (1974)
Stocker, J.W., Marchalonis, J.J., Harris, A.W.: J. Exp. Med. 139, 785-790 (1974)
Swan, D., Aviv, H., Leder, P.: Proc. Nat. Acad. Sci. 69, 1967-1971 (1972)
Takakura, K., Mason, W.B., Hollander, V.P.: Cancer Res. 26, 596-599 (1966)
Thompson, J., Van Furth, R.: J. Exp. Med. 131, 429 (1970)
Thompson, J., Van Furth, R.: J. Exp. Med. 137, 10-21 (1973)
Toivanen, P., Toivanen, A., Good, R.: J. Immunol. 109, 1058-1070 (1972a)
Toivanen, P., Toivanen, A., Linna, T.J., Good, R.A.: J. Immunol. 109 1071-1080 (1972b)
Trentin, J.J., Wolf, N., Cheng, V., Fahlber, W., Weiss, D., Bonhag, R.: J. Immunol. 98, 1326-1337 (1967)
Trowbridge, I.S., Hyman, R.: Cell 6, 279-289 (1975)
Vitetta, E.S., Boyse, E.A., Uhr, J.W.: Europ. J. Immunol. 3, 446-453 (1973)
Vitetta, E.S. Uhr, J.W.: Biochim. Biophys. Acta 415, 253-271 (1975)
Wang, A.C., Wilson, S.K., Hooper, J.E., Fundenberg, H.H., Nisonoff, A.: Proc. Nat. Acad. Sci. 66, 337-343 (1970)
Warner, N.L.: Immunogenetics 2, 1-20 (1975)

Warner, N.L., Harris, A.W., Gutman, G.A.: In: Membrane Receptor of
   Lymphocytes. Seligmann, M., Preudhomme, J.L., Kourilsky, F.M. (eds.).
   Amsterdam: North Holland 1975
Wivel, N.A., Potter, M.: Unpublished data
Zauderer, M., Askonas, B.A.: Nature (Lond.) <u>260</u>, 611-613 (1976)

Macrophages

H. Fischer

The organisers of this Symposium have asked me to give a short pladoyer
for the macrophage and its position in the immune system. I would like
to open this case with two statements:

1. The specificity of the immune response is a sole function of committed lymphocytes. Macrophages do not play an instructional role.

2. Phagocytosis was the primordial mechanism of defence; the specific
elements, the lymphocytes, came later.

If you accept these two statements and put them together it is legitimate to look at the role of the macrophage in evolutionary terms: macrophages did not develop to help lymphocytes but rather the reverse is
true. Specific immune induction developed in order to help the macrophage to recognise, eliminate and destroy foreign noxious agents.

Starting with these considerations I will now briefly explain the mononuclear phagocytic system, then discuss briefly the role of the macrophage in immune induction and finally the way in which lymphocytes
help macrophages.

The Mononuclear Phagocytic System

Studies by Gowans, Cohn, van Furth and others have made it possible
to define the so-called mononuclear phagocytic system in a more generic
way, superior to the classically described "Reticuloendothelial System"
(1-3). Stem cells and promonocytes reside in the bone marrow, from
which monocytes are released into the blood stream. Monocytes represent the relatively immature transport form of macrophages; in various
organs, depending on the conditions of the local environment, they
mature into "tissue macrophages", which differ in their morphology and
function. They may also be found ubiquitously in connective tissue as
relatively immature "histiocytes" (Table 1). Common to all of these
cell types is their relatively long life-span, capability of phagocytosis, pinocytosis and furthermore synthesis and liberation of material to the environment through secretion and exocytosis.

Time will not permit me to go into macrophage physiology. However, I
would like to discuss two alternatives: in one instance foreign antigenic material is rapidly phagocytosed and undergoes complete degradation, the other extreme being that antigen adheres longer to the out-
side, is internally only partially degraded and eventually reappears
extracellularly via exocytosis. One can deduce from these extremes
that macrophages may act in immune responses in different ways. Depending on the assay conditions this role may appear suppressive or
enhancing. What happens in the intact animal and in tissues is hard
to analyze. However, in culture the various phenomena can be reproducibly analyzed. In vitro systems have also permitted us to gain insight into the role of macrophages during the different phases of immune reactions.

Table 1. Mononuclear phagocyte system

      Promonocyte   (bone marrow)
            ↓
      Monocyte   (blood)
           ↓
    Macrophage   (tissues)
        highly phagocytic

connective tissue (histiocyte)
liver (Kupffer cell)
lung (alveolar macrophage)
spleen (free and fixed macrophage, sinusoidal lining cell)
lymph node (free and fixed macrophage)
bone marrow (macrophages, sinusoidal lining cell)
serous cavity (peritoneal macrophage)
bone tissue (osteoclast)
nervous system (microglia?)

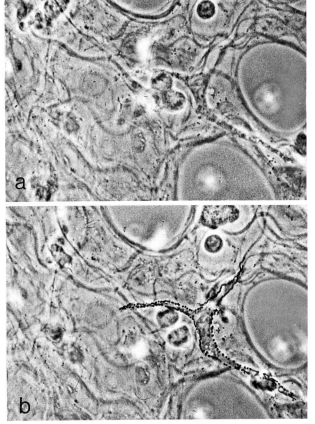

Fig. 1a and b. Lymphocyte undergoing division in close contact with dendrite of omental macrophage (taken from Ref. 5). (a) before mitosis; (b) after mitosis

Role of Macrophages in Immune Induction

It has been shown using various antigens in in vitro systems that those antigens which are avidly bound to macrophages are good immunogens.

Table 2. From Ritter et al. (1975)

Stimulator	C3H	
DBA/2 cell	C.r. Thymocytes	
1000 r	3H-Thym. incorp.	specific 51 Cr release
Spleen	+ + + +	76 %
Lymph node	+ +	18 %
c.r. Thymocyte	∅	∅
Thymoc + Mac	+ + +	44 %
Thymoc + 1 γ LPS	+ + +	48 %
Thymoc + AMS + 1 γ LPS	∅	∅

Those soluble antigens which are not bound to macrophages, on the other hand, lead to the development of tolerance (4).

Induction of Antibody Formation

Years ago we studied the induction of humoral antibodies using the mouse omentum and time lapse microcinematography. These studies, carried out in collaboration with Drs. Ax, Malchow, and Lohmann-Matthes, showed us that motile lymphocytes attracted by and attached to antigen-loaded macrophages undergo division and differentiation (Fig. 1). Time lapse analysis has revealed that small lymphocytes have extensive encounters with macrophages only if the latter are stimulated by antigen. Blast transformation and mitosis only occurred when lymphocytes were arrested in close contact with macrophage dendrites. Within three days lymphocyte proliferation led to the formation of antibody-secreting plasma cells (5, 6). The omentum observation was meant to give a visible example of how macrophages act during the induction of antibody formation. For further information see Reference 7.

Induction of Cell-Mediated Immunity

In order to demonstrate the involvement of macrophages in cell-mediated immunity, I will again choose experiments done in our laboratory. Drs. J. Ritter, Sonntag, Marie-Luise Lohmann-Matthes and myself have been interested in the development of specific lymphocyte-cytotoxicity in the course of mixed lymphocyte reactions (8).

As can be seen in Table 2, cortisone-resistant (c.r.) thymocytes from C3H mice can be effectively stimulated by irradiation DBA/2 stimulator cells from spleen and later become highly cytotoxic.

Irradiated DBA/2 lymph node cells are less effective, and cortisone-resistant thymocytes have no stimulatory effect at all. However, when macrophages — either from DBA/2 or of allogeneic origin — are present from the very beginning, good stimulation and development of cytotoxicity are observed.

In another series of experiments it turned out that very small concentrations of lipopolysaccharide (LPS) had a similar effect to the addition of macrophages. LPS is not a T cell stimulator and a possible

explanation for its effect was thought to be the activation of the few
macrophages present in a population of cortisone-resistant thymocytes.
That this speculative idea is indeed true was shown with the help of
antimacrophage serum and complement.

While our experiments demonstrated in a convincing manner the necessity
of macrophages for stimulation and cytotoxicity of thymocytes, they do
not give us any clue as to the mechanism. In this respect experiments
by Rosenthal et al. are most illuminating (9). Using purified cells
from guinea pigs, PPD as soluble antigen, and pulse labeling with
tritiated thymidine, they showed that T lymphocytes in the absence of
macrophages were not stimulated by addition of PPD as soluble antigen.
Only in the presence of macrophages pretreated with the antigen did
stimulation occur. A prerequisite for stimulation is a rather long-
lasting macrophage-lymphocyte contact.

Further studies then showed that cell contact only sufficed and led to
a good stimulation if histocompatible macrophages were loaded with antigen. With semiallogeneic macrophages, on the other hand, the stimulation amounts to only 50 %, while allogeneic macrophages preexposed to
the antigen PPD failed to stimulate lymphocytes at all.

If these findings are as true for men as for mice, they tell us that
a competent lymphocyte, in order to be stimulated by a soluble antigen,
has to recognize antigen plus a neighboring HLA or IR gene product on
the macrophage surface.

Lymphocyte-Macrophage Interaction

If I now turn to the important question of how lymphocytes help the
macrophages, I am entering a field which in recent years has found wide
attention: that of the lymphokins.

In pioneering experiments Mackaness (10, 11) has shown that macrophages
from animals with chronic infections (for instance, listeriosis) are
highly active. Not only do they kill the inducing bacteria, but in a
rather non-specific way they are also able to kill unrelated germs,
parasites and tumor cells. From in vitro experiments done in many laboratories it has become clear that sensitized T lymphocytes secrete
mediators, some of which active macrophages. I will show the effects
of a rather crude preparation of migration inhibition factor (MIF)
functions (12; Table 3).

This table is taken from a recent article by David, and I have pur-
posely omitted data not directly related to macrophage function. We
are here at the very beginning of a fascinating new chapter of biochemistry. As purification and characterization of the lymphocyte
mediators is progressing rapidly, one may also predict that their mode
of action will soon be unraveled (13).

This goal can however only be achieved if well-defined unstimulated
macrophages are available, where all desired biochemical parameters
of activations can be followed in a reproducible manner.

I would like to describe such a system now. As you will remember,
macrophages originate from stem cells in the bone marrow.

When culturing bone marrow under appropriate conditions one is able
to end up with an extremely pure population of non-stimulated slightly

Table 3. Effect of MIF-rich Sephadex fractions on macrophages (from David, 1975)

Increased adherence
Increased ruffled membrane movement
Increased glucose oxidation
Increased number of granules

Enhanced bacteriostasis
Enhanced tumoricidal activity

adherent macrophages. These macrophages and their precursors can be harvested and subcultured at various stages of maturation and reproducibly be activated by lymphokins and other stimulants like endotoxins (14).

In closing my case for the macrophage, I would like to summarize. Phagocytes, the representatives of the "archetype" of defence mechanisms, are essential in humoral as well as in cell-mediated specific immune reactions. They serve in the early recognition and are most important in the effector phase. What would be the value of a sophisticated surveillance system that could not dispose of the unwanted material?

References

1. Furth, R. van (ed.): Mononuclear Phagocytes. Oxford and Edinburgh: Blackwell Sci. Publ. 1970
2. Furth, R. van (ed.): Phagocytes in Immunity, Infection and Pathology. Oxford: Blackwell Sci. Publ. 1975
3. Wagner, H.W., Hahn, H. (eds.): Workshop Conference Hoechst, Vol. II. Excerpta Medica Amsterdam. New York: American Elsevier Publ. Co. 1974
4. Unanue, E.R.: J. Immunol. 105 (1970)
5. Matthes, M.-L., Ax, W., Fischer, H.: In: Cell Interactions and Receptor Antibodies in Immune Responses. Mäkelä, O., Cross, A., Kosunen, T.U. (eds.). London, New York: Academic Press 1971
6. Matthes, M.-L., Ax, W., Fischer, H.: Z. Ges. Exp. Med. 154, 253-264 (1971)
7. Unanue, E.R., Calderon, J.: Federation Proc. 34, 1737-1742 (1975)
8. Ritter, J., Lohmann-Matthes, M.-L., Sonntag, H.G., Fischer, H.: Cell. Immunol. 16, 153-161 (1975)
9. Rosenthal, A.S., Lipsky, P.E., Shevach, E.M.: Federation Proc. 34, 1743-1748 (1975)
10. Mackaness, G.G.: J. Exp. Med. 129, 973-984 (1969)
11. Mackaness, G.B., Lagrange, P.H., Miller, T.E., Ishibashi, T.: In: Activation of Macrophages. Workshop Conference Hoechst, Vol. II. (Wagner, H.W., Hahn, H. (eds.). Excerpta Medica, Amsterdam. New York: American Elsevier Publ. Co. 1974
12. David, J.R.: Federation Proc. 34, 1730-1736 (1975)
13. Lohmann-Matthes, M.-L., Fischer, H.: Transpl. Rev. 17, 151-171 (1973)
14. Meerpohl, H.G., Lohmann-Matthes, M.-L., Fischer, H.: Europ. J. Immunol. 6, 213-217 (1976)

The Major Histocompatibility Complex
and Cellular Recognition

The Major Histocompatibility Complex and Its Biological Function

E. D. Albert

The Concept of a Major Histocompatibility Complex

It was one of the very early findings in transplantation research that
the survival of grafted organs was largely dependent on the degree of
genetic similarity between donor and recipient. Selective breeding in
connection with transplantation experiments in mice has led to the identification of a number of independent genetic systems, all of which
play a role in the rejection of allografts (Histocompatibility Systems =
H-Systems). In all mammalian species studied so far one has been able
to isolate one system, which is of paramount importance for transplantation. This system in each species is now generally referred to as
the major histocompatibility complex (MHC). The current knowledge about
the MHC resulted from widely differing approaches depending on the population structure of the species investigated: in rodents the availability of inbred strains and the possibility of selective breeding
formed a major advantage in the analysis of the biological function
of the chromosomal region. In the outbred species such as dog, rhesus
monkey and man, one has to resort to the laws of population genetics.
In man, two unrelated individuals have to go back for very many generations to come to the latest common ancestor chromosome in relation
to the MHC. In this long time very many recombinations must have occurred,
separating even very closely linked genes. Thus it must be expected
that in the unrelated population of an outbred species most or all
theoretically possible combinations in cis-position of alleles of
closely linked loci (= haplotypes) do in fact exist. This is in contrast to the situation in the laboratory mouse, where only a relatively
limited number of haplotpyes (i.e. original haplotypes and recombinant
haplotypes) is available for testing. It was therefore a logical consequence that investigators of the mouse MHC tried to widen the scope
by investigating wild mice while on the other side investigators of
the human system attempted to reduce the polymorphisms by searching
for "inbred" individuals homozygous at the MHC. In spite of basically
different methods of analysis a very remarkable degree of analogy has
been observed for the MHCs in different mammalian species.

The Components of the MHC

The genetic region of the MHC has been subdivided in recent years into
a number of genetically separate (by recombination) but functionally
related systems. In the following the components of the MHC will be
reviewed with special reference to the H-2 system of the mouse and the
HLA-system in man.

The "Classical" Transplantation Antigens Such as H-2 K and H-2 D in Mice and HLA-A, B and C in Man

The antigens coded for by these loci are located on the cell surface
of all nucleated cells and on platelets. They are generally detected

182

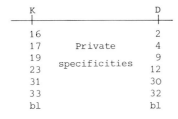

Fig. 1. The SC loci of H2 and their private specificities

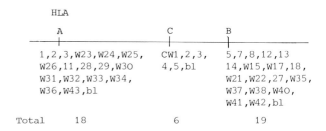

Fig. 2. The SD loci and alleles of the human MHC

in a complement dependent microlymphocytotoxicity test using sera from multiparous women or sera obtained through planned immunization. In a operational, oversimplified and convenient nomenclature these antigens are generally referred to as serologically defined (SD) antigens. Each of the closely linked loci - H2-K and H2-D in the mouse and HLA-A, B and C in man - governs the expression of a series of allelic specificities, which are in the mouse sometimes referred to as "private specificities". A schematic picture is given in Figure 1 for H2-K and D and in Figure 2 for HLA-A, B and C.

There is evidence for the existence of at least one additional SD locus in the mouse, whose exact map position and - ironically - serologic definition is still somewhat unclear (Wagner, pers. comm.). Serological studies as well as absorption-inhibition experiments and the molecular redistribution of SD-antigens on the cell surface have demonstrated that each "antigen" is composed of a number of highly associated antigenic factors (14) which correspond to separate antigenic features on the same molecule. This in turn suggests the existence of multiple mutational sites within the gene responsible for the production of the entire "antigen"-molecule.

The genetic unit of inheritance is the haplotype which comprises the combination of all alleles in cis position of the different MHC-loci, so that the polymorphism of the MHC is best described by the matrix of haplotype frequencies, which can be determined by counting in family data and by estimation from the phenotype data using a formula given by Piazza in 1975 (20).

The analysis of HLA-A,B haplotype frequencies has shown that there are marked differences in frequencies in different populations and that within one population there are certain haplotypes much more common than one would expect on the basis of the gene frequencies of the antigens forming the haplotype. In the Caucasian population per example the haplotype HLA-A1, B8 is found with a frequency of 0.079. Since the HLA-A and B loci are separable by recombination (approximately 0.85 % of all meiotic divisions per generation) (25) one would expect that after a large number of generations (which are available in the evolution of the human species) two alleles of two linked loci would behave as independent characters, so that the frequency of their joint occurence on a haplotype would be equal to the product of their mutal gene frequencies. In our example of HLA-A1, B8 this product is 0.015 which is highly significantly different from the observed haplotype frequency of 0.079. The prevalence of the Haplotype HLA-A1, B8 is reflected by a significant positive association of HLA-A1 and B8 in the phenotype data of a random Caucasian population. This phenomenon of linkage disequilibrium is wide spread throughout the entire MHC. The

Fig. 3. LD and SD systems in man   Fig. 4. LD and SD systems in mice

reasons for this deviation are not known. It is certainly very appealing
to speculate that selection has favored certain haplotypic combinations.
On the other hand, there is no doubt, that migration and mixture of
different inbred populations as well as random genetic drift produce
linkage disequilibrium; the relative contributons, however, of these
simultaneously acting different mechanisms are not clear.

Lymphocyte Defined (LD) Antigens

Another in vitro test of histocompatibility is based on the mixture of
lymphocytes (= immunocompetent cells) under culture conditions (mixed
leucocyte culture = MLC). Depending on the genetic disparity of the
two lymphocyte populations a marked blast transformation (= stimulation)
can be observed after several days of culture and quantitated by the
uptake of ^{3}H-thymidine. The two-way reactivity of cell populations -
A and B - (A B) is disected into two one-way reactions by blocking the
blast transformation of one cell population with Mitomicin C (subscript
m) or irradiation (A + Bm and B + Am). MLC-nonreactivity between two
cell populations in a well controlled experiment is generally taken
as evidence for identity of LD-antigens. Family studies in man and
segregation studies in mice have shown that MLC-reactivity is governed
by the MHC, while recombinations have demonstrated that SD antigens
and LD antigens are coded for by distinct but closely linked loci.
Both in the MHC of man and mice there is one LD locus, for which disparity causes strong MLC stimulation, and at least one more LD locus
responsible for a rather low grade stimulation. The relative map positions of the LD loci in man and in mice are shown in Figure 3 and 4
respectively.

MLC-Tests in the unrelated human population revealed a very low incidence of mutual nonstimulation (i.e. identity for LD antigens) indicating a high degree of polymorphism for the LD system. The identification of the LD alleles in man has become possible through the use
of reference cells, which were shown by family analysis to be homozygous
for the LD antigens (17): A test cell heterozygous for the LD antigen
A (A/X) shows a one-way-nonreactivity in MLC with the (A/A) homozygous
reference cell (A/X ⟶ A/A; A/A ⟶ A/X), where the LD-antigen X of
A/X causes stimulation in the A/A homozygous cell, while A/A is not
able to stimulate A/X because A is not foreign to A/X. This rather
schematical description does not mention a number of technical and/or
genetic problems involved in LD typing. Nevertheless, it has been possible to establish several internationally recognised LD-antigens (HLA-
D according to the new nomenclature) and to test the population distribution as well as the relationship of LD antigens to the HLA-A, and B
alleles. Strong linkage disequilibrium has been observed between HLA-B
antigens and HLA-A, B haplotpyes on one side and the HLA-D alleles on
the other side (1).

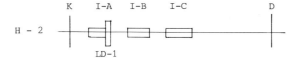

Fig. 5. The IR region in mice

Immune Response Genes and Immune Response Region Associated (Ia) Antigens

The findings of McDevitt and Benacerraf (16) that the immune response directed against a wide variety of thymus-dependent, synthetic and natural antigens is controlled by genes closely linked with the MHC in many species has demonstrated, that this genetic region may be of major importance for basic immunology. Analyses of intra-H-2 recombinant strains have mapped most of the H2-linked immune response (IR) genes between H2-K and D close to, but separated from H-2K. Recombinations which occurred inside the IR region have divided this region into at least three "subregions" I-A, I-B, and I-C. The LD-1 locus is very closely associated or included in the I-A subregion (4).

In an attempt to characterize the gene products of the IR-genes, antibodies were produced by cross-immunization of inbred mouse strains identical for H-2 K and D and differing only for the IR region. The antigens defined by such antisera were termed Immune-response-region-associated (Ia) antigens and have been found to be expressed predominantly on B-cells and macrophages (4). The Ia antigens are associated with, but clearly distinct from the Fc receptor on the cell membrane, as shown by blocking experiments (31). Anti Ia antisera have been found to block specifically the MLC stimulation, which suggests that at least some Ia antigens could be sterically associated or even identical with LD antigens (18). In addition, Ia antibodies can at least partially block the immune response against foreign red blood cells by splenic lymphocytes cultured in vitro (8). Nevertheless, it must be stressed that in the IR region there is ample space for many genes whose function may or may not be associated with immune response and MLC. In rhesus monkeys a series of presumably allelic Ia-like specificities is governed by a locus, which is far separated from the LD locus in that species (22). The Ia antigens can be distinguished from SD antigens (H2-K, H-2D, HLA-A, B anc C) by the restricted tissue distribution (B-cell antigens), by the independent molecular redistribution on the cell surface and by the lack of association with β-2 microglobulin, which is characteristic for "SD" antigens. Following these criteria, Ia-like antisera have also been identified in man. Some of these Ia-like antigens are very highly associated with HLA-D antigens (LD antigens) in the random population (23).

MHC-Linked Disease Susceptibility Genes

The first information about MHC-linked disease susceptibility was provided by Lilly et al. (15), who showed that all the strains carrying the H-2^k or H-2^d haplotypes are susceptible to gross leukemia virus, while all strains carrying H-2^b or H-2j are resistant. Consequently, susceptibility was shown to be linked with H-2 by classical linkage analysis in backcross experiments. Interestingly, resistance to gross leucemia virus is dominant over susceptibility. The gene for susceptibility(rgv-1) has been mapped in the close proximity of the H-2K locus. In more recent experiments, the susceptibility to autoimmune thyroiditis has been found to be H-2 linked (30). In contrast to the relatively scarce data in mice, there is an increasing amount of information about the association of HLA antigens with disease. For instance it was shown that patients with Coeliac disease had a frequency of over 80 % for HLA-B8 which compares with 20 % for the normal population. Such an association between HLA-B8 and Coeliac disease suggests either a direct involvement of the antigen in the pathogenesis or - much more likely - the presence of a susceptibility gene, which is closely linked to the HLA-B locus and which displays strong linkage disequilibrium with HLA-B8, resulting in the association observed in the patient population. In recent years a growing number of human diseases, including Psoriasis, Multiple Sklerosis, Ankylosing Spondylitis, Nyasthenia gravis, Addisons disease, Graves disease, chronic autoaggressive hepatitis and juvenile diabetes mellitus to name only the outstanding examples, were found to be associated with HLA. These diseases have a number of common features:

1. Evidence for familial incidence

2. Immunological (autoimmune) mechanisms are suspected to play a role in the pathogenesis

3. The associations are almost exclusively with antigens of the HLA-B locus

4. The susceptibility rather acts as a dominant character.

Since it may be assumed by analogy from the mouse, that the region adjacent to the HLA-B locus should represent the human equivalent of the human IR region, the following working hypothesis may be formulated: Disease susceptibility is determined by genes which may represent pathologic immune response genes, which upon the impact of environmental factors such as viral or bacterial infection or contact with certain antigens, produce a wrongly directed and/or overshooting immune response leading to the observed autoimmune diseases. Unfortunately, no convincing evidence for association of HLA with malignant diseases in man has been presented so far. It should be expected, however, that in the case of malignancy there should be rather a lack of immune surveillance and therefore a recessive mode of inheritance of the susceptibility.

Genes Controlling Complement Components

It has been demonstrated by Demant et al. in 1971 that in mice there is a H-2-linked gene which controls the complement level in mouse serum. This gene could be mapped in the Ss-Slp region between H-2K and H-2D. Recently, Lachmann et al. (13), and Curman et al. (5) reported that the long known Ss protein represents the C4 component of complement. In man there is also evidence for the linkage of complement genes with the MHC: The polymorphism of the Bf-protein, which functions as the proactivator of the C3-Component has been shown to be linked with HLA (3) and could be located between the HLA-B and D loci (2). This polymorphism has four relatively frequent alleles, some of which reveal significant linkage disequilibrium with HLA-B alleles. The deficiency of the C2 complement component, which in the homozygous state presents with a lupus like syndrome, was also found to be linked with HLA (9). Rittner et al. (21) presented a family, in which a deficiency for C4 also segregates with HLA. The map position for the C2 and C4 genes is not yet known. Thus there is evidence that several components of the complement cascade are coded for somewhere within the human MHC. The combined knowledge about the MHC in man and mice is given in Figures 6 and 7 respectively.

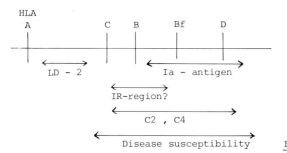

Fig. 6. The MHC in man

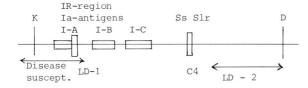

Fig. 7. The MHC in mice

The Biological Function of the MHC Genes

The investigations of the MHC in many different mammalian species indicate that this group of genes has remained remarkably stabile during
evolution. This leads to the suspicion that these functionally related
genes could have been held together by selective forces. The linear
order of the genes within the MHC does not seem to be of major importance since there are clear differences between - for example - mouse,
rhesus monkey and man. This is compatible with the fact that selection
generally acts upon the phenotype rather than the genotype.

If one accepts the assumption that the components of the MHC are forming a functional unit, then the question immediately arises: what is
really this physiological function and is the enormous polymorphism
of this system related to this function?

Clearly, a satisfactory answer cannot be given at present. Nevertheless,
most investigators in this field are convinced that the biological
significance of the MHC goes far beyond the artificial systems of transplantation immunity, which may however serve as a model for the immune
processes directed at cell surface bound antigens: Disparity for LD
antigens is responsible for proliferation of cells as a prerequisite
for the development of cytotoxic T cells, which are sensitized by and
act upon the SD antigens of the target cells.

Certainly the inclusion of immune response genes within the MHC is of
major biological importance, although it is still unclear just what
the gene products of IR-genes are and how they function. The fact that
Ia antigens are coded for in the IR-region suggests that some Ia antigens could be associated or even identical with IR-gene product. It
has been shown that antigen-specific T-cell factors, which are not reactive with antiimmunoglobulin but contain gene products of the H-2
haplotype, are effective in "helping" (19, 28) and in "suppressing"
(26, 27) antibody production. In the former case the factor described
by Taussig et al. (28) has been shown to contain Ia antigen by immuno-

adsorption experiments (29). Thus H-2 and more specifically Ia antigens are intimately involved in the regulation of the immune response.

More intriguing information has been gathered by the study of histocompatibility requirements in T-B cell cooperation (10, 11), in the cooperation between T cells and macrophages (24) and in the cell mediated lysis of cells infected with lymphocytic choriomeningitis (LMC) virus (32, 33). In all these systems it is necessary that the cells involved share at least one MHC haplotype or part of it. The genetic requirements have been narrowed down to the Ia antigens for the T-B cooperation and to the SD antigens for cell mediated lysis. This has led to the concept of "immunological surveillance against altered self components by sensitized T-cells" (33), where it is assumed that LMC virus binds to the SD antigen, causing confirmational changes in the SD part of the SD-LCM complex, which is recognised as foreign (altered self) and forms the target for lysis.

Doherty and Zinkernagel (7) went on to demonstrate that there are LCM-immune T-cell clones directed at the altered self characteristic of each SD antigen in the infected animal, so that F1 animals heterozygous at the MHC would have twice as many immune T-cell clones as their homozygous parents. This would easily provide for a selective advantage of MHC heterozygotes, which could account for the production and maintenance of the enormous polymorphism observed in the MHC systems.

In conclusion, the MHC genes code for molecules which are somehow involved in antigen recognition, in cell-interaction, in regulation of the immune response by helper and suppressor effects and in receptor functions for viruses. Furthermore, MHC-coded molecules serve as target structures of cell-mediated cytolysis, and they may be forming the substrate for self-nonself-recognition. Finally, some of the MHC genes govern essential components of a very important immunological effector mechanism: the complement system. All these genes are sufficiently similar in function so that they could conceivably have arisen from one gene by duplication and mutation.

Acknowledgments. This work was supported by SFB 37 and by Forschungsgemeinschaft DFG A1 92/9

References

1. Albert, E.D., Mempel, W., Grosse-Wild, H.: Transpl. Proc. V, No. 4, 1551-1553 (1973)
2. Albert, E.D., Rittner, Ch., Grosse-Wild, H., Netzel, B., Scholz, S.: In: Histocompatibility Testing. Copenhagen: Munksgaard 1975, p. 941-944
3. Allen, F.H., Jr.: Vox Sang. 27, 382-384 (1974)
4. Benacerraf, B., Katz, D.H.: In: Immunogenetics and Immunodeficiency. Benacerraf, B. (ed.). Lancaster: MTP 1975, pp. 117-178
5. Curman, B., Östberg, L., Sandberg, L., Malmheden-Eriksson, I., Stalenheim, G., Rask, L., Peterson, P.A.: Nature (Lond.) 258, 243-245 (1975)
6. Dermant, P., Capkova, J., Hinzova, E., Voracova, B.: Proc. Nat. Acad. Sci. 70, 863-864 (1973)
7. Doherty, P., Zinkernagel, R.M.: Nature (Lond.) 256, 50-52 (1975)
8. Frelinger, J.A., Niederhuber, J.E., Shreffler, D.C.: Science 188, 268 (1975)
9. Fu, S.M., Kunkel, H.G., Brusman, H.P., Allen, F.H.Jr., Fotino, M.: J. Exp. Med. 140, 1108 (1974)

188

10. Katz, D.H., Hamaoka, T., Benacerraf, B.: J. Exp. Med. 137, 1405 (1973)
11. Kindred, B., Shreffler, D.C.: J. Immunol. 109, 940 (1972)
12. Klein, J.: Biology of the mouse histocompatibility-2 complex. New York-Heidelberg-Berlin: Springer 1975
13. Lachmann, P.J., Grennan, D., Martin, A.: Nature (Lond.) 258, 242-243 (1975)
14. Legrand, L., Dausset, J.: Histocompatibility Testing 1972, pp. 441-453
15. Lilly, F., Boyse, E.A., Old, L.J.: Lancet 2, 1207 (1964)
16. McDevitt, H.O., Benacerraf, B.: Advan. Immunol. 11, 31-74 (1969)
17. Mempel, W., Grosse-Wilde,H., Albert, E.D., Thierfelder, S.: Transpl. Proc. 5, 401-408 (1973)
18. Meo, T., David, C.S., Rijnbeek, A.M., Nabholz, M., Miggiano, V., Shreffler, D.C.: Transpl. Proc. 7, 127 (1975)
19. Munro, A.J., Taussig, M.J., Campbell, R., Williams, H., Lawson, Y.: J. Exp. Med. 140, 1579 (1974)
20. Piazza, A.: In: Histocompatibility Testing 1975, pp. 923-927
21. Rittner, Ch., Hauptmann, G., Grosse-Wilde, H., Grosshans, E., Tongio, M.M., Mayer, S.: In: Histocompatibility Testing 1975, pp. 945-954
22. Roger, V.H., van Vreeswijk, W., Dorf, M.E., Balner, H.: Tissue Antigens, 1976 in press
23. van Rood, J.J., van Leeuwen, A., Parlevliet, J., Termijtelen, A., Keuning, J.J.: In: Histocompatibility Testing 1975, pp. 629-636
24. Rosenthal, A.S., Shevach, E.M.: J. Exp. Med. 138, 1194 (1973)
25. Svejgaard, A., Bratlie, A., Hedin, P.J., Høgman, C., Jersild, C., Kissmeyer-Nielsen, F., Lindblom, B., Lindblom, A., Löw, B., Messeter, L., Møller, E., Sandberg, L., Staub-Nielsen, L., Thorsby, E.: Tissue Antigens 1, 81-88 (1971)
26. Tada, T., Takemori, T.: J. Exp. Med. 140, 239 (1974)
27. Takemori, T., Tada, T.: J. Exp. Med. 140, 253 (1974)
28. Taussig, M.J., Mozes, E., Isac, R.: J. Exp. Med. 140, 301 (1974)
29. Taussig, M.J., Munro, A.: Proc. Ninth leucocyte culture conference. New York: Academic Press 1975, p. 791
30. Vladutiu, A.O., Rose, N.R.: Science 174, 1137-1139 (1971)
31. Wernet, P., Rieber, E.P., Winchester, R.J., Kunkel, H.G.: In: Histocompatibility Testing 1975, pp. 647-650
32. Zinkernagel, R.M., Doherty, P.C.: Nature (Lond.) 248, 701-702 (1974)
33. Zinkernagel, R.M., Doherty, P.C.: Nature (Lond.) 251, 547-548 (1974)

Specificity of Immune Response Control by H-Linked Ir Genes

E. Rüde and E. Günther

The processes required for the induction of an immune response are
quite complex, and it is obvious that genetic regulation may take place
at different levels. Among the genetic factors determining the ability
of an individual to respond to an antigen dominant immune response
(Ir) genes that are closely linked to the genes of the major histocompatibility complex (MHC) (H-linked Ir genes) play an important role
(Benacerraf and McDevitt, 1972; Benacerraf and Katz, 1975). The presence of such an H-linked Ir gene in an individual is one of the pre-
requisites for the recognition of a particular antigen as an immunogen.
H-linked Ir genes have been first recognized in studies on the immunogenicity of synthetic polypeptide antigens of limited heterogeneity
in mice (McDevitt and Tyan, 1968). The same type of control has in the
meantime been demonstrated for a large variety of natural antigens as
well. Furthermore, H-linked Ir genes have been identified in several
other species, such as guinea pigs (Ellman et al., 1970), rats (Günther
et al., 1972; Würzburg, 1971), rhesus monkeys (Dorf et al., 1974), and
chickens (Günther et al., 1974; Karakoz et al., 1974).

The mechanism of Ir gene function is of both, theoretical and practical
interest because these genes control functions that are essential for
the stimulation of lymphocytes by (thymus-dependent) antigens, and
because they may participate in determining susceptibility or resistance to certain autoimmune, infectious, and neoplastic diseases
(Svejgaard et al., 1975).

One of the major characteristics of H-linked Ir genes is their high
degree of specificity with respect to a certain antigen. This is exemplified by the immune response to a variety of linear and branched
synthetic polypeptide antigens. The structure of some of the branched
antigens originally developed by Sela and coworkers (1962) is shown
schematically in Figure 1. These antigens are built on a backbone of
poly-L-lysine to which poly-D, L-alanine or poly-L-proline side chains
are attached. The side chains can then be further elongated by short
copolymers composed of either L-tyrosine, L-histidine, or L-phenyl-
alanine combined with L-glutamic acid residues. The antibody response
of different rat strains to such antigens is shown in Table 1. A typical pattern of high and low responsiveness to each antigen as measured
by the production of antibodies was observed which depended on the H-1
haplotypes of the particular strains (Table 1). It should be noted
that most of the strains used were H-1 congenic i.e. they differed
from each other only by the genes of the major histocompatibility system termed H-1 in the rat. Thus the gene(s) determining high or low
responsiveness must be linked to the genes of the H-1 system.

It is of particular interest that the closely related antigens (T,G)-
A--L, (H,G)-A--L, and (Phe,G)-A--L gave rise to different patterns
(Günther et al., 1972; Günther et al., 1973), indicating that these
polypeptides are under separate control and it appears that the specificity of this control must be mainly determined by the aromatic amino
acid residues in the otherwise very similar polypeptides. However, the
antibodies produced to these antigens cross-react extensively. This

190

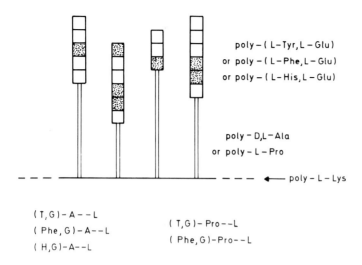

poly – (L–Tyr,L – Glu)
or poly – (L–Phe,L – Glu)
or poly – (L–His,L – Glu)

poly – D,L–Ala
or poly – L – Pro

◄── poly – L – Lys

(T,G)– A – – L
(Phe,G)– A – – L
(H,G)– A – – L

(T,G)– Pro – – L
(Phe,G)– Pro – – L

Fig. 1. Schematic representation of branched synthetic polypeptide antigens prepared by polymerization of amino acids

Table 1. Response patterns of rats of different H-1 haplotypes to branched and linear synthetic polypeptides[a]

H - 1 haplotype	Secondary antibody response to:								
	(T,G)-A--L	(H,G)-A--L	(Phe,G)-A--L	Pro--L	(T,G)-Pro--L	(Phe,G)-Pro--L	$G^{52}L^{33}T^{15}$	$G^{50}A^{50}$	$G^{50}T^{50}$
a	h	l	h	l	l	h	h	l	h
b	l	l			h		l	h	l
c	m h	h	h				h	h	h
d	l	l	h						
e	l	h	h		l				
f	l m	l	h	h	h	h			
l	l	m	h				l	h	l
n	l	l	l	m	m	m	l	l	l
w	l	h	l	m	m	l	m	h	l
Reference	1	2	2	3	3	3	4	5	5

[a]Classification of antibody responses as high (h), medium (m) and low (l) is relative and is valid only for the particular test-antigen used.

References: 1 Günther et al. (1972); 2 Günther et al. (1973); 3 Günther et al. (1976); 4 Kunz et al. (1974); 5 Armerding et al. (1974)

Table 2. Cross-absorption of anti-(H,G)-A--L antibodies by (T,G)-A--L and
(Phe,G)-A--L bound to Sepharose

Antisera from rat strain	Anti-(H,G)-A--L titers[a] after absorption with:			
	(H,G)-A--L-Sepharose Sepharose		(T,G)-A--L-Sepharose	(Phe,G)-A--L-Sepharose
LEW (med. resp.)	125	10	4	4
L.WP (high resp.)	640	6	18	4
(LEW x L.BDV)F1 (high resp.)	780	13	76	37

[a]Determined using the homologous (H,G)-A--L as test-antigen.

is demonstrated in Table 2 for anti-(H,G)-A--L antibodies which can be
absorbed to a large extent by (T,G)-A--L and (Phe,G)-A--L bound to
sepharose. Thus, the specificity of the anti-polypeptide antibodies
is rather similar, whereas the specificity of genetic control that
decides whether antibodies are formed at all is different.

The linear copolymer $G^{50}T^{50}$ exhibited a response pattern (Armerding et
al., 1974) very similar to that of (T,G)-A--L. This would be an agreement with the assumption that in (T,G)-A--L the (T,G)-copolymers also
play a major role in determining the specificity of the control.

On the other hand, the influence of the core structure becomes apparent
when the poly-D,L-alanine side chains of (T,G)-A--L were replaced by
poly-L-proline (Günther et al., 1976). The resulting (T,G)-Pro--L
elicited a response pattern completely different from that of (T,G)-A--
L. In this case the Pro--L core appears to be the dominant region for
the specificity of Ir gene control since in the strains tested Pro--L
itself induced the same antibody response as (T,G)-Pro--L. This even
applies to the specificity of the antibodies produced which were mainly
anti-Pro--L antibodies, while the (T,G)-portion seemed of little importance. In contrast, (Phe,G)-Pro--L exhibited a response pattern in the
four H-1 congenic rat strains tested up to now, which was rather similar to that of (Phe,G)-A--L but different from Pro--L. This indicates
a relatively greater importance of the (Phe,G)-copolymers as compared
to the (T,G)-portion of (T,G)-Pro--L. These examples demonstrate that
no predictions can so far be made concerning the relative contributions
of different portions of the branched polypeptide antigens to the Ir
gene controlled recognition.

In mice the response to each of about 25 different antigens both natural and synthetic was found to be under H-linked Ir gene control and
most of the antigens exhibited different response patterns in strains
representing 12 different major histocompatibility (H-2) haplotypes
(for review see Benacerraf and Katz, 1975a). This again shows that the
Ir genes involved are distinct and suggests that their function must
be related to the specific recognition of antigen.

With respect to the relationship of structure and specificity of Ir genes
the results obtained in mice with synthetic polypeptide antigens, some
of them closely related, are again most informative (Table 3). The classical data obtained with the branched antigens of the (T,G)-A--L type

Table 3. Response patterns of mice of different H-2 haplotypes to branched and linear polypeptides[a]

H-1 haplotype	Antibody response to															
	(T,G)-A--L	(H,G)-A--L	(Phe,G)-A--L	$G_{40}A_{60}$	$G_{60}A_{40}$	$G_{90}A_{10}$	$G_{50}T_{50}$	$G_{90}T_{10}$	$G_{60}A_{30}T_{10}$	$G_{35}L_{35}A_{30}$	$G_{54}L_{36}A_{10}$	$G_{57}L_{38}A_{5}$	$G_{57}L_{38}Pro_{5}$	$G_{57}L_{38}T_{5}$	$G_{51}L_{34}T_{15}$	$G_{53}L_{36}Phe_{11}$
b	h	l	h	h	h	l	l	l	h	h	l	l	l	l	l	l
d	m	m	h	h	h	l	l	l	h	h	m h	m h	l	h	h	h
f	l	l	h		h				h	h	h	h	l	l	h	l
j	l	m	h		l				l	l	h	m h		h	h	h
k	l	h	h	h	h	l	l	l	h	h	m h	m h	l	l	l	l
p	l	l	h	l	l	l	l	l	l	l	l	l m	l	l	l	h
q	l	l	h	l	l	l	l	l	l	l	m	l l m	l	l	m	h
r	l	l	h	h	h		l	l	h	h	m	m	l	h	h	h
s	l	l	l	h	h	l	l	l	l	h	m h	h	m h	l	l	l
u					h[5]				h[5]				l	h	h	h
v					h[5]				h[5]				l	l	l	l
Reference	1,2	1,2	1,2	3	3	3	3	3	4	5	6	6	5	7	7	7,8

[a] Classification of antibody responses as high (h), medium (m), and low or non-responsive (l) is relative and valid only for the particular antigen.

References: 1 McDevitt and Chinitz (1969); 2 Grumet and McDevitt (1972); 3 Merryman and Maurer (1976); 4 Merryman et al. (1975); 5 Benacerraf and Katz (1975a); 6 Maurer and Merryman (1974); 7 Dorf et al. (1976); 8 Dorf and Benacerraf (1975)

(McDevitt and Chinitz, 1969) need no further comment since they are essentially the same as already discussed for rats. It should be mentioned, however, that in contrast to rats the response to the Pro--L portion of (T,G)-Pro--L and (Phe,G)-Pro--L was reported to be controlled by an Ir gene termed Ir-3 not linked to H-2 (Mozes et al. ,1969). These antigens are therefore not included in the table.

A situation very similar to that encountered with the (T,G)-A--L series of polypeptides can also be observed with the linear copolymers $G^{57}L^{38}A^5$, $G^{57}L^{38}T^5$, $G^{57}L^{38}Pro^5$, and $G^{53}L^{36}Phe^{11}$. These antigens differ from each other only by the third amino acid. Although present in relatively small amount (5-11 % of the total number of residues) this amino acid determines the specificity of the response patterns which are distinct for each antigen. Extensive cross-reaction of the antibodies produced against some of these copolymers as in the case of (T,G)-A--L, (H,G)-A--L, and (Phe,G)-A--L has been reported (Benacerraf and Katz, 1975a). This is also true for the copolymers $G^{60}A^{40}$ and $G^{60}A^{30}T^{10}$. In this case the response patterns of the two antigens is the same except for mice of the $H-2^S$ haplotype which respond to $G^{60}A^{40}$ but not to $G^{60}A^{30}T^{10}$. Thus, the introduction of relatively few tyrosine residues into $G^{60}A^{40}$ has eliminated responsiveness. Since $G^{60}A^{30}T^{10}$ was found to stimulate suppressor T cells in $H-2^S$ mice it has been speculated that the tyrosine containing peptides which were added to $G^{60}A^{40}$ could be responsible for this activation of suppressor T cells which in turn could be responsible for low responsiveness (Benacerraf and Katz, 1975a).

In addition to these examples stressing the specificity of Ir gene control there are a few related but not identical polypeptides such as $G^{35}L^{35}A^{30}$ and $G^{60}A^{40}$ or $G^{60}L^{30}A^{10}$ and $G^{57}L^{38}A^5$ which in the strains tested exhibited the same or a very similar pattern. This can, of course, be explained by the recognition of structures common to each of these antigens by the same Ir gene. Similar results obtained with more defined branched polypeptides will be discussed below.

In view of the high degree of specificity exhibited by Ir gene function it is of particular importance that there is strong experimental evidence indicating that this specificity is not expressed at the level of the antibodies but at another level of immunological specificity (Benacerraf and McDevitt, 1972). It has already been discussed that related antigens which are under separate genetic control may nevertheless induce the production of highly cross-reactive antibodies. Furthermore, low responder animals can also produce high concentrations of antibodies against the respective antigen if this antigen is presented to the animal not as such but as a conjugate with a protein carrier (Green et al., 1966). In certain cases it could in addition be shown by electrofocusing analysis that the antibodies of low responders are not different from those of the high responders (Melchers et al., 1973). Finally, there is no evidence that major histocompatibility genes are genetically linked to known immunoglobulin structural genes.

Although much could be learned about the specificity of H-linked Ir genes by using linear and branched polypeptides prepared by copolymerization techniques these antigens having essentially random amino acid sequences are still too complex for the further characterization of the structures responsible for genetically controlled recognition. In particular, it would be of interest to know more about the relationship between the antigenic structures which are responsible for the specificity of genetic control and the determinants which are recognized by T and B lymphocytes, respectively. For this purpose synthetic antigens of known amino acid sequence can be prepared and it can be

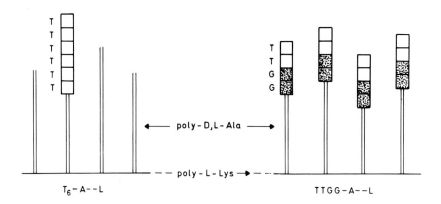

<u>Fig. 2.</u> Schematic representation of branched polypeptide antigens with defined oligopeptide side chains

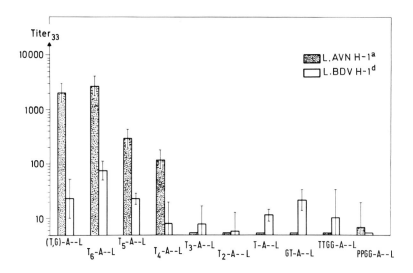

<u>Fig. 3.</u> Secondary antibody responses of two H-1 congenic rat strains to branched polypeptides with defined oliogopeptide side chains. Antibody titers were measured by a radioimmune coprecipitation assay with the homologous antigens

tested whether the pattern of genetic control is changed by defined modifications of these antigens.

In our own studies branched synthetic antigens analogous to (T,G)-A--L but carrying defined side chains were used to test for the fine specificity of Ir gene control in rats and mice (Rüde and Günther, 1974; Rüde and Günther, in prep.). As already discussed, it appears that the specificity of control to (T,G)-A--L, (H,G)-A--L, and (Phe,G)-A--L is mainly determined by the aromatic amino acid residues. With respect to (T,G)-A--L the random (T,G)-sequences were, therefore, replaced by defined oligopeptides containing only tyrosine or certain sequences of tyrosine and glutamic acid residues (Fig. 2).

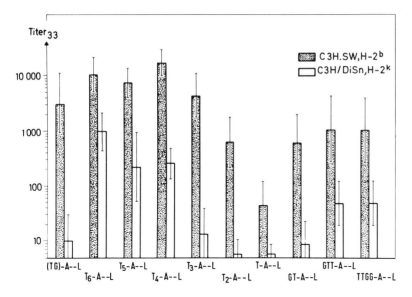

Fig. 4. Secondary antibody responses of two H-2 congenic mouse strains to branched polypeptides with defined oligopeptide side chains. Antibody titers were measured by a radioimmune coprecipitation assay with the homologous antigen

The antibody response to several of the resulting polypeptides in two H-1 congenic rat strains which are high and low responders to (T,G)-A--L, respectively, is shown in Figure 3. It can be seen that only T_6-, T_5-, and T_4-A--L gave a high/low responder split concordant with that of (T,G)-A--L while some of the other polypeptides which were all poorly immunogenic even tended to show a reverse pattern. In several other rat strains of different H-1 types the response pattern to T_6-A--L was also found to be identical to that of (T,G)-A--L. Thus, these polypeptides may be under the control of the same Ir gene and oligotyrosine peptides in combination with poly-D,L-alanine residues of the side chains may be the structures which define the specificity of the Ir-(T,G)-A--L gene(s).

In mice the results obtained with these analogs of (T,G)-A--L were clearly different (Fig. 4). $H-2^b$ mice which are high responders to (T,G)-A--L reacted with significantly higher antibody titers than $H-2^k$ low responder mice to all of the polypeptides tested. However, considerable quantitative differences existed in the high/low responder split between antigens carrying various defined oligopeptides. It is remarkable that some of the polypeptides such as TTGG-A--L, T_2-A--L, or GT-A--L differ considerably in structure but nevertheless gave a very similar high/low responder split. This structural difference is, for instance, reflected in the very poor cross-reactivity of antibodies produced against T_2-A--L and GT-A--L indicating that the major determinants of these two antigens recognized by B cells differ.

Similar studies have been carried out by Mozes et al. (1974) and by Seaver (personal communication). A summary of the data is given in Table 4. Apart from the first group of analogs which behaved as (T,G)-A--L some other tyrosine and glutamic acid containing peptides on A--L backbone were reported to exhibit a response pattern that was different from that of (T,G)-A--L. In summary these results indicate that Ir gene control is highly specific since it distinguishes between

Table 4. Specificity of Ir gene control in mice to defined analogs of (T,G)-A--L

	Response pattern similar to (T,G)-A--L
T_2-A--L[1]	yes
T_3-A--L[2]	yes
T_4-A--L[1]	yes
T_5-A--L[2]	yes
T_6-A--L[1]	yes
GT-A--L[1,3]	yes
GTT-A--L[2]	yes
TTGG-A--L[2,4]	yes
TG-A--L[3]	no
TGTG-A--L[4]	no
GTTG-A--L[4]	no

References: 1 Rüde and Günther (1974); 2 Rüde and Günther, in prep.; 3 Seaver, pers. commun.; 4 Mozes et al. (1974); Schwartz et al. (1975)

different oligopeptides bound to A--L. Only some of these elicited a response pattern concordant to that of (T,G)-A--L. On the other hand, if one assumes that control to the concordant group of antigens and to (T,G)-A--L is exerted by the same Ir gene the specificity of this gene would be relatively broad since it would include peptides as different as GT-A--L and TTGG-A--L. Alternatively, several closely linked genes with different specificity could control responsiveness to this first group of antigens. The response to (T,G)-A--L would then be the resultant of several separately controlled immune responses. The finding that in certain F_1 hybrids of rats high responsiveness to (H,G)-A--L is dependent on the complementation of at least two H-linked Ir genes (Günther and Rüde, 1975) would be in agreement with such an assumption. It is not yet known, however, whether such complementing genes are indeed specific for separate structures or determinants of the antigen molecule.

An alternative approach to study the fine specificity of Ir gene control and to identify the responsible structures of an antigen molecule is to use natural protein antigens of known sequence and tertiary structure which are under Ir gene control and for which variants from various species with limited amino acid replacements are available. Examples are lysozyme and insulin.

The antibody response of mice to hen egg white lysozyme was found by Hill and Sercarz (1975) to be under control of an H-2 linked Ir gene. From the response pattern of C57BL/6 mice to various gallinaceous lysozymes of known amino acid sequences it was suggested that the recognition of these molecules depends on a region comprising the amino acid residues No. 99 to 103.

Results on the Ir gene control of antibody production to insulins and to hapten conjugates of insulin are shown in a somewhat simplified form

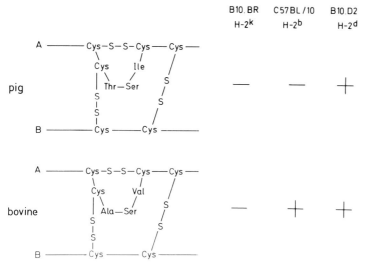

Fig. 5. Schematic representation of pig and bovine insulin and responsiveness of mice of three H-2 haplotypes to these antigens (according to Keck, 1975a, b)

in Figure 5 (Keck, 1975a, b). Mice of the $H-2^k$ haplotype were non-responders to both pig and bovine insulin while $H-2^d$ mice responded to both. In $H-2^b$ mice only bovine insulin but not pig insulin induced antibody formation. Since the two insulins differ from each other by only two amino acids in a small and rigid peptide loop and since the loop peptide is the same in pig and mouse insulins it can be concluded that these amino acids (Ala and Val) must be responsible for the Ir gene controlled recognition of bovine insulin in $H-2^b$ mice. It cannot be excluded, however, that another area of bovine insulin which in addition to the loop differs from the two mouse insulins may also contribute to recognition. This must be the case in $H-2^d$ mice where pig and bovine insulins are equally immunogenic and the loop peptide of bovine insulin cannot be exclusively responsible for immunogenicity. These studies with lysozymes and insulins again stress the high degree of specificity of the recognition process controlled by Ir genes and demonstrate that also natural antigens are suitable to identify small amino acid sequences responsible for this specificity.

For a better understanding of Ir gene specificity it would, of course, be important to know more about the mechanism of Ir gene function. A variety of experimental data suggest that H-linked Ir genes control processes that are associated with the specific stimulation of T lymphocytes by antigen (which requires the participation of macrophages) and/or the cooperation of these cells with B lymphocytes (Mozes, 1975). This explains why Ir gene control can be observed both by antibody production to thymus dependent antigens requiring the cooperation of B with T lymphocytes and, on the other hand, in cellular immune reactions that are dependent on the function of T lymphocytes. This is illustrated in Figure 6 for the antibody response and for the proliferative secondary response of lymphocytes of three rat strains (high, medium, and low responding) to (H,G)-A--L. For the latter test which is generally considered to measure T cell function lymph node cells of preimmunized animals were cultured in vitro in the presence of dif-

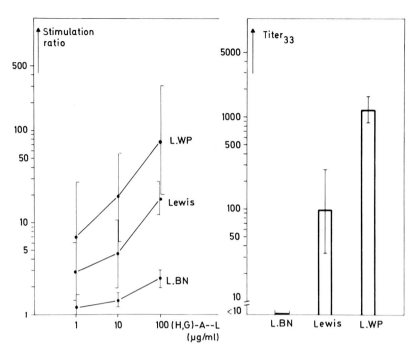

Fig. 6. Ir gene controlled immune responsiveness in rats to (H,G)-A--L. Correlation between (H,G)-A--L induced secondary lymphocyte proliferation (left) and antibody production (right). Lymphocyte proliferation was measured by ^{3}H-thymidine uptake and is expressed as stimulation ratio relative to the control cultures without antigen. Antibody titers were determined by a radioimmune coprecipitation assay

ferent concentrations of antigen. Cell proliferation was measured by the uptake of tritiated thymidine. It can be seen that responsiveness is equally expressed in lymphocyte proliferation and antibody production.

Specificity of Ir gene control at the cellular level was studied using this in vitro assay by testing the cross-stimulation between the polypeptides (H,G)-A--L, (Phe,G)-A--L and (T,G)-A--L (Fig. 7). These antigens show extensive cross-reaction of the respective antibodies but are under separate Ir gene control as discussed before. (H,G)-A--L and (Phe,G)-A--L primed lymph node cells were stimulated only by the homologous but not by any of the other antigens tested. The same was true for (T,G)-A--L primed cells with respect to (H,G)-A--L or A--L. However, these cells were cross-stimulated by (Phe,G)-A--L to the same extent as with the homologous (T,G)-A--L (Günther and Rüde, 1976). Thus, there is an unilateral cross-reaction at the cellular (probably T cell) level between (T,G)-A--L and (Phe,G)-A--L. This may not be too difficult to accept because the phenylalanine residues differ from the tyrosine residues only be the lack of the phenolic hydroxyl group. However, in order to explain these results on a functional level it would be important to know more about the relationship between the products of Ir genes and the receptors of T cells for antigen. Recently, evidence has been obtained for the expression of the variable region of immunoglobulins on T-cells, presumably as part of the T-cell receptor (Binz et al., 1975). However, as mentioned before, V-region

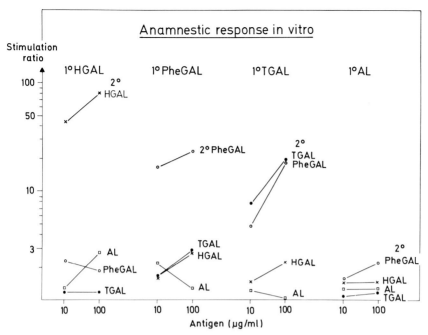

Fig. 7. Cross-stimulation between the branched synthetic polypeptides (H,G)-A--L, (Phe,G)-A--L, (T,G)-A--L, and A--L on the cellular level. After priming in vivo (1°) of the respective high responder rats lymph node cells were cultured in vitro in the presence of antigen. Cell proliferation was measured by ³H-thymidine uptake which is expressed as stimulation ratio relative to the control cultures without antigen

genes of immunoglobulins are not identical with H-linked Ir genes. The above cross-stimulation could, therefore, occur at the level of the T-cell receptor which might be distinct from the Ir gene product. But stimulation may in addition depend on the expression of the respective Ir genes on the cells (T cells and/or macrophages) participating in this stimulation. Alternatively, the unilateral cross-reaction could occur directly at the level of the Ir gene product itself. It is interesting to note that (T,G)-A--L high responding strains are also high responders to (Phe,G)-A--L.

The high degree of Ir gene specificity could be most easily explained by the interaction of the antigen with a specific receptor which is at least partially a product of the Ir gene. However, in addition to the problems just mentioned there is the further complication that in certain cases high responsiveness to an antigen depends on the interaction of two separate Ir genes both of which may be specific and may have different functions in different cells (Munro and Taussig, 1975). Furthermore, in certain types of low responders the respective antigen was found to induce T cells which can actively suppress the antibody response (Gershon et al., 1973). This activity is also specific and controlled by genes linked to the genes of the major histocompatibility system (Benacerraf et al., 1975).

The specificity of Ir gene function is one aspect of the mechanism by which Ir genes regulate immune responsiveness and, therefore, how antigen stimulates lymphocytes.

Acknowledgment. This work was supported by the Deutsche Forschungsgemeinschaft.

References

Armerding, D., Katz, D.H., Benacerraf, B.: Immunogenetics 1, 340-351 (1974)

Benacerraf, B., McDevitt, H.O.: Science 173, 273-279 (1972)

Benacerraf, B., Kapp, J.A., Debré, P., Pierce, C.W., de la Croix, F.: Transpl. Rev. 26, 21-38 (1975)

Benacerraf, B., Katz, D.H.: Advan. Cancer Res. 21, 121-173 (1975b)

Benacerraf, B., Katz, D.H.: In: Immunogenetics and Immunodeficiency, Benacerraf, B. (ed.). Lancaster: MTP 1975a, pp. 117-177

Binz, H., Kimura, A., Wigzell, H.: Scand. J. Immunol. 4, 413-420 (1975)

Dorf, M.E., Balner, H., de Groot, H.L., Benacerraf, B.: Transpl. Proc. 6, 119-123 (1974)

Dorf, M.E., Benacerraf, B.: Proc. Nat. Acad. Sci. 72, 3671-3675 (1975)

Dorf, M.E., Maurer, P.H., Merryman, C.F., Benacerraf, B.: J. Exp. Med. 143, 889-896 (1976)

Ellman, L., Green, I., Martin, W.J., Benacerraf, B.: Proc. Nat. Acad. Sci. 66, 322-328 (1970)

Gershon, R.K., Maurer, P.J., Merryman, C.F.: Proc. Nat. Acad. Sci. 70, 250-254 (1973)

Green, I., Paul, W.E., Benacerraf, B.: J. Exp. Med. 123, 859-879 (1966)

Grumet, F.C., McDevitt, H.O.: Transplantation 13, 171-173 (1972)

Günther, E., Balcarova, J., Hala, K., Rüde, E., Hraba, T.: Europ. J. Immunol. 4, 548-553 (1974)

Günther, E., Mozes, E., Rüde, E., Sela, M.: J. Immunol. in press (1976)

Günther, E., Rüde, E.: J. Immunol. 115, 1387-1393 (1975)

Günther, E., Rüde, E.: Immunogenetics 3, 261-269 (1976)

Günther, E., Rüde, E., Meyer-Delius, M., Stark, O.: Transpl. Proc. 5, 1467-1469 (1973)

Günther, E., Rüde, E., Stark, O.: Europ. J. Immunol. 2, 151-155 (1972)

Hill, S.W., Sercarz, E.E.: Europ. J. Immunol. 5, 317-324 (1975)

Karakoz, I., Krejci, J., Hala, K., Blaszczyk, B., Hraba, T., Pekarek, J.: Europ. J. Immunol. 4, 545-548 (1974)

Keck, K.: Nature 254, 78-79 (1975a)

Keck, K.: Eur. J. Immunol. 5, 801-807 (1975b)

Kunz, H.W., Gill III, T.J., Borland, B.: J. Immunogenetics 1, 277-287 (1974)

Maurer, P.H., Merryman, C.F.: Immunogenetics 1, 174-183 (1974)

McDevitt, H.O., Chinitz, A.: Science 163, 1207-1208 (1969)

McDevitt, H.O., Tyan, M.L.: J. Exp. Med. 128, 1-11 (1968)

Melchers, I., Rajewsky, K., Shreffler, D.C.: Europ. J. Immunol. 3, 754-761 (1973)

Merryman, C.F., Maurer, P.H., Stimpfling, J.H.: Immunogenetics 2, 441-448 (1975)

Merryman, C.F., Maurer, P.H.: J. Immunol. 116, 739-742 (1976)

Mozes, E.: Immunogenetics 2, 397-410 (1975)

Mozes, E., McDevitt, H.O., Jaton, J.C., Sela, M.: J. Exp. Med. 130, 493-504 (1969)

Mozes, E., Schwartz, M., Sela, M.: J. Exp. Med. 140, 349-355 (1974)

Munro, H.J., Taussig, M.J.: Nature (Lond.) 256, 103-106 (1975)

Rüde, E., Günther, E.: In: Progress in Immunology II, Brent, L., Holborow, J. (Eds.). North-Holland: 1974, Vol. II, pp. 223-233

Schwartz, M., Mozes, E., Sela, M.: Europ. J. Immunol. 5, 866-871 (1975)

Sela, M., Fuchs, S., Arnon, R.: Biochem. J. 85, 223-235 (1962)
Svejgaard, A., Platz, P., Ryder, L.P., Staub-Nielsen, L., Thomson, M.:
  Transpl. Rev. 22, 3-43 (1975)
Würzburg, U.: Europ. J. Immunol. 1, 496-497 (1971)

The Structure of Products of the Major Histocompatibility Complex in Man

J. L. Strominger, R. E. Humphreys, J. F. Kaufman, D. L. Mann, P. Parham, R. Robb, T. Springer, and C. Terhorst

Introduction
_____

The major histocompatibility complex (MHC) is a genetic region which encodes the major barrier to transplantation in man and in all other species which have been studied. It was first identified through mouse-breeding studies which led to the identification of the H-2 region, the second transplantation region discovered in these studies and the strongest barrier to transplantation in this species. There are at least 15 genetic loci which determine the ability of mice to exchange grafts. Little is known about the minor loci. HLA is the human analog of the H-2 region, and correspondingly the strongest of the transplantation barriers in man. The existence of other loci greatly increases the complexity of our ultimate understanding of the phenomenon of graft acceptance. The sixth human chromosome (or seventeenth mouse chromosome) contains genetic regions HLA-A, HLA-B and HLA-C (analogous to the H2-K and H2-D loci in the mouse). There are multiple alleles at each of these loci. The existence of this polymorphism and its meaning for the natural function of this region (which certainly did not evolve to prevent the exchange of surgical grafts) poses the most intriguing questions about this system. This paper will deal mainly with the products of the HLA-A and HLA-B loci. Little is known about the products of the HLA-C locus at present. The products of the HLA-D locus are defined by the mixed leucocyte culture (MLC) and are probably the human analog(s) of the mouse Ia antigen(s); some information about the probable product of this locus will also be presented.

The products of the HLA-A and HLA-B genes are measured using specific alloantisera, the major source of which is pregnancy antisera, i.e., sera from women who have been multiply pregnant and who have been immunized against paternal antigens. Indeed, one of the most interesting, incompletely solved problems in the field is the reason why pregnant females do not reject the commonest graft in the species, i.e. the fetus in utero. Pregnancy alloantisera in the presence of complement lyse Cr^{51}-labeled lymphocytes of appropriate specificity. Soluble antigen is measured by its ability to inhibit this lysis by combination with antibody.

Purification of HLA Antigens from Human Lymphocytes
_____

A number of attempts had been made to purify HLA antigens from human lymphocytes, notably by Sanderson and Batchelor (1) who used splenic lymphocytes. Very small amounts were obtained from this source. Cultured lymphoblasts appeared to be a preferable source, both because they were available in potentially larger amounts than human spleens and because an experiment with a single human spleen cannot be repeated. However, cultured lymphoblasts are available in reproducible supply and do not seem to have any altered HLA specificities. Remarkably,

Table 1. Purification of papain-solubilized HLA antigens from 100 gm of JY cells

Purification step	Protein (mg)	Total number of inhibitory units recovered		Specific activity (inhibitory units/mg)		Purifi- cation
		HLA-A2	HLA-B7	HLA-A2	HLA-B7	
1. Cell membrane	1760	640,000	280,000	360	159	1
2. Papain digestion	1480	504,000	248,000	340	167	1
3. DE-52 batchwise dilution	360	956,000	288,000	2,560	800	~6
4. Sephadex G-150 chromatography	26	780,000	236,000	30,000	9,100	~70
5. DE-52 chromatography	4.1 (HLA-A2) 4.9 (HLA-B7)	900,000	216,000	220,000	44,000	~300

these lymphoblasts also contained far more HLA antigens than splenic or peripheral blood lymphocytes. This was shown by absorption experiments with HLA antisera using three types of cells, all from the same individual (2). Peripheral blood lymphocytes from RH had a very low absorptive capacity for these antisera. PHA-stimulated lymphocytes had a greatly enhanced capacity and RH lymphocytes transformed by EBV and growing continuously in cultures had an enormously enhanced absorptive capacity for HLA antisera, i.e. the representation of HLA antigens on the surface of the cultured lymphocyte transformed by EBV was in the range of 20-50-fold greater than peripheral blood lymphocytes; all four of the HLA specificities of RH cells were similarly affected. This enhanced representation was specific in that other membrane markers, such as 5'-nucleotidase or radioiodinatable surface protein, were increased in the transformed cell only 2-3-fold, the same as the increase in surface area. One interpretation is that the virus itself induced or enhanced expression of the antigen in some way. Alternatively, the virus may have selected for transformation a subpopulation of B lymphocytes which already had an enhanced representation. The explanation is not known but the fact that the HLA antigens are so much more densely represented on the cultured lymphoblasts has made possible their isolation in relatively large amounts.

Preparation of the HLA Antigens after Papain Solubilization

Two principle methods have been used for solubilization of HLA antigens. These are: (a) treatment with papain and (b) solubilization with detergent. Preparation of HLA from the cell line JY after papain solubilization is shown in Table 1. The use of papain to solubilize HLA antigens goes back to the pioneering studies of Shimada and Nathenson (3) who first solubilized H-2 antigens in this way. The procedure for purification of HLA has been improved considerably over that previously reported (4) and yields of about 4-5 gm per 50 gm batch can now be obtained. The procedure for isolation is not difficult, requiring only four steps. The most interesting feature is that only about 70-fold purification was needed to obtain pure HLA antigen from cell membranes, i.e., something in the order of 1-2 % of the total membrane protein in the cultured lymphoblast is HLA antigen. That is a very large representation of a single protein on the lymphocyte membrane surface. The cells used were homozygous for HLA-A2 at the first histocompatibility locus and for HLA-B7 at the second. HLA-A2 has a charge difference which distinguishes it from most of the other specificities. It is readily separated on DEAE-cellulose chromatography from HLA-B7 at the last step of purification. Separating the allelic specificities from each other is one of the biggest problems in this field. It is relatively easy to obtain pure HLA antigens, i.e., a mixture of the four specificities. Of course, the most interesting part of the chemistry requires that they be separated. Starting with doubly homozygous cell lines greatly reduces the problem. Several such lines have been started from homozygous individuals in the Indiana Amish community, an inbred religious sect, one of which is the JY cell line. HLA-A2 and a mixture of HLA-B7,12 have previously been obtained from the cell line RPMI 4265 (4).

Our first interesting finding was that these antigens contain two subunits (5, 6): a heavy chain which is glycoprotein and a light chain which is now known to be β_2-microglobulin (a protein first isolated from human urine). SDS gels of the HLA-A2 antigen preparation showed the heavy chain to have a molecular weight of 34,000 and the light chain to have a molecular weight of 12,000 (β_2-microglobulin). SDS gels of the HLA-B7,12 mixture showed a doublet at 34,000 molecular weight in addition to β_2-microglobulin. One of the doublet glycoproteins may be HLA-B7 and the other HLA-B12. However, since these gels are denaturing gels, the glycoproteins could not be recovered to prove that point.

Despite the apparent purity of the HLA antigen preparations, isoelectric focusing revealed considerable heterogeneity (4). For example, in the HLA-A2 preparation at least four bands with HLA-A2 antigenic activity were seen. The most interesting possible interpretation of this heterogeneity was that there was heterogeneity in the amino acid sequence and, therefore, that some kind of V region might exist in the HLA antigens. However, the heterogeneity turned out to be due to variability in the number of sialic acid residues on the molecule (7). The HLA-A2 preparation was treated with neuraminidase as a function of time. The initial preparation contained a species with two sialic acid residues as the major component but there were also species with three and species with one sialic acid residue. As the result of treatment, a preparation which was sialic acid-free was obtained and all the heterogeneity disappeared. The same result was obtained for other HLA preparations.

Are these antigen preparations really pure? To approach that question, the HLA antigen preparations were labeled by reductive methylation with formaldehyde and sodium borohydride (8). About two methyl groups/mol were introduced on the ϵ-amino groups of lysine of the molecule. This treatment did not alter immunological activity at all. Essentially 100 % of these labeled preparations of HLA-A2 antigens formed a specific complex with HLA-A2 antisera. No significant complexation was observed with normal serum or with specificity controls (HLA antisera with specificities other than HLA-A2).

However, when the HLA-B7,12 preparation or another HLA antigen preparation containing HLA-A3, AW25, B12 and B27 were used, only about 70 % of the total antigen could be complexed (8). With the HLA-B7,12 preparation, about 40 % complexation was obtained with HLA-B7 antiserum, 30 % with HLA-B12 antiserum and a total of 70 % with a mixture of antisera. What does that mean? The residual 30 % of material could be the product of the HLA-C locus or other unidentified HLA antigens, denatured antigens, or some other unrelated material co-purifying with HLA.

<u>Table 2.</u> Amino acid composition of papain solubilized HLA2 and HLA7+12 from cell line RPMI 4265 and papain-solubilized HLA2 and HLA7 from cell line JY in mol/100 mol. Each analysis was done in duplicate for 24, 48 and 72 h

Amino acid	HLA2_{JY}	HLA2₄₂₆₅	HLA7_{JY}	HLA7+12₄₂₆₅
Asp[a]	7.9	7.9	9.7	9.4
Thr[b]	7.6	7.5	6.2	6.9
Ser[b]	5.2	5.2	5.2	5.0
Glu[a]	13.9	13.9	14.8	14.4
Pro	4.3	4.5	5.4	5.4
Gly	7.9	7.6	8.0	6.9
Ala	8.8	8.5	8.0	7.8
Val[c]	6.3	6.4	5.0	5.0
Met	1.3	1.5	1.1	1.0
Ile[c]	1.6	1.8	2.6	2.9
Leu	6.2	6.3	6.4	6.7
Tyr[b]	4.8	4.7	4.5	5.3
Phe	2.9	2.9	2.3	2.4
Lys	4.2	4.3	3.7	3.6
His	4.8	4.5	3.5	3.3
Arg	7.6	7.8	8.8	9.3
CMCys	1.5	1.6	1.6	1.7
Trp[d]	3.2	3.2	3.1	3.2

[a]Ammonia not determined. [b] Extrapolated zero-time values. [c]72 h value only.
[d]Determined spectrophotometrically.

Another evidence of purity is the single common amino terminal group found in the HLA antigen preparations (8). The five preparations available all had glycine as the N-terminal amino acid of the heavy chain. An isoleucine residue was also found; it is the N-terminal residue of β_2-microglobulin.

With confidence that these preparations were pure, the heavy and light chains were reduced and alkylated by treatment with iodoacetic acid and then separated by gel filtration. Amino acid analysis of the heavy chain has been carried out and sequence studies initiated. The analyses of four preparations of the heavy chains of HLA antigens obtained from RPMI 4265 cells and JY cells are shown in Table 2. First of all, there were no significant differences between the heavy chain of HLA-A2 from JY and HLA-A2 from RPMI 4265 cells. Very small differences between HLA-A7 and the HLA-B7,12 mixture were found. However, on the order of 20 to 30 amino acid differences between HLA-A2 and HLA-B7 may be estimated. The degree of relatedness of these proteins can be examined from their amino acid analyses by a statistical method described by Marchalonis and Weltman (9). In this method, none of several thousand unrelated proteins had SΔQ values of less than 100. A SΔQ value of less than 50 was therefore considered to suggest a significant relatedness between the proteins. By this method, HLA-A2 from JY was identical to

Table 3. Relatedness among HLA antigens and β_2-microglobulin as determined from the amino acid compositions by the SΔQ method

Protein	HLA2 (4265)	HLA2 (JY)	HLA7 (JY)	HLA7+12 (4265)	β_2-M (human)	β_2-M (mouse)
HLA2$_{4265}$	O					
HLA2$_{JY}$	1	O				
HLA7$_{JY}$	13	16	O			
HLA7+12$_{4265}$	14	18	4	O		
β_2-M$_{human}$	172	183	183	173	O	
β_2-M$_{mouse}$	138	147	153	141	70	O

HLA-A2 from 4265 cells (Table 3). HLA-A2 was very closely related to HLA-B7 or to HLA-B7,12. However, the heavy chains of the various HLA antigen preparations and β_2-microglobulin were not related (SΔQ$\sim$170). This calculation carried out with various classes of immunoglobulin heavy chains (IgG, IgA, IgD and IgM) also resulted in relatively larger SΔQ values. The lowest SΔQ was obtained in comparison with IgD heavy chain but it was in the order of 70 (the same value as was obtained in comparison of human β_2-microglobulin with either mouse β_2-microglobulin or with the Fc fragment of Eu myeloma protein (an IgG)). There appears to be a relatedness of the heavy chain of the HLA antigens and the heavy chain of the IgD, but it may not be very extensive.

N-terminal sequence data (Table 4) (10) showed no differences between the two HLA-A2 antigen preparations. HLA-B7 and the HLA-B7,12 mixtures were very similar and the latter showed heterogeneity at only one position. However, there was only one difference in the first 25 amino acids between HLA-A2 and HLA-B7, although there may be only in the order of 20-30 amino acid differences in the whole molecule. There are four half cystine residues per heavy chain in the papain-derived product (see below). The heavy chain has a polypeptide molecular weight of about 29,000. If the heavy chain of HLA is homologous to the heavy chain of immunoglobulin then one would expect four half cystine residues in a molecule of 29,000 molecular weight. Using [^{3}H]carboxymethyl cysteine-labeled HLA antigens in sequence studies up to 40 residues, well past where one would expect the first cysteine residue, no significant counts were found. This is beyond the place that one might expect to find the first cysteine residue if there was strong homology to immunoglobulins.

There are several important points which can be made from these data.

1. The products of the two loci (HLA-A and HLA-B) although separated by a distance equivalent to several thousand genes are remarkably similar to each other and they must have arisen by gene duplication.

2. Allelic products at a single locus do not differ very much from each other, i.e., the immunological difference is not reflected in a very large difference in amino acid composition, or in N-terminal sequence.

3. Comparison of the sequence data for HLA-A and HLA-B with that obtained in several laboratories for mouse H2-D and H2-K antigens reveals

Table 4. Partial amino acid sequence analysis of heavy chains from HLA antigens: HLA-2 (JY and RPMI 4265), HLA-7,12 (RPMI 4265) and HLA-7 (JY). Partial sequence analysis of the N-terminal residues was obtained in the Beckman 890B Automatic Sequencer. The sequences of the two HLA-2 preparations were identical

| Step Number | Amino acid identified | | |
	HLA-2	HLA-7,12	HLA-7
1	Gly	Gly	Gly
2	Ser	Ser	Ser
3			
4	Ser	Ser	Ser
5	Met	Met	Met
6	Arg	Arg,Val	Arg
7	Tyr	Tyr	Tyr
8	Phe	Phe	Phe
9	Phe	Tyr	Tyr
10	Thr	Thr	Thr
11	Ser	Ala	Ser
12	Val	Val	Val
13	Ser	Ser	Ser
14	Arg	Arg	Arg
15	Pro	Pro	Pro
16	Gly	Gly	Gly
17			
18	Gly	Gly	Gly
19	Glu	Glu	Glu
20			
21			
22	Phe	Phe	Phe
23	Ile	Ile	Ile
24	Ala	Ala	
25	Val	Val	Val

strong homologies between the human and murine antigens. These similarities are discussed in other papers in this volume and have recently been reviewed editorially (11) in connection with recent N-terminal sequence data for additional HLA antigens (12, and see also 12a). The latter are strikingly similar to those presented here.

4. Several points which may indicate similarities between HLA antigen and immunoglobulins may be summarized: (a) Limited proteolysis by papain; (b) A two-chain structure with sequence homology between the light chain of HLA (β_2-microglobulin) and some heavy chain immunoglobulin domains (6, 13, 14, 15, 16). (c) However, from the compositional and sequence data available so far, there is little apparent homology between the heavy chain of HLA antigens and immunoglobulins. The cys-

Table 5. Purification of detergent solubilized HL-A from 150 g of J. Yoder cells

Purification step	mg	% Recovery of inhibitory units		Specific activity (inhibitory units/mg)		Fold purification
		HL-A2	HL-A7	HL-A2	HL-A7	
Detergent-solubilized membrane	750	100	100	330	140	(1)
Lectin column chromatography	111	85	83	1890	760	5.6x
Anti-β_2-micro-globulin column	–	63	59	–	–	
Bio-gel A-5m column	7	54	50	19000	7300	55x

teine content, is however, relevant. The heavy chain of papain-solubilized HLA antigens is equal to two immunoglobulin domains in size. Like immunoglobulins there are two cysteine residues in intrachain disulfide linkage for each polypeptide "domain" of approximately 14,000 daltons. There are four half cystines for each polypeptide size of about 29,000 daltons in papain-solubilized HLA antigens and these half cystines are all present in intrachain disulfide linkages. Knowledge of the linkage and distribution of these will therefore be of great interest and further data will greatly clarify our knowledge of the structure and evolution of these interesting proteins.

Preparation of HLA Antigens After Detergent Solubilization

Most membrane proteins are solubilized by detergents which have HLB (hydrophilic lipophilic balance) numbers in the range of 12-14. The HLB number is an empirical measure of a detergent's tendency to make oil-in-water or water-in-oil emulsions. The solubilization of HLA antigens appeared similar to the solubilization of bacterial membrane proteins except for the fact that a group of relatively hydrophilic Brij detergents appeared to be relatively selective in solubilizing HLA antigens (18).

The purification of detergent-soluble material using an anti-β_2-microglobulin immunoabsorbent column is summarized in Table 5 (19). An earlier procedure (17) yielded partially purified material. After membrane preparation and solubilization in detergent, the next steps are passage through a lectin affinity column, absorption on an anti-β_2-microglobulin affinity column, and subsequent elution with purified soluble β_2-microglobulin, and removal of the excess β_2-microglobulin on a Bio-Gel A-5m column. The purification required to obtain pure antigen was only about 50-fold over the detergent-solubilized membranes and the yields were on the order of 50 %. About 7 mg of HLA antigen were prepared from 150 g of cells. The anti-β_2-microglobulin column has also been used succesfully without the lectin column step (19). Alternatively, repeated agarose gel filtration after passage through the lectin column also yielded pure antigen and may be more applicable to large-scale work (18). The detergent-solubilized HLA antigens also contained two polypeptides, a heavy chain of 44,000 daltons and a light

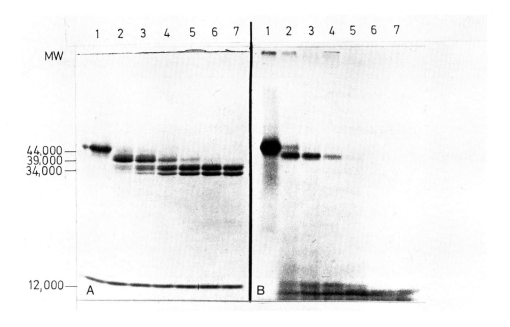

<u>Fig. 1A and B.</u> SDS slab gel electrophoresis of iodo-$[^{14}C]$acetate-labeled HLA anti-
gen. Antigen was digested with the indicated ratios of papain to protein: 1: none;
2: 1/3840; 3: 1/1920; 4: 1/960; 5: 1/480; 6: 1/240; 7: 1/120. Each aliquot was divided into halves and electrophoresed on two sides of an 11 % acrylamide SDS slab
gel. One half of the slab gel was stained (A); the other half was autoradiographed
(B)

chain of 12,000 daltons, thus showing that the previously observed
structure was not the result of proteolysis by papain.

The heavy chain of HLA antigens is therefore 10,000 daltons larger
than the heavy chain isolated from the papain-solubilized product; i.e.,
the papain cleaves a piece of 10,000 daltons from the molecule. Moreover, the detergent product has one or possibly two additional sulf-
hydryls present in easily reduced form (and therefore not in intrachain linkage), which are not found on the papain product. These easily
reduced sulfhydryl groups can be specifically labeled with $[^{14}C]$iodo-
acetic acid. The kinetics of papain cleavage of this material (Fig. 1)
clearly reveals that it proceeds in two steps (17, 20). An intermediate
of 39,000 daltons is first formed, followed by cleavage to a final
product of about 34,000 daltons. The easily reduced cysteines which
had been labeled are removed in two steps, one at the first step of
the cleavage and the remainder at the second step. The amino acid compositions of the 44,000, 39,000, and 34,000 dalton polypeptides have

Table 6. Amino acid composition, residues/molecule

Amino acid	Δ p44-p39	Δ p39-p34
CM-Cys	0.3	0.6
Asp	8	1
Thr[a]	5	0
Ser[a]	5	5
Glu	8	0
Pro	4	2
Gly	4	4
Ala	6	1
Val[c]	0	6
Met[b]	0	1
Ile[c]	0	6
Leu	0	7
Tyr[b]	2	0
Phe	2	3
His	0	0
Lys	2	2
Arg	4	0
Trp	n.d.	n.d.
Polarity[d]	63.7	20.7

Determinations are rounded to the nearest integer, except
for CM-Cys. n.d. = not determined.
[a]Extrapolated to zero time values. [b]24-h values. [c]72-h values.
[d]The polarity of p44 was 49.2; of p39, 46.7; and of p.34, 50.7.

been obtained. The difference between p44 and p39 is the composition
of the first peptide released, and the difference between the p39 and
p34 is the composition of the second peptide released (Table 6). The
first peptide released is extremely hydrophilic, much more hydrophilic
than the protein as a whole. The second peptide released is very hydrophobic and contain large numbers of leucine, isoleucine and valine
residues. The hydrophobic region is likely to be inserted in the membrane and the hydrophilic region is likely to be internal (although a
hair-pin structure is not excluded by the data available).

Do HLA Antigens Have a Four-Chain Structure?

Since immunoglobulins have a four-chain structure (i.e., two heavy
chains and two light chains), it seemed reasonable, in view of the
other indications of homology, to ask whether or not HLA antigens also
existed in this form. A number of observations in several laboratories
including our own (21-24) indicated that dimers and oligomers of the
two-chain unit could occur. What is the significance of these observation? Is the polymerization an artifact which occurs after isolation

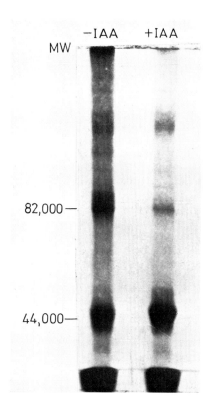

Fig. 2. SDS gel electrophoresis of HLA prepared in the presence and absence of
iodoacetamide. Aliquots of the preparation were electrophoresed on Laemmli 7.5 %
polyacrylamide SDS gels in the absence of ME

of the HLA antigen, i.e., did heavy chains in the preparation become
disulfide-linked to each other during isolation? Two kinds of experiment to examine that possibility have been carried out (25). If, indeed, the antigens are present in a tetrameric structure in the membrane before solubilization, then a chemical cross-linking reagent
should cross-link the chains in various ways, forming at least heavy-
chain dimers and light-heavy dimers. The only product obtained with
cross-linking reagents was a dimer containing a light chain and a
heavy chain. A dimer containing two heavy chains was not formed. In
another set of experiments cells were treated with iodoacetamide to
block all the free SH groups before isolating the antigens. Under these
conditions little or no oligomer was present in the isolated HLA antigens (Fig. 2) (20). Both of these experiments seem to suggest that HLA
antigens do not exist in the membrane as disulfide-linked oligomeric
forms. The possibility remains that they may nevertheless be associated
as tetramers or higher oligomers by non-covalent interactions (for a
discussion, see Ref. 25).

212

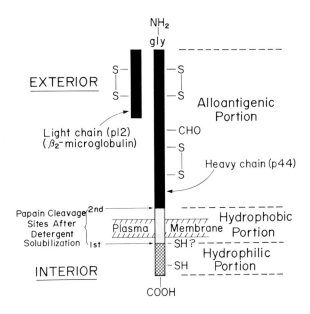

Fig. 3. Arrangement of HLA-A,B antigen in the membrane with the hydrophilic peptide hypothetically shown inside the cell membrane. The position of β_2-microglobulin relative to the alloantigenic portion of the heavy chain is not known. The possible presence and position of a second sulfhydryl, shown near the first papain cleavage site, is unclear. It may be present on only some HLA specificities, or the exact position of papain cleavage relative to it may vary

Summary

From these data a model summarizing schematically our present knowledge of the structure and orientation of the HLA antigenic molecule in the lymphocyte membrane is shown in Figure 3. It seems likely that the heavy chain spans the membrane with the hydrophobic region inserted in the membrane and the hydrophilic C-terminus inside the cell. This C-terminal region bears one, possibly two, SH residues which have the potential for forming interchain disulfides. Whether or not these are actually formed physiologically remains an interesting question. There is the attractive possibility that whatever the physiological function(s) of HLA antigens are, these molecules provide structurally the potential for signaling from outside the cell to inside the cell because they span the membrane. It is even conceivable that this function might be expressed via opening and closing of disulfide bridges.

B Cell-Specific Antigens from Human Lymphocytes

A second type of molecule which also has a two-chain structure and may be structurally related to HLA antigens has also been isolated. These materials are probably the products of the HLA-D genes and may be analogous to mouse Ia antigens. At an early stage of purification the HLA antigen preparations (obtained after papain solubilization) all contained impurities in varying amounts with molecular weights of 70,000; 30,000; 23,000 and 13,500. The latter was distinguishable from β_2-microglobulin. Detergent solubilized preparations contained analogous materials with molecular weights of 105,000; 34,000; 29,000 and 16,000.

Papain-Solubilized B-Cell Antigens

These materials from a papain preparation were separated by careful gel filtration. In addition to the peak of HLA antigen, three additional

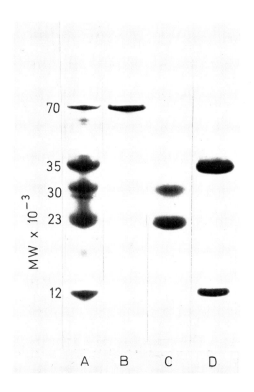

Fig. 4A-D. SDS-polyacrylamide gels of lymphocyte membrane proteins. (A) HLA antigen prepared with DTT preactivated papain containing "contaminants" of 70,000; 30,000 and 23,000 daltons. (B) Purified 70,000 daltons component. (C) Purified complex of 23,000 and 30,000 daltons components. (D) Purified HLA antigen containing 35,000 and 12,000 daltons components

peaks of protein were obtained which calibrated on the gel column at molecular weights of 50-70,000; 135,000, and in the excluded volume. The 50-70,000 molecular weight material was composed of two polypeptides with molecular weights of 23,000 and 30,000. The material which calibrated at a molecular weight of 135,000 on the Sephadex column was composed of apparently identical polypeptides of molecular weight 70,000 each. The material of very high molecular weight in the excluded volume of the column contained a single polypeptide of molecular weight 13,500, apparently highly aggregated. The purity of some of these preparations is illustrated by SDS gels (Fig. 4) (26).

Rabbits were immunized with all these preparations. The properties of the antisera which were obtained are extemely interesting. In this section we will focus on antiserum prepared to the material which contains polypeptides of 23,000 and 30,000 molecular weight. It had the following interesting properties:

Lysis of Peripheral Blood Lymphocytes and Lymphoblast Lines. Only a fraction of peripheral blood lymphocytes were lysed by the antiserum in complement-mediated cytotoxicity assays (Fig. 5). However, two B-cell lymphocyte lines, RH-1 and IM-1, were totally lysed at antiserum dilutions of 1:2000. When peripheral blood lymphocytes were separated into B, T and null cells, the T cells were not lysed at all; the B cells were completely lysed and a fraction of the null cells was lysed. In separate experiments with the null cell population, the population lysed by anti-p23,30 serum was found to bear the EAC rosette receptor and to cause the ADCC reaction (27). The null cell population which

214

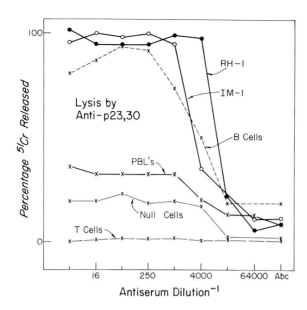

Fig. 5. Lysis of several purified populations of lymphocytes by anti-p23,30 rabbit
sera. The IM-1 and RH-1 are
lymphoblasts. Peripheral
blood lymphocytes (PBLs) from
one individual (RH) were separated into B, T, and Null
lymphocyte populations

was not lysed by the anti-p23,30 serum did not bear the EAC rosette
receptor and did not participate in the ADCC reaction. An anti-p23,30
serum blocked the ADCC reaction. By contrast, anti-β_2-microglobulin
serum lysed all of these populations of cells.

Peripheral blood lymphocytes of 40 individuals were separated into T
cells and B cells. The B cells of all of these individuals were lysed
by the sera and, at the low dilution used (1:10), some of the T cells
were also lysed. However, at 1:500 or 1:1000 the antiserum was absolutely specific for B cells of the separated populations. T-cell lines
and B-cell lines established from the same individuals were also examined. Three such pairs of lines were available. Again, in each case
only the B-cell line was lysed at high dilution (1:500). At low dilution some partial lysis of one of the T cell lines was observed.

Separation of Cells in the Fluorescence-Activated Cell Sorter. In the
Herzenberg fluorescence activated cell sorter the difference between
T, B and null cells was dramatically observed (26, 28). Cells were
treated with anti-p23,30 serum and then with FITC-conjugated goat anti-
rabbit IgG. The fluorescence-activated cell sorter yields data regarding both the number of cells and their relative fluorescence. A very
high fluorescence was obtained with B cells, but no reactions was observed with T cells. A subpopulation of null cells was shown to react
with the p23,30 antiserum.

Precipitation of Polypeptides from [35S]Methionine Internally Labeled
and Detergent-Solubilized Membranes. Another way of examining the specificity of these antisera is to radiolabel membranes and ask what
polypeptides are precipitated. This question was of special interest
with the p23,30 antiserum because it had been obtained by immunization
with p23,30 polypeptides prepared after papain solubilization. The native form of the p23,30 complex might be different in the membrane.
When radiolabeled membranes were solubilized in detergent and their
extract treated with p23,30 antiserum, three polypeptides were observed
in the precipitate with molecular weights of 39,000; 34,000 and 29,000.

Table 7. Inhibition by p23,30 of cytolysis of Amish cell lines by Amish antisera

Amish cell line	Cell source of p23,30 added	Amish antisera					
		35	76	192	289	590	RMB
KL	IM-1	+	-	-	+	-	+
	RPMI 4265	-	+	-	-	-	+
	JY	+	+	+	+	-	+
SL	IM-1	+	-	-	-	-	+
	RPMI 4265	-	+	-	-	-	+
	JY	+	+	+	+	-	+
PY	IM-1	+	-	-	+	-	+
	RPMI 4265	+	+	-	-	-	+
	JY	+	+	+	+	+	+
JY	IM-1	+	-	-	+	-	+
	RMPI	-	+	-	-	-	+
	JY	+	+	+	+	-	+

Apparently the p23,30 polypeptides originated from these. The precursor product relationship among these has not yet been elucidated.

Many HLA antisera contain additional antibodies specific for polypeptides other than HLA (obtained by double antibody precipitation from radiolabeled membranes). DAL (an HLA-A27 antiserum), MWS (a W28 antiserum) and BC (an HLA-A3 antiserum) all brought down a small amount of the 30,000 daltons polypeptide. BEL (an HLA-A27 antiserum) is very interesting because it contains an antibody directed against a 70,000 dalton component as well as an antibody directed against the 30,000 daltons component.

Using a rabbit anti-β_2-microglobulin serum only the 44,000 daltons heavy chain of the HLA antigens and the 12,000 daltons β_2-microglobulins were precipitated. No polypeptide corresponding to p39,34,29 were observed. At least as defined by this type of experiment, no polypeptide other than that of 44,000 daltons is associated with β_2-microglobulin in the human lymphocyte membrane.

Lysis of Various Amish Cell Lines by Amish Antisera and Its Inhibition by p23,30 Antigens. The p23,30 antigen has been obtained from three different cell lines: IM-1 (presently available as the purest of the preparations), RPMI 4265 and JY. JY is a member of the Indiana Amish community, a highly inbred human population, as mentioned previously. A number of cytolytic antisera from multiparous women of this community do not contain HLA antibodies but apparently have antibodies directed against other lymphocyte membrane components (29). These sera were used in cytotoxicity assays against four cell lines, also derived from the Amish population (Table 7) (26). The antisera used lysed all four of the cell lines. Several patterns are evident. First of all, JY p23,30 antigen inhibited lysis of the JY cell line by all of the antisera; the same was true of JY p23,30 antigens as an inhibitor of lysis of cells of his relative, PY. An interesting pair of antisera are 35 and 76. Antiserum 35, which lysed all four cell lines, was blocked by the IM-1 p23,30 antigen in each case. Antiserum 76, however, was not inhibited by the IM-1 p23,30 antigen but it was inhibited by p23,30 from RPMI 4265 cells. All of the data suggest that the p23,30 antigens

from IM-1, 4265, and JY cells are alloantigens; some of them inhibit
some of the Amish alloantisera and others inhibit other Amish alloantisera. Antiserum RMB is interesting because it was inhibited by all
three of the p23,30 antigens. Possibly it recognized a determinant
common to all of them.

Use of p23,30 Rabbit Antisera in Purification of p23,30 Antigens. An
interesting use of this antiserum is in following the p23,30 complex
in a crude mixture, e.g., the separation of solubilized membrane proteins on a gel filtration column. The p23,30 complex, detected by inhibition of cytolysis, immediately preceded the HLA antigens. The
p23,30 antiserum might also be useful in the purification of these
materials by immunoabsorbent chromatography.

Similarities to Mouse Ia Antigens and Antisera. Several similarities
of this antiserum to mouse Ia antisera, include the fact that it
blocks the MLC reaction of human lymphocytes and their Fc receptors.
The molecular weights of the two polypeptides of the antigen complex
are also similar to those reported for mouse Ia antigens.

Detergent-Solubilized B-Cell Antigens

Similar materials have been separated from HLA antigens after solubilization of lymphocyte membranes with detergent (30). This material
was present in a complex with a molecular weight of 55,000 which could
be separated into two polypeptides with molecular weight of 29,000
and 34,000 in a 1/1 ratio. Rabbit antiserum prepared against the p55
complex (i.e., p29,34) had properties very similar to that prepared
against p23,30.

Amino acid analysis of separated p29 and p34 revealed that these polypeptides were strikingly similar, and they yielded very low SΔQ values
when compared by the statistical method (9). Similarly, low values
were obtained when p44, the heavy chain of the HLA-A and HLA-B antigens
was compared to p29 and p34 (Table 8). These data suggest not only
that the p29 and p34 polypeptides are similar, but that there may be
similarities of structure also to p44. Further comparison were made
by radioiodinating the tyrosine residues of these polypeptides with
lactoperoxidase and then preparing tyrosine-labeled tryptic peptide
maps of the three polypeptides. P29 and p34 shared six major tyrosyl-
tryptic peptides and they each also contained unique peptides. The
number of peptides obtained was in agreement with the amino acid analysis showing that p29 and p34 contain eight and seven tyrosine residues
respectively. A large number of the shared tyrosyl-tryptic peptides
suggest that these peptides are highly homologous in structure. However, this homology does not extend to the amino terminus because N-
terminal analyses of the two polypeptides (Table 9) (30) showed little
similarity. P44 contained many unique tyrosyl-tryptic peptides not
found in either p29 or p34, but it also contained one peptide which
was in the same position in two-dimensional peptide maps as one of the
labeled peptides shared by p29 and p34. These data are further suggestive of a low degree of homology between p44 and p29 or p34. However,
much more extensive data will be required to prove this point.

It is clear from the present studies that gene duplication has been
involved in the evolution of the heavy chain of the HLA-A and HLA-B
antigens and also in the evolution of the two chains of the B-cell
specific alloantigens. At least one of the latter polypeptides is a
product of the HLA-D locus, though it is not clear whether or not the
other polypeptides are also specified by genes in this region. If

Table 8. SΔQ values

	p12	p29	p34
p29	103		
p34	94	16	
p44	225	27	43

Table 9. N-Terminal sequences of human HLA-linked B cell antigens (cell line RPMI 4265) (from Springer et al. (30))

	1	5	10	15
p29	GLY-ASP-THR-PRO-	-	-PHE-LEU-GLU-GLN-VAL-	
p34	ILE-LYS-GLU-GLU(ARG)	VAL-ILE-	-GLN-ALA-GLU-PHE-TYR-LEU-SER-	

further study confirms the suggestion that p44 and p29,34 are structurally related, then the intriguing question might be posed as to
whether many of the genes in the major histocompatibility complex on
the sixth human chromosome (many of which serve some kind of immunological or cell recognition function) could have arisen from duplication
of one or a few primitive ancestral genes. It is hoped that further
structural studies will help to shed light on the possible biological
functions of these intriguing molecules as well as on their evolutionary relationships.

Acknowledgments. This research was supported by grants AI O9576 and AI 1O736 from
the National Institutes of Health, and by grant BMS 574-O9718 from the National
Science Foundation to the Protein Chemistry Facility at Harvard University.

References

1. Sanderson, A.R., Batchelor, J.R.: Nature (Lond.) 219, 184 (1968)
2. McCune, J.M., Humphreys, R.E., Yocum, R.R., Strominger, J.L.: Proc.
   Nat. Acad. Sci. 72, 3206 (1975)
3. Shimada, A., Nathenson, S.G.: Biochem. Biophys. Res. Comm. 29, 828
   (1967)
4. Turner, M.J., Cresswell, P., Parham, P., Mann, D.L., Sanderson, A.R.,
   Strominger, J.L.: J. Biol. Chem. 250, 4512 (1975)
5. Cresswell, P., Turner, M.J., Strominger, J.L.: Proc. Nat. Acad.
   Sci. 71, 2123 (1973)
6. Cresswell, P., Springer, T., Strominger, J.L., Turner, M.J., Grey,
   J.M., Kubo, R.T.: Proc. Nat. Acad. Sci. 71, 2123 (1974)
7. Parham, P., Humphreys, R.E., Turner, M.J., Strominger, J.L.: Proc.
   Nat. Acad. Sci. 71, 3998 (1974)
8. Parham, P., Terhorst, C., Herrmann, H., Humphreys, R.E., Waterfield,
   M.D., Strominger, J.L.: Proc. Nat. Acad. Sci. 72, 1594 (1975)
9. Marchalonis, J.J., Weltman, J.K.: Comp. Biochem. Physiol. 38, 609
   (1971)

218

10. Terhorst, C., Parham, P., Mann, D.L., Strominger, J.L.: Proc. Nat. Acad. Sci. 73, 910 (1976)
11. Howard, J.C.: Nature (Lond.) 261, 189 (1976)
12. Bridgen, J., Snary, D., Crumpton, M.J., Barnstable, C., Goodfellow, P., Bodmer, W.F.: Nature (Lond.) 261, 200 (1976)
12a. Appella, E., Tanigaki, N., Faiwell, T., Pressman, D.: Biochem. Biophys. Res. Comm. 71, 286 (1976)
13. Grey, H.M., Kubo, R.T., Colon, S.M., Poulik, M.D., Cresswell, P., Springer, T., Turner, M., Strominger, J.L.: J. Exp. Med. 138, 1608 (1973)
14. Nakamuro, K., Tanigaka, N., Pressman, D.: Proc. Nat. Acad. Sci. 70, 2863 (1973)
15. Peterson, P.A., Rask, L., Lindblom, J.B.: Proc. Nat. Acad. Sci. 71, 35 (1974)
16. Cunningham, B.A., Berggard, I.: Transplant. Rev. 21, 3 (1974)
17. Springer, T.A., Strominger, J.L., Mann, D.: Proc. Nat. Acad. Sci. 71, 1539 (1974)
18. Springer, T.A., Mann, D.L., DeFranco, A.L., Strominger, J.L.: J. Biol. Chem. in press (1976)
19. Robb, R., Strominger, J.L.: J. Biol. Chem. 251, 5427 (1976)
20. Springer, T.A., Strominger, J.L.: Proc. Nat. Acad. Sci. 73, 2481 (1976)
21. Strominger, J.L., Cresswell, P., Grey, H., Humphreys, R.E., Mann, D., McCune, J., Parham, P., Robb, R., Sanderson, A.R., Springer, T., Terhorst, C., Turner, M.J.: Transplant. Rev. 21, 126 (1974)
22. Strominger, J.L., Humphreys, R.E., McCune, J.M., Parham, P., Robb, R., Springer, T., Terhorst, C.: Federation 35, 1177 (1976)
23. Cresswell, P., Dawson, J.R.: J. Immunol. 114, 523 (1975)
24. Peterson, P.A., Rask, L., Sege, K., Klareskog, L., Anundi, H., Ostberg, L.: Proc. Nat. Acad. Sci. 72, 1612 (1975)
25. Springer, T.A., Robb, R.J., Terhorst, C., Strominger, J.L.: J. Biol. Chem. in press (1976)
26. Humphreys, R.E., McCune, J., Chess, L., Herrmann, H., Malenka, D., Mann, D., Parham, P., Schlossmann, S., Strominger, J.L.: J. Exp. Med. 144, 98 (1976)
27. Chess, L., Evans, R., Humphreys, R.E., Strominger, J.L., Schlossmann, S.F.: J. Exp. Med. 144, 113 (1976)
28. Schlossmann, S.F., Chess, L., Humphreys, R.E., Strominger, J.L.: Proc. Nat. Acad. Sci. 73, 1288-1292 (1976)
29. Mann, D.L., Abelson, L., Harris, S.D., Amos, D.B.: J. Exp. Med. 142, 84 (1975)
30. Springer, T.A., Kaufman, J.F., Terhorst, C., Strominger, J.L.: Nature (Lond.) in press (1976)

Discussion

Dr. Rajewsky: Since you are comparing the products of the MHC with immunoglobulin: have you found antibody-like heterogeneity for any of the products?

Dr. Strominger: There are two ways to answer that question. The first thing to say is that the similarity to immunoglobulins would be that the HLA antigens correspond only to the constant region domains. That is, if one added a V region of 12,000 to each of the chains of HLA antigens, then one would have something which looked much more like an immunoglobulin. The HLA antigens would correspond to immunoglobulins minus V regions. The second point is that, although the 34,000 dalton polypeptide of the D locus product may be homogenous, the 29,000 dalton product is heterogeneous on isoelectric focusing and urea-Laemmli gel electrophoresis. It remains to be clarified further what that heterogeneity means.

Dr. Rajewsky: It may be useful to say again that the latter product would correspond to an I region product in mice, wouldn't it? And in the I region the Ir genes etc. are located.

The Structure and Evolution of Transplantation Antigens

L. Hood and J. Silver

Introduction

Our laboratory has been interested for the past ten years in the strategies complex multigenic families, such as antibodies, employed for
dealing with extremely large amounts of information (Hood and Prahl,
1971; Hood, 1973; Hood et al., 1975a). These include strategies for
information storage, information expression and information evolution.
Recent studies of the immune system, carried out at the biochemical,
genetic, and cellular levels, show that the families of genes coding
for antibody molecules display unusual evolutionary features that are
shared by other multigene families as diverse in character as the ribosomal RNA genes, the histone genes, the tRNA genes and even the simple
sequences coding for DNA satellites (for review see Hood et al., 1975a).
Thus we define a multigene family as a group of nucleotide sequences
or genes that exhibits four properties: multiplicity, close linkage,
sequence homology and related or overlapping phenotypic functions.
Multigene families can be divided into three general categories by
several criteria (Hood et al., 1975a). The simple-sequence family en-
compasses the DNA satellites. The multiplication family consists of
those (nearly) identical gene families whose gene products are required
in large quantities during certain stages of development or the cell
cycle (e.g. ribosomal, tRNA and histone genes). An informational family
has individual members that can differ markedly in sequence from one
another, although all are homologous and obviously share an ancient
ancestry. Informational families can be relatively simple (e.g., the
β-like genes of the hemoglobins) or complex (e.g. most families of
antibody variable (V) region genes).

The complex informational multigene families are of particular interest.
The only well documented example of this category, antibody V genes,
employs novel mechanisms for information generation, expression and
evolution (see Hood et al., 1975a). One might argue that the antibody
system is unusual and peculiar and that these strategies are unique to
it. We prefer to believe that evolution develops sophisticated information-handling strategies once and then employs these mechanisms in
a variety of different systems with a common evolutionary origin. Hence
our long-term goal has been to find and characterize additional complex
multigene families to determine whether they share basic information-
handling mechanisms with the antibody gene families. Our basic approach,
as with our studies on antibodies, has been to use primary amino acid
sequence information to discern basic genetic and evolutionary mechanisms in new complex multigene families (see Hood et al., 1975a).

The major histocompatibility complex of mammals is a genetic region
which encodes a variety of cell surface molecules, all of which seem
interrelated with the immune system. In the mouse, the major histocompatibility complex, or the H-2 complex, can be divided by recombina-
tional studies into four major regions — K, I, S and D — that code for
three classes of gene products; (1) the classical transplantation antigens K and D (Shreffler and David, 1975); (2) the I region gene prod-

ucts into which the immune response genes map (McDevitt et al., 1974); and (3) the S region gene products which appear to code for certain complement components found in the serum (McDevitt, 1976; Meo et al., 1975). By simplistic genetic calculations, the H-2 complex may have sufficient DNA to code for up to 2000 structural proteins of molecular weight of about 20,000 (Klein, 1975b). Two of these categories, the I and S region genes, appear to be potential multigene families. The other category, the transplantation antigens, appears by all known criteria to be encoded by just two genes (i.e. K and D), although both exhibit an unprecedented degree of polymorphism (Klein, 1974).

Because relatively little chemistry has been carried out on cell surface molecules, we decided to investigate first the simpler transplantation antigen system. The K and the D gene products are found on almost all tissues of the mouse, although in highest concentration in lymphoid tissues (Klein, 1975b). The H-2K and H-2D cell surface antigens are hydrophobic glycoproteins of molecular weight 45,000 (Nathenson and Cullen, 1974). These molecules are noncovalently associated with β_2-microglobulin (Silver and Hood, 1974a), a polypeptide homologous to the constant region domains of immunoglobulin molecules (Smithies and Poulik, 1972). We report in this paper the partial amino acid sequences of two K and two D alleles determined in our laboratory (Silver and Hood, 1976) and compare these with the similar data obtained from several different laboratories on mouse transplantation antigens (Ewenstein et al., 1976; Henning et al., 1976; Vitetta et al., 1976) and human transplantation antigens from the HL-A complex (Terhorst et al., 1976).

The Isolation and Chemical Characterization of H-2 Molecules

Isolation

Three properties of the cell surface antigens coded by the H-2 complex dictate the types of approaches that can be employed in the isolation of these molecules: (1) they are hydrophobic glycoproteins; (2) they are intermingled with many other cell surface molecules; and (3) they are present in limited quantities even on those cells with the highest concentration of these molecules. The basic approach employed for the isolation of cell surface antigens has been to (1) radiolabel the cell surface molecules of lymphocytes (e.g. spleen cells), (2) dissolve lymphocytes and their membranes in a nonionic detergent, (3) precipitate the H-2 gene product with highly specific alloantisera, and (4) fractionate and isolate these molecules by molecular weight sieving in SDS polyacrylamide gels. These techniques have been developed in a number of different laboratories and are well discussed in the literature (see Nathenson and Cullen, 1974). This approach yields two gene products, the transplantation antigen or heavy chain (molecular weight 45,000) and the noncovalently associated β_2-microglobulin or light chain (molecular weight 12,000). Each was analyzed by microsequence analysis.

Microsequence Analysis

A mouse spleen has approximately 10^8 cells. A spleen lymphocyte may have as many as 5×10^5 K and D molecules on its cell surface (Hämmerling and Eggers, 1970). Accordingly, a single spleen should theoretically yield 5×10^{13} molecules or $\sim$ 100 picomoles of K or D gene prod-

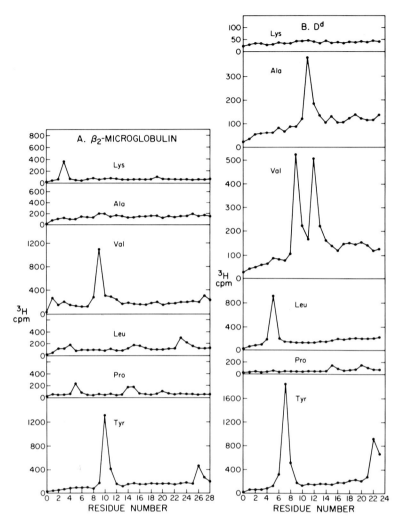

Fig. 1. (A) Amino acid sequence data from the β_2-microglobulin of mouse. (B) Amino acid sequence data from the D^d gene product. The amount of radioactivity associated with each of six incorporated amino acids is plotted against residue number (see text)

uct. This estimate is high because the yield will not be 100 % nor are all the cells in the spleen lymphocytes. Conventional amino acid sequence analysis on the automated sequenator employs 100 to 800 nanomoles of sample. Since we are dealing with 1,000 times less material, the need to employ microsequencing techniques is obvious.

We have spent the past few years developing microsequencing techniques (Silver and Hood, 1974b; Silver and Hood, 1975a; Silver and Hood, 1975b). These may be summarized as follows. Spleens from individual mice are removed and the cells are cultured in vitro for 4-6 hours with groups of tritiated amino acids. The K and D gene products and β_2-microglobulin are isolated by indirect immunoprecipitation, eluted from SDS polyacrylamide gels, concentrated and sequenced in the presence of carrier ovalbumin (which has a blocked N-terminal group) as previously described (Silver and Hood, 1975b). The tritiated phenylthiohydantoin (PTH) amino acid residues were resolved by thin layer chromatography in the presence of the appropriate unlabeled PTH-amino acids and counted.

The sequence data obtained for the β_2-microglobulin and one typical D gene product are presented in Figure 1. The data display several characteristics typical of conventional automated sequence analysis. There is a gradually rising background due to random hydrolysis of the protein. Sequence residues are characterized by a sharp rise in radioactivity associated with a particular amino acid followed by a more gradual decline. The latter phenomenon, known as "lag" is the result of an incomplete Edman reaction and tends to increase throughout the run. We believe the amino acid residue present in lowest yield, lysine at position 19, is real in that three separate analyses of the β_2-microglobulin molecule have each given this residue with virtually no background noise in preceding or successive residues. Since each amino acid is incorporated to a different extent, the data for each amino acid must be treated independently. For those residues that appear two or more times (e.g. tyrosine, valine, proline) an average repetitive yield can be calculated. The repetitive yields for these three residues ranged between 89-94 % which is similar to the repetitive yields obtained from conventional runs. Furthermore, these high repetitive yields suggest that a single major component is being sequenced. We should be able to dectect a 10-20 % minor contaminant. Obviously, this statement is qualified by our inability to detect polypeptides with a blocked N-terminus or heterogeneity in residues that were not labeled.

Let us now consider the genetic and evolutionary implications of the partial amino acid sequence data that have been obtained on gene products coded by the major histocompatibility locus of mouse and man.

Genetic and Evolutionary Implications of Partial N-Terminal Amino Acid Sequences of the Transplantation Antigens

The Light Chain (12,000 Daltons) is the Mouse β_2-Microglobulin

All eleven residues that we were able to identify are identical to the corresponding residues of the mouse β_2-microglobulin analyzed by conventional macrosequence analysis (E. Appella, personal communication). This identity demonstrates unequivocally that the small polypeptide associated with the 45,000 molecular weight components is the mouse β_2-microglobulin. More important, the analysis of this partial amino acid sequence at the level of a hundred picomoles or less demonstrates the feasibility and reliability of extending these microsequencing methods to the analysis of other cell surface molecules available in limited quantities.

The Heavy Chain (45,000 Daltons) Represents an H-2K or H-2D Gene Product

The heavy chain can no longer react with specific alloantisera after isolation from SDS polyacrylamide gels, presumably because the SDS has denatured the polypeptide. However, three lines of evidence convince us that this component is the K or D gene product. First, a papain fragment 37,000 in molecular weight can be isolated from spleen cells and retains the serological reactivity of the native molecule (Shimada and Nathenson, 1969). The N-terminal sequence of the papain fragment is identical to that of the SDS purified component from the same cells (Henning et al., 1976; Ewenstein et al., 1976). If this serological activity is contained in a minor component, the component must cleave with papain in a fashion identical to the putative major non H-2 component. We feel this possibility is unlikely. Second, control indirect immunoprecipitation experiments using specific alloantisera and radioactivity labeled lymphocytes that lack the corresponding H-2K or H-2D specificities yield very little ^3H protein in the 45,000 molecular weight range (Nathenson and Cullen, 1974; Silver et al., 1975). Thus, very little radiolabeled protein in the 45,000 molecular weight range is isolated by this procedure unless the specific antigen is present in the immunoprecipitation mixture. Finally, it is possible that a small contaminant is being picked up by this procedure and sequenced by our sensitive techniques. This possibility is unlikely because each of the two K and two D gene products is different from all the rest. Therefore, if a contaminant is present, it must vary in accordance with the different allelic gene products. In addition, there is a remarkable degree of concordance obtained from four different laboratories which are sequencing the K and D gene products. Furthermore, the mouse and human transplantation antigens, isolated by completely different procedures, are homologous, as will be discussed subsequently. Thus the serological controls and the presence of amino acid sequence variability correlated with the differing alleles provide strong support for the supposition that the heavy (45,00 dalton) polypeptides are the H-2K or H-2D gene products.

The K and D Gene Products are Homologous to One Another and Probably have Descended from a Common Ancestral Gene

The preliminary amino acid sequence results from four different laboratories on various K and D gene products are presented in Figure 2 (Ewenstein et al., 1976; Henning et al., 1976; Silver and Hood, 1976; Vitetta et al., 1976). Also given in Figure 2 are data obtained from the LA and four gene products of the human HL-A system (Terhorst et al., 1976). The dashes, as well as the identified residues in the mouse transplantation antigens, represent important data. Since a limited number of amino acids were labeled, a dash indicates that the corresponding residue is not one of the labeled residues. Accordingly, residue positions represented by dashes have information and can be used in homology comparisons.

Of the 12-16 residue positions that are identifiable in each K or D molecule (Fig. 2), six are identical in all four proteins (residues 6, 7, 8, 11, 12, 15). In addition, the K and D molecules share residues at several other positions (e.g. positions 1, 2, 3, 5, 9, 14, 17, 21, 22). Individual K and D molecules show 63-85 % homology with one another over the positions that can be compared. These observations strongly support earlier suppositions that the K and D loci are homologous to one another and must have come from a common ancestral gene (Shreffler et al., 1971; Klein and Schreffler, 1971; Brown et al., 1974; Murphy and Shreffler, 1975).

The K and D Products of the H-2 Complex are Homologous to the LA and Four Gene Products of the HL-A Complex

The gene products from the four region were isolated from a human

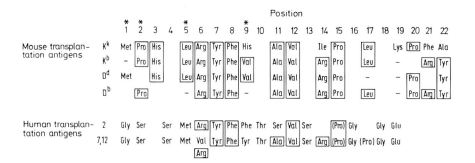

Fig. 2. Partial N-terminal sequences of mouse and human transplantation antigens.
*: species-associated residues; Dash: absence of a particular residue found at that
position in other mouse chains; +: position 9 in the D^b molecule is not valine (histidine was not tested); Blank: no sequence information is available at that position;
Boxes around the mouse residues: positions of identity; Boxes around the human residues: identity with their mouse counterparts; Parentheses: uncertainty in residue
assignment. The 2 and 7, 12 specificities are gene products from the human four and
LA regions, respectively. The 7 and 12 molecules could not be separated from one another and were sequenced as a mixture. Mouse data from Ewenstein et al., 1976;
Henning et al., 1976; Silver and Hood, 1976; and Vitetta et al., 1976. Human data
from Terhorst et al., 1976

cell line heterozygous (i.e. 7 and 12) at this locus. One allelic product from the four locus (e.g. 7) differs from a second (e.g. 12) by a
single valine-arginine interchange at position 7. The LA gene product
has identical residues at 13/16 and 14/16 residues (81-88 % homology)
when compared with the two "four" gene products. Hence the four and
LA gene products are homologous to one another and their corresponding
genes must have descended from a common ancestral gene.

Of the thirteen amino acid sequence residues that can be compared among
the mouse H-2 and human HL-A gene products, four positions are identical (positions 7, 11, 12 and 15) and three positions share at least
one residue (positions 6, 11 and 14) (Fig. 2). Thus the three human
HL-A gene products demonstrate 44-67 % homology with their mouse counterparts over those positions that can be compared. This homology
strongly supports the supposition that the human and murine transplantation antigens have descended from a common ancestral gene. This indicates that the major histocompatibility complexes of the mouse and
man are homologous to one another in structure and presumably should
exhibit a similar spectrum of biological functions as well as employ
similar genetic and evolutionary strategies.

The K and D Gene Products do not Appear to be Homologous to Immunoglobulins Based on Limited Sequence Comparisons

It has been proposed that the immunoglobulin gene families descended from the major transplantation complex. This postulated evolutionary relationship is based on several general features shared by both
systems - their extreme polymorphism, their cell surface location
and their role in regulating the immune response (Bodmer, 1972;
Gally and Edelman, 1972). In addition, the K and D gene products
are noncovalently associated with β₂-microglobulin, a molecule homo-

logous to the constant domains of immunoglobulins (see Smithies
and Poulik, 1972) and preliminary structural studies have suggested
the presence of immunoglobulin-like domains in the K and D molecules
(Peterson et al., 1975). The limited sequence data presented in Figure 2
show questionable, if any, sequence homology with the various immunoglobulin domains or the β_2-microglobulins. There are a number of possible explanations for this apparent lack of homolgy.
(1) The amino acid sequence data are insufficient to determine sequence homology. (2) The K and D genes are not evolutionarily related to immunoglobulin genes. (3) The N-terminal portions of the K and D molecules may
be particularly variable, perhaps reflecting some important associated
function; other portions of these molecules may show homologies with
immunoglobulins. (4) The transplantation antigens did share a common
ancestor with immunoglobulins; however, they have diverged sufficiently from immunoglobulins to mask any obvious sequence relationships.
This is true of the V and C regions of immunoglobulins (Poljak et al.,
1973). The association of the K and D products with an immunoglobulin-
like molecule suggests that conformational homology may exist. Perhaps
the contact residues between the two chains will be structurally homologous to their immunoglobulin counterparts from V_L and V_H domains.
Indeed, the six contact residues for the V_L regions (positions 35, 37,
42, 43, 86 and 99) are strikingly conserved in all light chains as are
their seven V_H region counterparts (positions 37, 39, 43, 45, 47, 95
and 108) (Poljak et al., 1975). These postulated homologies can be
tested by additional sequence data. Whatever the case may be, the data
currently available do not reveal any striking homology relationship
between the transplantation antigens and immunoglobulins at the level
of primary amino acid sequence.

The H-2 and HL-A Transplantation Antigens Exhibit Species-Associated Residues

Both the K and D transplantation antigens exhibit common amino acid
residues at certain positions which distinguish these polypeptides from
the LA and four transplantation antigens of man. These are termed
species-associated residues[1].
For example, at positions 1, 2, 5 and 9, two or three of the mouse
transplantation antigens share residues that are different from those
of the human transplantation antigens at the corresponding positions
(Fig. 2). In this regard, it will be important to examine additional
human and mouse transplantation antigens, particularly some from wild
mice, to verify the presence of extensive species-associated residues.
However, the fact that four gene products from the mouse and three
from two unrelated humans exhibit species-associated residues suggests
to us they will be found in most human and mouse transplantation antigens. Indeed, preliminary sequence data on a fifth mouse transplantation antigen (K^b) supports the presence of species-associated residues at positions 2 and 5 (the other two positions were not examined)
(Henning et al., 1976).

The presence of species-associated residues for both the K and D gene
products of the mouse on the one hand and the LA and four gene products
of man on the other suggests that these two loci arose in the mouse

---

[1] A species-associated residue need not be present in every member of a polymorphic
population. For example, if the heterogeneous immunoglobulin V_κ regions of the
guinea pig had aspartic acid and glutamic acid in a 9/1 ratio at position 1 and
the rabbit had alanine and isoleucine in a 9/1 ratio, then aspartic acid-glutamic
acid and alanine-isoleucine would, respectively, be species-associated residues for
guinea pigs and rabbits.

and human evolutionary lines <u>after</u> their divergence. The alternative
hypothesis that gene duplication occurred in the common ancestor before
speciation, as suggested by the observation that most vertebrate species analyzed to date have multiple transplantation antigens, would
require extensive identical (parallel) mutations in two pairs of genes
(i.e. K and D on the one hand and LA and four on the other) to generate
the mouse and human species-associated residues. Indeed, duplication
of some genes coding for the transplantation antigens <u>after</u> speciation
is already required to explain the observation that the HL-A complex
of man contains three regions coding for transplantation antigens (LA,
four, AJ) (van Rood et al., 1976), whereas the H-2 complex appears to
contain only two (K and D). Thus, different species may have undergone
different numbers of gene duplication events to produce one, two, three
or more loci coding for classical transplantation antigens. Accordingly,
these genetic loci may be part of a multigenic system that can undergo
gene expansion and concentration, presumably by homologous-but-unequal
crossing over (see Hood et al., 1975a).

The Two K Allelic Products Differ from One Another by Multiple Amino Acid Residues as do the Two D Allelic Products

The K gene products differ by six out of sixteen residues and the D
products by four out of fourteen residues (residues 1, 9, 14, 19, 21,
22 and 2, 5, 9, 12, respectively - see Fig.2. These constitute 29-38 %
sequence differences over the limited regions examined. Furthermore,
two of three K substitutions with identified residue alternatives
(e.g. positions 9, 14, and 21) differ by two base substitutions in
the genetic code dictionary - further emphasizing the evolutionary
separation of these "alleles". These differences are consistent with
the multiple serological specificities which differentiate the K (and D)
alleles from one another and with the multiple differences noted in
peptide maps. To our knowledge these amino acid differences constitute
some of the largest ever reported for two "alleles" if, indeed, differences of this magnitude are found throughout the entire molecule.

We have recently suggested that alleles (or allotypes) can be divided
into two categories (Gutman et al., 1975). Alternative forms of <u>simple</u>
allotypes segregate in a Mendelian fashion in mating studies and differ by one or a few amino acid substitutions (e.g. the Inv marker of
the human κ chain (Terry et al., 1969)). In contrast, alternative forms
of <u>complex allotypes</u> differ by multiple amino acid residues and generally segregate in a Mendelian fashion (e.g. the group a and group b
allotypes of the rabbit (see Mage et al., 1973)). By these definitions
the K and D alleles, based on limited sequence data, are complex allotypes. The importance of the distinction between simple and complex
allotypes lies in the very different types of genetic or evolutionary
mechanisms implied (see Gutman et al., 1975). Simple allotypes are
probably coded by alternative alleles at a single structural locus.
In contrast, complex allotypes may be explained by one of three general genetic models (Fig. 3).

<u>Classical Allelic Model.</u> The K (or D) alleles may have evolved by the
divergence of alleles at a single genetic locus (Fig. 3a). Because of
the extensive differences among these alleles, it is tempting to postulate that when mice evolved as a species, multiple alleles already
differing significantly in sequence were incorporated into the new
evolutionary line. However, this does not appear to be the case, at
least with the gene products examined to date, because the presence

228

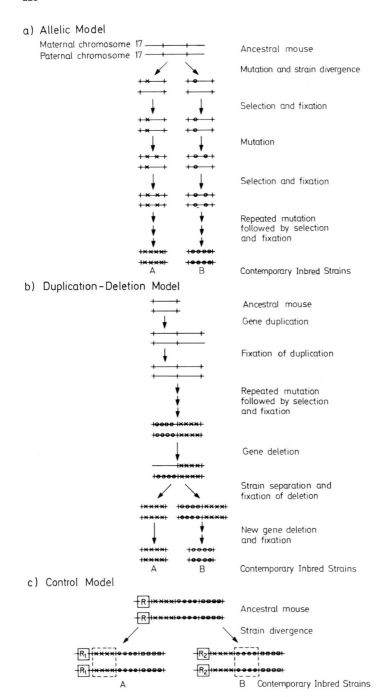

a) Allelic Model

Maternal chromosome 17 — Ancestral mouse
Paternal chromosome 17 —

Mutation and strain divergence

Selection and fixation

Mutation

Selection and fixation

Repeated mutation
followed by selection
and fixation

A   B   Contemporary Inbred Strains

b) Duplication-Deletion Model

Ancestral mouse

Gene duplication

Fixation of duplication

Repeated mutation
followed by selection
and fixation

Gene deletion

Strain separation and
fixation of deletion

New gene deletion
and fixation

A   B   Contemporary Inbred Strains

c) Control Model

Ancestral mouse

Strain divergence

A   B   Contemporary Inbred Strains

Fig. 3a-c. Three genetic and evolutionary models for complex allotypes (see text).
X and O: divergence of genes that exhibiti Mendelian behavior in contemporary inbred
strains of mice, A and B. In the control model, dotted boxes indicate distinct genes
that are expressed in different inbred strains, regulated by control mechanism R_1
and R_2

of species-associated residues suggests that the genes coding for the mouse and human transplantation antigens diverged from common ancestral genes after speciation (so as to avoid multiple identical or parallel mutations). If these allelic gene products differ by 30-40 % of their amino acid sequence throughout their entire length, then these poly-peptides may differ by as many as ∿100 residues. The allelic model assumes that a variant (single base substitution) gene arises by muta-tion and is fixed in the population. This new gene must then incur a second mutation that is once again fixed and this entire process must be repeated 100 times.

The structure of the wild mouse population may favor the rapid diver-gence of alleles (Klein, 1974). Wild mice breed in small demes (groups) with a single dominant male and sharp geographical boundaries. There is little, if any, migration among demes. However, young mice are con-tinually migrating from the old deme to establish new and subsequently independent demes. In these small isolated populations, variants can be fixed either by natural selection or by nonrandom fluctuations in a very limited pool of chromosomes. Since the fluctuations in gene frequencies are extreme in small populations, the chances for fixation (or elimination) of an allele are greatly enhanced over the chances in a large population (Hood et al., 1975b). However, alleles of gene products for other systems in the mouse (e.g. cytochrome c, the hemo-globins, etc.) do not show the same degree of diversity exhibited by the transplantation antigens (Dayhoff, 1972). Hence one must assume that natural selection plays an important role in generating this diversity.

The question arises as to whether mice as a species have had sufficient evolutionary time for their K (and D) alleles to have evolved to be so different. In this regard, it will be interesting to determine whether the transplantation antigens of guinea pigs and rats have species-associated residues when compared to those of mice. If they do, the time span over which the allelic differences can be produced is further reduced. For example, if the rat and mouse transplantation antigens have species-associated residues, their corresponding "alleles" must have arisen after speciation some 10 million years ago. In contrast, the C_K immunoglobulin regions of man and mouse, two species which di-verged about 75 million years ago, differ by only 40 % of their amino acid sequence (Dayhoff, 1972). Finally, it should be pointed out that there is no precedent for alleles that differ by as much as 30-40 % of their amino acid sequence (Gutman et al., 1975).

Duplication-Deletion Model. This model suggests that the transplanta-tion genes of the mouse were duplicated and that many differences were fixed in these duplicated genes (Fig. 3b). Later in the evolution of the mouse line, crossing-over events deleted most of these genes so individual H-2 chromosomes of mice had just a single K (and D) gene. The presence of species-associated residues suggests that this gene expansion and contraction process must have occurred after the di-vergence of man and mouse. This model would permit many more mutations to be fixed as opposed to the allelic model since it would ease the requirements for stringent natural selection. That is, one K gene may be freed to accept many substitutions while the second is temporarily the functional transplantation antigen. Indeed, this is presumably how one gene of a duplicated pair evolves to assume a new function (Ohno, 1970). There are precedents for the evolution, via a crossing-over event, of alleles that differ by multiple substitutions (Gilman, 1972; Huisman, 1975).

Control Model. The transplantation "alleles" may be closely linked duplicated genes with a control mechanism that permits the duplicated genes to be expressed so as to mimic a Mendelian pattern of genetic segregation (Fig. 3c). This model suggests that each mouse chromosome 17 has genes coding for many of the K and D "alleles" the mouse species can express. The genetic polymorphism may reflect a control mechanism which operates at the chromosomal level to determine which of the K and which of the D alleles is expressed by a particular chromosome 17. This theory postulates that the chromosomal mechanism for commitment (to the expression of one K and one D allele) is transmitted through the germ line and is very stable. Perhaps this mechanism is mediated by the translocation of a genetic element as may occur in maize (Mc-Klintock, 1956) and immunoglobulin genes (Gally and Edelman, 1972). The high rate of mutation at the K locus ($\sim$5 x 10^{-4} mutations/generation) in contrast to other mammalian genetic systems (Klein, 1975a) may be a reflection of an alteration in this control mechanism so that different K (or D) genes are expressed. Thus the "mutant" K (or D) genes may reflect a switch in the control mechanism to express another closely-linked gene rather than the actual mutation of a structural gene. In this regard, it will be extremely important to determine whether the "mutant" gene product differs from its "parental" gene product by one or multiple amino acid substitutions. The former would be consistent with the mutation of a structural gene, whereas the latter would favor the control mechanism concept.[2]

There are possible precedents for this type of control mechanism. The group a and b allotypes of rabbit immunoglobulins are complex allotypes of the V_H and C_κ regions, respectively (i.e. their allelic products exhibit multiple amino acid differences). Some individual rabbits can express at low levels a "wrong" V_H allotype — one that should not be present in the individual according to the genotype of his parents (Mudgett et al., 1975). Indeed, one unusual rabbit expressed three group a and three group b allotpyes (Strosberg et al., 1974). Both these observations suggest that the complex allotypes of the rabbit are, at least in part, closely linked genes regulated by a control mechanism that generally permits the complex allotypes to be expressed in a Mendelian fashion (see Farnsworth et al., 1976, for a discussion of this point). Accordingly, the complex allotypes of the rabbit may employ a control gene mechanism similar to that depicted in Figure 3c. This model has also been elegantly discussed by Bodmer (1973).

There are several distinctions and similarities among these three models that need to be emphasized. The duplication-deletion model and the control model view the K (and D) allelic as having evolved as duplicated and closely linked genes, while the classical allelic model assumes that these alleles evolved as alternative forms of a single structural gene. However, the duplication-deletion model and the classical allelic model contend that a single structural gene with multiple allelic forms encodes the contemporary K (and D) gene products.

The classical allelic model would be eliminated as an explanation for the complex K (and D) allotypes if it is unequivocally demonstrated (e.g. by structural analysis) that a mouse can produce H-2 alleles it should not have inherited from its parents (i.e. "wrong" alleles). This model would also find it difficult to explain K (or D) mutants

---

[2] Preliminary peptide may analysis suggests that two H-2 mutants differ from their corresponding wild type gene products by one or a few peptide differences (Nathenson et al., 1976). However, this analysis only examined a small portion of the molecules (many peptides are insoluble) and did not determine whether the different peptides contained multiple amino acid substitutions. The important unanswered question is whether the mutant and wild type alleles differ by one or multiple base substitutions.

that differed by multiple amino acid residues from the wild-type gene product. The control gene model could be verified if it could be demonstrated (e.g. by nucleic acid hybridization) that each mouse has many genes coding for the K (or D) gene products. Furthermore, if this model is correct, perhaps experimental conditions could be found which might cause the control mechanism to "switch" to the expression of different gene products [(e.g. certain in vitro conditions (Rivat et al., 1973) and stress states (Bosma and Bosma, 1974) appear to cause "wrong" complex allotypes of immunoglobulin to be expressed)]. The duplication-deletion model would predict that an occasional mouse may have chromosomes with "hybrid" K (or D) genes or with multiple K (or D) genes. Because of the species-associated residues which distinguish the mouse K and D gene products from their human counterparts, each of these theories must assume the K and D genes (or clusters of genes) arose subsequent to the human-mouse speciation.

It is intriguing to note that the only two allelic products of a single human HL-A region examined, 7 and 12, differ far less from one another than do their mouse counterparts (e.g., ∿5 % vs. ∿35 %, see Fig. 2). Indeed, these human alleles will still differ from one another by multiple residues, but it is not apparent why there should be such a striking difference in the apparent extent of diversity among the polymorphic forms of the human and mouse transplantation systems.

The K Products as a Class Cannot be Distinguished from the D Products as a Class on the Basis of Limited Sequence Data

The K gene products do not appear to be significantly more closely related to one another than to the D gene products. For example, the D^d gene product shows 79 % homology with K^b, 63 % homology with K^k, and 69 % homology with its allelic counterpart, the D^b gene product (Fig. 2). Likewise, no amino acids are restricted to only the K or the D gene products (Fig. 2). This lack of "D-ness" of "K-ness" is perhaps the most surprising observation in these data.

These amino acid sequence observations are in apparent conflict with the serological data on the K and D gene products. In general, multiple public serological specificities are shared by gene products of a single region (K or D). For example, public specificities 7, 8, 11, 25, 37, 38, 39, 45, 46 and 47 are found only on K molecules, whereas specificities 6, 13, 41, 43, 44 and 49 are found only on D molecules. Thus the serological data suggest that there is "K-ness" and "D-ness" in the respective molecules (see Klein, 1975b).

How can the apparent paradox posed by the serological and sequence data be resolved? There are three general explanations, the first two of which are trivial. (1) Our limited partial sequence data may be misleading in this regard; this may be the most likely possibility. (2) The N-terminal regions of these molecules may not have any important function or they may have a function compatible with great variability. Thus they are capable of accepting many mutations. Other regions of these molecules may show "K-ness" and "D-ness". These trivial explanations can be tested by additional amino acid sequence data.

Let us consider, however, the genetic and evolutionary implications of a third possibility, namely that subsequent amino acid sequence data will demonstrate a lack of K-ness or D-ness in at least a portion of these molecules. This observation coupled with two others, species-associated residues and the shared public specificities of allelic products, place severe constraints on possible genetic and evolutionary models of these regions. None of the models for complex allotypes discussed above can easily explain the lack of K-ness or D-ness. In the allelic model, if the allelic forms of the K or D genes arose subsequent to the divergence of man and mouse, the alleles of one region should be more closely related to one another than to those of the second region and, accordingly, display K-ness and D-ness (Fig. 3a). The duplication-deletion model would require that the K and D genes form a single gene (or cluster) before reduction to a series of alleles at two structural loci in order to explain the lack of D-ness or K-ness (Fig. 3b). The same would be true of the control gene model (Fig. 3c). However, even if the K and D genes formed a single gene family, this would be inconsistent with the fact that many public specificities are found only on K (or D) gene products.

One possible resolution of the paradox caused by the amino acid sequence and serological data would be the possibility that the K and D proteins are actually coded by two genes, analogous to the V and C genes of immunoglobulins. The two C genes (C_D and C_K) might contain the public (shared) specificities, whereas the V genes for both K and D molecules might be drawn from a common gene family (cluster). Thus the family of V genes could evolve to have species-associated residues by crossing over just as appears to be the case with the families of antibody V genes (Hood et al., 1975a). Each V gene would have its own private specificity (analogous to an idiotype). There would be a lack of K-ness or D-ness among the V genes since they are evolving in a common multigene family. Accordingly, if subsequent sequence data confirm the lack of K-ness or D-ness in a major portion of the transplantation antigens, an explanation formally analogous to that given above will be necessary.

Summary

The preliminary amino acid sequence data on the transplantation antigens place us in a position similar to that for immunoglobulin structure in 1965. Provocative hypotheses about the genetic organization and evolution of the genes coding for the transplantation antigens have been raised. Indeed, the K and D "genes" appear to have certain properties shared by the multigenic antibody system — they undergo gene expansion and contraction; they are complex allotypes; and the apparent lack of D-ness or K-ness raise the possibility they have V and C regions. New microsequencing techniques should permit a detailed analysis of these gene products and lead to an eventual choice among the alternative hypotheses now proposed. A study of the multigenic I region gene products has already commenced in a number of laboratories. These data have made it apparent that the H-2 complex is a fascinating and complicated chromosomal region which will continue for some time to intrigue immunologists, geneticists, biochemists and cell biologists.

Acknowledgments. This work was supported by Grant No. PCM71-00770 from the National Science Foundation and Grant No. GM 06965 from the National Institutes of Health. J.S. has an Established Investigatorship Award from the American Heart Association. L.H. has an NIH Research Career Development Award. Portions of this paper were adapted from a recent review (Silver, J., Hood, L.: Contemporary Topics in Molecular Immunology, in press, 1976).

References

Bodmer, W.: Nature (Lond.) 237, 139 (1972)
Bodmer, W.: Transpl. Proc. V, 1471 (1973)
Bosma, M.J., Bosma, G.: J. Exp. Med. 139, 512 (1974)
Brown, J.L., Kato, K., Silver, J., Nathenson, S.G.: Biochem. 13, 3174 (1974)
Dayhoff, M.: The Atlas of Protein Sequence and Structure. Nat. Biomed. Res. Found. Silver Spring, Md. (1972)
Ewenstein, B.M., Freed, J.H., Mole, L.E., Nathenson, S.G.: Proc. Nat. Acad. Sci. 73, 915 (1976)
Farnsworth, V., Fleishmann, R., Rodkey, S., Hood, L.: Proc. Nat. Acad. Sci. in press (1976)
Gally, J., Edelmann, G.M.: Ann. Rev. Genet. 6, 1 (1972)
Gilman, J.G.: Science 178, 873 (1972)
Gutman, G., Loh, E., Hood, L.: Proc. Nat. Acad. Sci. 72, 4046 (1975)
Hämmerling, U., Eggers, H.J.: Europ. J. Biochem. 17, 95 (1970)
Henning, R., Milner, R.J., Reske, K., Cunningham, B.A., Edelman, G.M.: Proc. Nat. Acad. Sci. 73, 118 (1976)
Hood, L.: In: Stadler Genet. Symp. 5, 73 (1973)
Hood, L., Campbell, J.H., Elgin, S.R.C.: Ann. Rev. Genet. 9, 305 (1975a)
Hood, L., Prahl, J.: Advan. Immunol. 14, 291 (1971)
Hood, L., Wilson, J., Wood, W.B.: Molecular Biology of Eucaryotic Cells. Menlo Park: W.A. Benjamin, Inc. 1975b, Chap. 7
Huisman, T.H.J.: Ann. N.Y. Acad. Sci. 241, 549 (1975)
Klein, J.: Ann. Rev. Genet. 8, 63 (1974)
Klein, J.: Contemp. Topics in Immunobiol. in press (1975a)
Klein, J.: Biology of the Mouse Histocompatibility-2 Complex, Principles of immunogenetics applied to a single system. New York: Springer 1975b
Klein, J., Shreffler, D.C.: Transpl. Rev. 6, 3 (1971)
Mage, R., Lieberman, R., Potter, M., Terry, W.D.: In: The Antigens. Sela, M. (ed.). New York, London: Academic Press 1973, Vol. I, p. 300
McDevitt, H.O.: Federation Proc. in press (1976)
McDevitt, H.O., Bechtol, K.B., Hämmerling, G.J., Lonai, P., Delovitch, T.L.: In: The Immune System: Genes, Receptors, Signals. Sercarz, Williamson, A.R., Fox, C.F. (eds.). New York: Academic Press 1974, p. 597
McKlintock, B.: Cold Spring Harbor Symp., Quant. Biol. 21, 197 (1956)
Meo, T., Krasteff, T., Shreffler, D.: Proc. Nat. Acad. Sci. 72, 4536 (1975)
Mudgett, M., Fraser, B.A., Kindt, T.J.: J. Exp. Med. 141, 1448 (1975)
Murphy, D., Shreffler, D.C.: J. Exp. Med. 141, 374 (1975)
Nathenson, S.G., Brown, J.K., Ewenstein, D.M., Rajan, T.V., Freed, T.H., Sears, D.W., Mole, L.E., Scharff, M.D.: In: The Role of Products of the Histocompatibility Gene Complex in Immune Responses. Katz, D.H., Benacerraf, B. (eds.). New York: Academic Press 1976, in press
Ohno, S.: Evolution by Gene Duplication. New York: Springer 1970
Peterson, P.A., Rask, L., Sege, K., Klaresky, L., Anundi, H., Östberg: Proc. Nat. Acad. Sci. 72, 1612 (1975)
Poljak, R., Amzel, L., Avey, H., Chen, B., Phizackerley, R., Saul, F.: Proc. Nat. Acad. Sci. 70, 3305 (1973)
Poljak, R., Amzel, L.M., Chen, B.L., Phizackerley, R.P., Saul, F.: Immunogenetics 2, 393 (1975)
Rivat, L., Gilbert, D., Ropartz, C.: Immunology 24, 1041 (1973)
Rood, J.L. van, van Leuween, A., Termijtelen, A., Keuuing, J.J.: In: The Role of Products of the Histocompatibility Gene Complex in Immune Responses. Katz, D.H., Benacerraf, B. (eds.). New York: Academic Press, in press (1976)

Shimada, A., Nathenson, S.G.: Biochemistry 8, 4048 (1969)
Shreffler, D.C., David, C.S.: Advan. Immun. 20, 125 (1975)
Shreffler, D.C., David, C.S., Passmore, H.C., Klein, J.: Transpl.
    Proc. 3, 176 (1971)
Silver, J., Hood, L.: Nature (Lond.) 249, 764 (1974a)
Silver, J., Hood, L.: Anal. Biochem. 60, 285 (1974b)
Silver, J., Hood, L.: Nature (Lond.) 256, 63 (1975a)
Silver, J., Hood, L.: Anal. Biochem. 67, 392 (1975b)
Silver, J., Hood, L.E.: Proc. Nat. Acad. Sci. 73, 599 (1976)
Silver, J., Sibley, C., Morand, P., Hood, L.: Transpl. Proc. 7, 201
    (1975)
Smithies, O., Poulik, M.D.: Proc. Nat. Acad. Sci. 69, 2914 (1972)
Strosberg, A.D., Hamers-Casterman, C., Van der Lou, W., Hamers, R.:
    J. Immunol. 113, 1313 (1974)
Terhorst, C., Parham, P., Mann, D.L., Strominger, J.L.: Proc. Nat.
    Acad. Sci. 73, 910 (1976)
Terry, W.D., Hood, L., Steinberg, A.G.: Proc. Nat. Acad. Sci. 63, 71
    (1969)
Vitetta, E.S., Capra, J.D., Klapper, D.G., Klein, J., Uhr, J.W.: Proc.
    Nat. Acad. Sci. 73, 905 (1976)

Chemical Structure and Biological Activities of Murine Histocompatibility Antigens

R. Henning, J. W. Schrader, R. J. Milner, K. Reske, J. A. Ziffer,
B. A. Cunningham, and G. M. Edelman

Histocompatibility antigens are glycoproteins found on the surface
membranes of vertebrate cells. The significance of these antigens in
the rejection of allografts between individuals has been well established (1), and the genetics of these polymorphic systems have been
studied in detail in the mouse (H-2) and in man (HL-A). The segment
of the 17th chromosome of the mouse containing the major histocompatibility complex is shown in Figure 1. Despite all the information available regarding these molecules, the physiological function of H-2 and
HL-A antigen is not known. A number of studies has shown that histocompatibility antigens are involved in cell-mediated lysis of xeno-
or allogeneic cells (2, 3). Recently it has been observed that murine
histocompatibility antigens participate in the lysis of syngeneic
chemically modified cells (4), virally infected cells (5) and neoplastic cells (6, 7, 8). These data support the notion that histocompatibility antigens play an important role in the elimination of antigenically abnormal cells, and thus, they may form a highly complicated
system of membrane proteins that serve as signals to distinguish "self"
from "non-self".

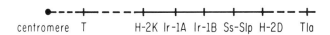

centromere   T   H-2K Ir-1A Ir-1B Ss-Slp H-2D   Tla

Fig. 1. Outline of the murine 17th chromosome showing the H-2 complex bounded by the
H-2D and H-2K loci in relation to the T locus and the thymus leukemia antigen (Tla)
locus. Immune response genes (Ir-1A, Ir-1B) and genes controlling the level of a serum
protein (Ss) and its sex-linked variant are (Slp) located within the H-2 gene complex

At present this hypothesis is under intensive investigation in many
laboratories. Understanding the molecular basis of the genetic polymorphism and the possible evolutionary relationships of these proteins
to other protein systems requires knowledge of the subunit structure
of these molecules, the amino acid sequence of their polypeptide
chains, and the distribution and number of antigenic sites per molecule. The basis for the molecular mechanisms underlying the function
of these molecules also depends upon knowing the subunit structure
and the cell surface orientation of these antigens.

We have begun a detailed analysis of the chemical structure and some
of the biological activities of murine histocompatibility (H-2) antigens. We describe here the molecular anatomy of H-2 antigens on the
cell surface and their physicochemical properties in solution. Partial
amino acid sequences of H-2 heavy chains were compared in four different H-2 gene products in order to obtain some knowledge of the nature of the polymorphism and the evolution of these molecules. We propose a working model of H-2 antigens in detergent solution and on the
cell surface. This model includes the arrangement of the polypeptide
chains, the location of interactions between chains, and the orienta-

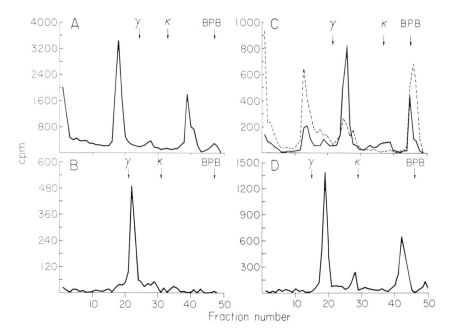

Fig. 2. SDS-polyacrylamide gel electrophoresis of immune precipitates of NP 40 solu-bilized H-2Kᵇ antigens from ¹²⁵I-labeled C57BL/1OJ spleen cells or EL4 cells. (A) Unreduced immune precipitate (10 % polyacrylamide gel). (B) Reelectrophoresis of ma-terial eluted from fractions 18-20 of the gel shown in panel A after reduction with mercaptoethanol (10 % gel). (C) Unreduced immune precipitates of NP 40 extracts of untreated EL4 cells (---) and of EL4 cells alkylated with idioacetamide prior to detergent extraction (——) (8.5 % gel). (D) Immune precipitate from papain-treated NP 40 extract after reduction with mercaptoethanol (12 % gel). γ, MOPC 21 heavy chain; κ, MOPC 21 light chain; BPB: bromophenol blue

tion of these molecules on the cell surface. The similarities found among different H-2 heavy chains allow some conclusions about the ori-gin and the evolution of H-2 antigens, and the differences found in the sequences reflect the diversity among different gene products. This diversity can also be shown by peptide mapping techniques de-monstrating the extreme polymorphism of this system (19).

It has been found that the same class of cytotoxic lymphocytes is in-volved in the lysis of allogeneic cells and the destruction of syn-geneic but antigenically abnormal cells, e.g. chemically modified, virally infected, or neoplastic cells (5-8). It has been proposed (10) that the target of the cytotoxic T lymphocytes that lyse virally in-fected syngeneic cells is, in fact, a virally modified H-2 antigen. The nature of this modification, however, has not been determined.

Our experiments indicate that the lysis of tumor cells by syngeneic or H-2 compatible cytotoxic lymphocytes involves the participation of H-2 antigens on the tumor cells. This finding suggests that cytotoxic lymphocytes may recognize foreign and H-2 antigens simultaneously, or recognize the H-2 antigens and the tumor or viral antigens by two in-dependent T cell receptors. Our preliminary observations favor the idea that physical interactions between H-2 and viral antigens take place on the cell surface, suggesting that H-2 antigens form a hybrid

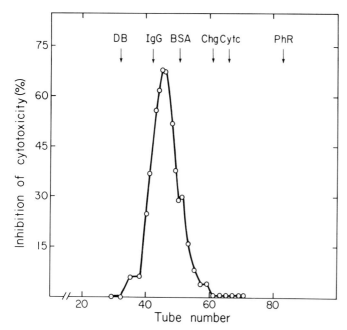

Fig. 3. Chromatography of detergent solubilized H-2 antigens on Sephadex G 200.
1 ml of 0.5 % DOC extract of a membrane fraction of C57BL/1 0J spleens was loaded
on a calibrated Sephadex G 200 column (1.2 x 75 cm). Eluant: 0.01 M Tris-HCl buffer,
pH 8.0, 0.15 M NaCl, 0.5 % DOC, 5 ml/h. Assay: anti-H-2K[b] on ^{51}Cr-loaded C57BL/1 0J
spleen lymphocytes. DB: dextran blue; IgG: MOPC 21 immunoglobulin; BSA: bovine
serum albumin; Chg: chymotrypsinogen; Cytc: cytochrome c; PhR: phenol red

antigen with foreign antigens. These results support the hypothesis
that H-2 antigens play a key role in immune surveillance.

Subunit Structure in Detergent Solution

The molecular weight of detergent solubilized H-2 antigens was determined under dissociating conditions by SDS-polyacrylamide gel electrophoresis in the presence and absence of reducing agents. H-2 antigens
were solubilized from cell membrane fractions using 0.5 % Nonidet P 40
(NP 40) and were isolated by immune precipitation using alloantisera
directed against only one H-2 gene product. The gel electrophoretic
patterns run in the absence of reducing agents (Fig. 2A) showed two
species of molecular weight 92,000 and 12,000. Occasionally a 46,000
dalton component was also observed. Elution, reduction, and reelectrophoresis of the 92,000 molecular weight component from the gel gave
only one component with a molecular weight of 46,000 (Fig. 2B), suggesting that the 92,000 dalton component is a disulfide linked dimer
of H-2 heavy chains. In accord with this result, reduction of immune
precipitates with 5 % β-mercaptoethanol resulted in the disappearance
of the heavy chain dimer (MW 92,000), the appearance of a monomer (MW
46,000), and no change in the 12,000 dalton light chain (β2-microglo-
bulin).

238

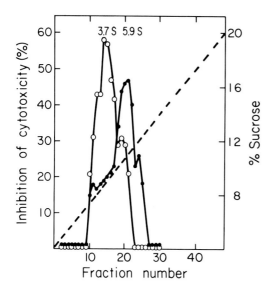

Fig. 4. Sucrose density gradient centrifugation of detergent (o) and papain (●) solubilized H-2 antigens. A 0.5 % NP 40 extract from membranes of DBA/2J spleens was treated with papain, mixed with an untreated extract from C57BL/1OJ spleen cell membranees, loaded onto a 4-2O % sucrose gradient in O.O1 M Tris-HCl, pH 8.O, O,15 M NaCl, 0.5 % DOC, and centrifuged for 25 h at 35,OOO rpm (5°) in an SW 39 Beckman rotor. Fractions (6 drops/fraction) were collected by tube puncture. Each was fraction assayed for H-2K^d and H-2K^b by inhibition of cytotoxicity of anti-H-2K^d and anti-H-2K^b on DBA/2J and C57BL/1OJ target cells respectively. (---) % sucrose. (From Ref. 19)

Molecular weight determinations were also performed under other conditions. Gel exclusion chromatography in 0.5 % deoxycholate (DOC) of NP 4O or DOC solubilized spleen cell or lymphoma cell membranes gave one major component with H-2 antigenic activity. This component had a molecular weight of approximately 120,000 (Fig. 3). Ultracentrifugation of detergent solubilized H-2 antigens on sucrose density gradients in 0.5 % DOC gave a major peak of antigenic activity at 5.9 S (Fig. 4) and some minor components at lower S values. The diffusion coefficient was calculated from the partition coefficients obtained by gel chromatography on calibrated Sephadex G 200 columns. The molecular weight, calculated from the diffusion and the sedimentation coefficients using the Svedberg equation was 116,000. This value is consistent with the hypothesis that intact H-2 antigens in detergent solution are composed of two heavy chains and two light chains (Fig. 5).

Recent data from other laboratories have suggested that human and murine histocompatibility antigens consist of two disulfide linked heavy chains and two non-covalently associated light chains (β_2-microglobulin) (11-15). The amino acid sequence of β_2-microglobulin has been shown to be homologous to the constant regions of immunoglobulin light and heavy chains (16-18), raising the possibility of an evolutionary relationship between the immunoglobulins and the histocompatibility antigens. In both the mouse and the human system, higher as well as lower molecular weight species were seen in nonionic detergents (9, 11) and the simultaneous presence of heavy chain dimers and monomers was observed in the absence of reducing agents (9). Our studies confirmed these reports. In addition, however, we found that the murine histocompatibility antigens exist as disulfide linked dimers only in detergent solution (19) and not on the cell surface.

H-2 Antigens on the Cell Surface

Occasionally immune precipitates of detergent showed the simultaneous presence of H-2 heavy chain monomers and dimers on SDS-polyacrylamide

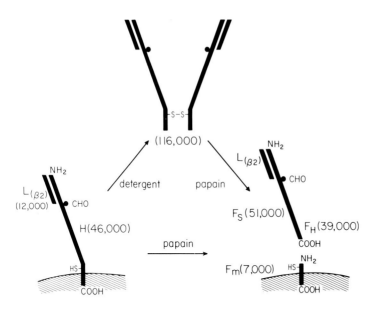

Fig. 5. Molecular model of H-2 antigens. Solubilization of H-2 antigens with detergents gives a structure containing 2 heavy chains and 2 light chains with at least
1 disulfide bond linking the 2 heavy chains. Papain treatment of cell surfaces or
detergent extracts results in fragments of identical size. H: H-2 heavy chain; L(β_2):
H-2 light chain (β_2-microglobulin); F_S: water-soluble fragment (F_H plus L) obtained
after papain treatment of cell surfaces or detergent extracts; F_H: fragment of H-2
heavy chain obtained after papain treatment; F_m: portion of H-2 heavy chain cleaved
from molecule by papain and apparently associated with cell membrane. Molecular
weights given in parentheses. (From Ref. 38)

gels in the absence of reducing agents (Fig. 2C). Similar observations
were reported previously by other laboratories (9, 20). These results
suggested that the H-2 heavy chain may not exist as a dimer on the
cell surface and that the disulfide bridge might be formed after extraction from the membrane. Treatment of cells or membrane fractions
with iodoacetamide prior to detergent extraction significantly increased the amount of H-2 heavy chain monomer on SDS-polyacrylamide
gels in the absence of reducing agents (Fig. 2C). These findings strongly suggest that the H-2 molecule exists predominantly as a monomer on
the cell surface. The transient occurrence of dimers, however, cannot
be totally excluded, and preliminary experiments indicate that some
dimers may exist. For example, oxidation of sulfhydryl groups with o-
phenantroline/$CuSO_4$ prior to alkylation of cells or membranes increased
the amount of heavy chain dimer, suggesting that some H-2 molecules
may be closely associated with each other as non-covalently bonded
dimers.

Fragments Obtained by Papain Cleavage

It has been shown that treatment of cells or detergent extracts of
cells with papain results in water soluble fragments of H-2 antigens
that contain antigenic acitvitiy (9). Immune precipitates of H-2 an-

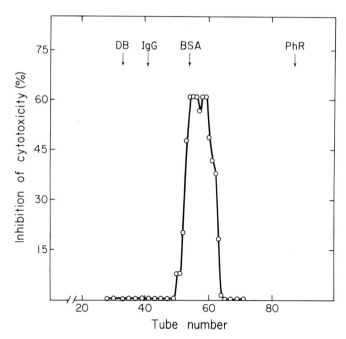

Fig. 6. Chromatography of papain solubilized H-2 antigens on Sephadex 200. 1 ml of a 0.5 % NP 40 extract of a membrane fraction of 10 BALB/c spleen cells was digested with papain and loaded on a column (1.2 x 75 cm) of Sephadex G 200. Eluant: 0.01 M Tris-HCl, pH 8.0, 4 ml/h. Assay: anti-H-2^d on ^{51}Cr-loaded BALB/c spleen lymphocytes. Markers: DB, IgG, BSA, and PhR as described in Fig. 3

tigens obtained by papain treatment of cells, membrane fractions or detergent extracts from membrane fractions were analyzed on SDS-poly-acrylamide gels. Two components of 39,000 and 12,000 molecular weight were detectable both in the presence and in the absence of reducing agents (Fig. 2D). Under nondissociating conditions and in the absence of detergent, gel chromatography (Fig. 6) and ultracentrifugation (Fig. 4) indicated that the molecular weight was approximately 49,000. These results suggest that the major fragment obtained by papain cleavage is monomeric and consists of a 39,000 dalton heavy chain fragment (F_H) and one non-covalently associated β_2-microglobulin molecule (Fig. 5). We have designated this monomeric fragment F_s for water-soluble fragment to distinguish it from that portion of the heavy chain, F_m, that putatively extends into the membrane. We have observed no F_s dimers after papain treatment, regardless of the ionic strength of the solutions. This observation is in agreement with results obtained by Peterson et al. (12) but is in contrast to the observations of Stro-minger et al. (11) who found a non-covalently associated dimer (MW 96,000) after papain treatment of HL-A antigen. Our results suggest that interchain disulfide bonds do not occur between F_s fragments but may occur in the F_m region.

Amino Terminal Amino Acid Sequences of Different H-2 Heavy Chains

To extend structural analyses to the more chemical level, we determined the amino acid sequences of several H-2 heavy chains using a radio-

Table 1. Amino acid sequences of H-2 heavy chain

		1				5					10					15
K^b	NH$_2$-	X	Pro	His	—	Leu	Arg	Tyr	Phe	Val	—	—	Val	—	Arg	—
$K^b F_H$	NH$_2$-	—	—	His	—	—	Arg	Tyr	—							
K^d	NH$_2$-	Met	X	His	—	X	Arg	Tyr	—	—	—	—	—	—	Arg	—
D^d	NH$_2$-	Met	X	His	—	Leu	Arg	Tyr	Phe	Val	—	—	Val	—	Arg	—
K^k	NH$_2$-	—	—	His	—	—	Arg	Tyr	—							

	16				20					25		
K^b	—	Leu	—	—	—	Arg	Tyr	—	—	—	—	Tyr
K^d	—	(Leu)	—	—	—	Arg	—	—	—	—	—	—
D^d	—	X	—	—	—	Arg	—	—	—	—	—	—

chemical approach. Different cell preparations were incubated with in-dividual ^{3}H-amino acids and the specifically labeled H-2 heavy chains or their papain fragments were isolated by preparative SDS-gel electro-phoresis. Sequence analysis (21) was performed manually by a procedure (19, 22) modified to obtain phenylthiohydantoin amino acids and auto-matically by a Beckman 890 C sequencer using the Quadrol double cleav-age program (Beckman Instruments).

To locate the region of the heavy chain that contained the papain cleav-age site, the partial amino terminal sequence of an intact K^b heavy chain was compared with that of its papain fragment F_H. For the three amino acids tested, no differences were found in the first eight posi-tions (Table 1). These results suggest that the F_H fragment contains the amino terminal part of the heavy chain and that the F_m fragment is derived from the carboxyl-terminal part of the polypeptide chain. These data have been confirmed recently by Ewenstein et al. (23) who showed that the Leu and Arg residues in H-2K^b and its papain fragment F_H are identical in positions 5, 6, 14, 17, and 21. Inasmuch as F_S can be obtained by direct papain treatment of the cell membrane, the amino terminal end of the H-2 heavy chain probably extends away from the cell surface (Fig. 5), and the carboxyl-terminal region (F_m) is probably associated with the plasma membrane. This orientation resem-bles that of other membrane glycoproteins such as glycophorin (24, 25) and immunoglobulin (26) and may be a general feature of all membrane proteins. It is possible that the F_m region of H-2 antigens has many features in common with similar portions of other membrane proteins including similarities in amino acid sequence.

Radiochemical sequence studies were also carried out on other H-2 heavy chains. The partial NH$_2$-terminal amino acid sequence of four different H-2 heavy chains are shown in Table 1. These sequences are identical with the exception of positions 1, 2, 5, and 17 where some chains lacked residues present in the other gene products. While the sequence data are not yet complete, two conclusions can now be drawn. First, the similarity in the sequences of K and D gene products sup-ports the hypothesis (27) that these genes evolved by duplication of an ancestral gene. Second, the antigenic polymorphism of the H-2 an-tigens is very likely reflected in the residues that differ in the various chains. The polymorphism probably extends to other positions yet to be determined, as it is unlikely that the short segment of the peptide chain shown in Table 1 is the only antigenic region expressing the polymorphic character of these molecules. There exist no data as yet to exclude the possibility that the antigenic sites are spread over the entire molecule. The sequence data shown in Table 1 have been confirmed very recently and extended by similar studies in three other laboratories (23, 28, 29). While the combined data are still too limited to allow any definitive conclusion about the postulated common origin (30, 31) of histocompatibility antigens and immunoglobulins, no convincing homology is as yet apparent.

Very recently, the partial NH_2-terminal amino acid sequences have been reported for papain fragments of three types of HL-A heavy chains (32). The number of residues common in H-2 heavy chains and in HL-A heavy chain fragments (6 out of 25 residues) reflects the similarity seen between these systems. The results also support the notion that the papain fragmentation of human histocompatibility antigens is similar to that observed in the murine system and suggest that the orientation of HL-A molecules on the cell surface is similar to that of H-2 antigens. Despite some earlier reports on the immunoglobulin-like structure of HL-A antigens (11), currently available sequence data for the HL-A antigens do not permit any definitive conclusions about possible homologies with immunoglobulins.

Biological Activities of Murine Histocompatibility Antigens

During an allograft reaction, a subclass of T lymphocytes that is capable of lysing specifically allogeneic target cells in vitro is generated (33). Similar cytotoxic T lymphocytes are generated in the in vitro destruction of virally infected (5) and neoplastic syngeneic cells (6-8). From these results, it has been suggested that one of the physiological functions of cytotoxic T lymphocytes is to eliminate antigenically abnormal cells, and thus this class of lymphocytes may be an important element in the immune surveillance system of vertebrate organisms.

Many observations suggest that there is a special relationship between cytotoxic T lymphocytes and H-2 antigens (3). Cytotoxic lymphocytes generated against virally infected syngeneic cells kill virally infected cells only if the target cells and the stimulated cells share the H-2K or the H-2D end region of the major histocompatibility antigen complex. In addition, antibodies against H-2 antigens inhibit the lysis of virally infected (34) or neoplastic cells (6, 7) by H-2 compatible cytotoxic T lymphocytes. While these experiments indicate that H-2 antigens may be involved, the molecular role of H-2 antigens in cell mediated lysis is unknown.

One possibility is that H-2 antigens interact with viral or tumor antigens on the cell surface to form a hybrid antigen. Alternatively, H-2 antigens and foreign antigens may undergo no physical interaction but may be independently required for the recognition and the elimination of antigenically abnormal cells. In order to distinguish between these possibilities, we have used a variety of approaches to examine the H-2 antigens on murine tumor cells.

Cytotoxic lymphocytes capable of lysing EL4 lymphoma cells (H-2^b) were generated using spleen cells from the parent C57BL/6J strain, and cells capable of lysing P388 lymphoma cells (H-2^d) were generated using lymphocytes from the parent strain DBA/2J or from the H-2 compatible strain BALB/c. In each case, the lysis of target cells by H-2 compatible cytotoxic lymphocytes was inhibited by alloantisera directed against their shared H-2 antigens (Fig. 7).

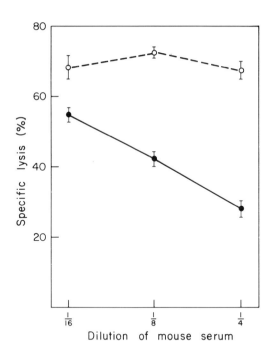

Fig. 7. Inhibition of the specific lysis of P388 tumor cells by H-2 compatible cytotoxic lymphocytes (BALB/c) in presence of anti-H-2 serum. Assays were carried out in presence of either anti-H-2^d serum (●) or normal mouse serum (o). (From Ref. 6)

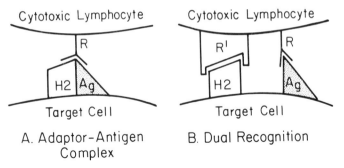

A. Adaptor-Antigen Complex   B. Dual Recognition

Fig. 8A and B. Diagrams of possible modes by which cytotoxic lymphocytes recognize target cells. H2: histocompatibility antigen; Ag: viral antigen; R: receptor on T cell; R': recognition unit for H-2. (From Ref. 6)

Control experiments demonstrated that the suppressive effect depended on the anti-H-2 activity of the alloantisera and was not due to the presence of anti-viral antibodies, which are detectable in most H-2 alloantisera (manuscript submitted for publication). Irrelevant H-2 alloantisera did not suppress the lysis of syngeneic tumor cells by cytotoxic T cells. Because each of the systems involved cells that were H-2 identical, inhibition of cell lysis might result from effects of the antisera on the target cells, the cytotoxic cells or both. Several lines of evidence indicated that the inhibition of cytotoxicity by the H-2 antisera was due to the blocking of H-2 antigens on the target cells rather than of those on cytotoxic lymphocytes (8).

244

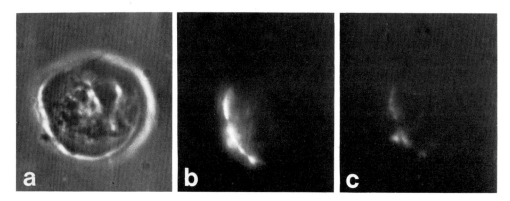

Fig. 9a-c. Co-capping of H-2 antigens and viral antigens on EL4 cells. H-2 antigens were capped using anti-H-2^b antiserum followed by Fl-goat anti-mouse Ig. The cells were treated with formaldehyde to prevent further redistribution of receptors and then were stained with rabbit anti-RLV, which had been absorbed twice with normal spleen cells, followed by TMR-goat anti-rabbit Ig. The same EL4 cell is shown. (a) phase contrast illumination; (b) fluorescence microscopy showing distribution of H-2 antigens as detected with Fl-goat anti-mouse Ig; (c) fluorescence microscopy showing distribution of viral antigens as detected with TMR-goat anti-rabbit Ig. (From Ref. 6)

These observations are compatible with the idea that H-2 antigens on the tumor cells are part of the target of the cytotoxic lymphocytes, and suggest that tumorigenesis may lead to chemical modification of H-2 antigens or that H-2 antigens and tumor antigens may be physically associated to form an antigenic complex (Fig. 8A). On the other hand, "dual recognition" may be required, i.e. cytotoxic lymphocytes must see the appropriate H-2 antigen with one set of receptors and the foreign antigens with a different set of receptors (Fig. 8B).

So far we have no proof for either of the two models, but we have obtained some indirect data that seem to favor the presence of a hybrid complex of H-2 antigens and tumor antigens on the surface of tumor cells. To test for such an interaction, we selected viral antigens as an example of tumor associated antigens and examined their relationship on the cell surface to the H-2 antigens on both EL4 and P388 tumor cells using an assay based on the known ability of polyvalent antisera to induce the redistribution of surface receptors to form "caps" on lymphocytes or tumor cells (36).

Co-Capping and Co-Patching of H-2 and Viral Antigens

Both EL4 and P388 tumor cells bear surface antigens that react with antisera against Rauscher leukemia virus (RLV). To test for a possible physical association between these antigens and the H-2 antigens on tumor cells, one type of antigen was first "capped" by reaction with specific antisera. The cap was detected using antibodies labeled with fluorescein. The effect of this capping on the other receptors was then tested by using a second specific antiserum and antibodies labeled with tetramethyl rhodamine. We found that the capping of H-2 antigens on the tumor cells induced co-capping of the viral antigens (Fig. 9), and capping of viral antigens resulted in a partial redistribution of H-2 antigens.

Capping is a highly cooperative process (36) and the co-capping of viral and H-2 antigens might be due to trapping effects. To exclude this possibility, we investigated whether the more limited process of patching of H-2 antigens by H-2 antisera would also induce patching of viral antigens. Experiments were identical to those described for co-capping but were carried out in the presence of an inhibitor of capping. The larger patches induced by anti-H-2^b serum on EL4 cells were also stained with anti-viral antibodies, indicating that some co-patching had occurred.

A number of control experiments were performed to ascertain the specificity of co-capping and co-patching. It is particularly important to exclude the influence of anti-viral antibody in H-2 antisera or H-2 antibody in anti-viral antisera. These possibilities were tested by absorption of the sera with various cells. Absorption of H-2^b antisera with H-2^b spleen cells eliminated capping and co-capping. On the other hand, absorption of H-2^b antisera with H-2^d spleen cells infected with Friend virus, which is closely related to Rauscher leukemia virus (35), did not affect the capping and co-capping results. Co-capping, however, was markedly reduced by absorption of the rabbit anti-RLV with H-2^d spleen cells infected with Friend virus, and not by absorption with normal H-2^b or H-2^d cells.

Similar results were obtained using a goat antiserum prepared against a purified component, gp69/71, of RLV and an H-2 antiserum monospecific for the H-2K^b locus. Indirect immune precipitates of NP 40 lysates of ^{125}I surface labeled EL4 cells were performed using these two sera, and the precipitates were subjected to SDS-gel electrophoresis. In each case only one component was detected on the gel, and the molecular weights of the radioactive peaks corresponded to H-2 antigens or gp69/71 as would be expected from the putative specificity of the antisera used.

All of these data support the idea that the H-2 antigens and some viral antigens may be capable of interacting on the cell surface. Many more experiments and controls must be performed, however, before this can be established as a general phenomenon. Although a number of attempts were made to rule out cross-reactions between the anti-H-2 and the anti-viral antisera, the results of these experiments must be viewed with great caution. All of the data so far provide only indirect evidence for a physical association between H-2 and viral antigens. Furthermore, cap formation results in a decrease in the threshold for the detection of antigens by immunofluorescence methods. Antigens which have been undectable in a diffuse distribution might become visible when concentrated in cap.

If the co-capping and co-patching data reflect an association between histocompatibility antigens and foreign antigens on cell surfaces, these experiments suggest how the H-2 antigens could form part of an antigenic structure containing both "self" and "non-self" (virus) components (Fig. 8). In this case the H-2 molecule would serve in a hybrid antigen as part of the target for the cell mediated lysis of syngeneic cells. The extreme polymorphism of H-2 molecules might either contribute to the variation in the immunogenicity of the "altered self" site (37), or it could be related to a binding site on the H-2 molecule for foreign antigens. The latter possibility seems less likely because H-2 antisera do appear to bind specifically to tumor and virally infected cells, suggesting that the H-2 antigens on the cells have all of the major specificities exposed. Verification of these ideas, however, must await a direct demonstration of the linkage between H-2 and viral antigens, a description of the binding mechanism, and an estimate of the range of foreign antigens that may be bound.

Acknowledgments. This work was supported by USPHS grants AM 04256, AI 09273, and AI 11378 from the NIH, and a Career Scientist Award to BAC from the Irma T. Hirschl Trust. JWS is supported by a C.J. Martin Travelling Fellowship of the National Health and Medical Research Council, Canberra, Australia. KR is a postdoctoral fellow of the Deutsche Forschungsgemeinschaft.

References

1. Gorer, P.A.: J. Pathol. Bacteriol. 47, 231-252 (1938)
2. Lindahl, K.F., Bach, F.H.: Nature (Lond.) 254, 607-609 (1975)
3. Alter, B.J., Schendel, O.J., Bach, M.L., Bach, F.H., Klein, J., Stimpfling, J.H.: J. Exp. Med. 137, 1303-1309 (1973)
4. Shearer, G.M., Rehn, T.G., Garbarino, C.A.: J. Exp. Med. 141, 1348-1364 (1975)
5. Zinkernagel, R.M., Doherty, P.C.: J. Exp. Med. 141, 1427-1436 (1975)
6. Schrader, J.W., Cunningham, B.A., Edelman, G.M.: Proc. Nat. Acad. Sci. 72, 5066-5070 (1975)
7. Germain, R.N., Dorf, M.E., Benacerraf, B.: J. Exp. Med. 142, 1023-1028 (1975)
8. Schrader, J.W., Edelman, G.M.: J. Exp. Med. 143, 601-614 (1976)
9. Schwartz, B.D., Kato, K., Cullen, S.E., Nathenson, S.G.: Biochemistry 12, 2157-2164 (1973)
10. Zinkernagel, R.F., Doherty, P.C.: Nature (Lond.) 251, 547-548 (1974)
11. Strominger, J.L., Cresswell, P., Grey, H., Humphreys, R.E., Mann, D., McCune, J., Parham, P., Robb, R., Sanderson, A.R., Springer, T.A., Terhorst, C., Turner, M.J.: Transpl. Rev. 21, 126-143 (1974)
12. Peterson, P.A., Rask, L., Sege, K., Klareskog, L., Anundi, H., Östberg, L.: Proc. Nat. Acad. Sci. 72, 1612-1616 (1975)
13. Grey, H.M., Kubo, R.T., Colon, S.M., Poulik, M.D., Cresswell, P., Springer, T., Turner, M., Strominger, J.L.: J. Exp. Med. 138, 1608-1612 (1973)
14. Nakamuro, K., Tanigaki, N., Pressman, D.: Proc. Nat. Acad. Sci. 70, 2863-2867 (1973)
15. Silver, J., Hood, L.: Nature (Lond.) 249, 764-765 (1974)
16. Peterson, P.A., Cunningham, B.A., Berggård, I., Edelman, G.M.: Proc. Nat. Acad. Sci. 72, 1697-1701 (1972)
17. Cunningham, B.A., Wang, J.L., Berggård, I., Peterson, P.A.: Biochemistry 12, 4811-4822 (1973)
18. Cunningham, B.A.: Federation Proc. 35, 1171-1176 (1976)
19. Henning, R., Milner, R.J., Reske, K., Cunningham, B.A., Edelman, G.M.: Proc. Nat. Acad. Sci. 73, 118-122 (1976)
20. Cresswell, P., Dawson, J.R.: J. Immunol. 114, 523-525 (1975)
21. Jacobs, J.W., Kemper, B., Niall, H.D. Habener, J.F., Potts, J.T., Jr.: Nature (Lond.) 249, 155-157 (1974)
22. Weiner, A.M., Platt, D., Weber, K.: J. Biol. Chem. 247, 3242-3251 (1972)
23. Ewenstein, B.M., Freed, J.H., Mole, L.E., Nathenson, S.G.: Proc. Nat. Acad. Sci. 73, 915-918 (1976)
24. Bretschner, M.S.: Nature (New Biol.) 231, 229-232 (1971)
25. Segrest, J.P., Kahane, I., Jackson, R.L., Marchesi, V.T.: Arch. Biochem. Biophys, 155, 167-183 (1973)
26. Fu, S.M., Kunkel, H.G.: J. Exp. Med. 140, 895-903 (1974)
27. Klein, J., Shreffler, D.C.: Transpl. Rev. 6, 3-29 (1971)
28. Silver, J., Hood, L.: Proc. Nat. Acad. Sci. 73, 599-603 (1976)
29. Vitetta, E.S., Capra, J.D., Klapper, D.G., Klein, J., Uhr, J.W.: 73, 905-909 (1976)
30. Burnet, F.M.: Nature (Lond.) 226, 123-126 (1970)

31. Gally, J.A., Edelman, G.M.: Ann. Rev. Genet. <u>6</u>, 1-46 (1972)
32. Terhorst, C., Parham, P., Mann, D.L., Strominger, J.L.: Proc. Nat. Acad. Sci. <u>73</u>, 910-914 (1976)
33. Cerottini, J.C., Brunner, K.T.: Advan. Immunol. <u>18</u>, 67-132 (1974)
34. Koszinowski, U., Ertl, H.: Nature (Lond.) <u>255</u>, 552-554 (1975)
35. Cerny, J., Essex, M.: Nature (Lond.) <u>251</u>, 742-744 (1974)
36. Edelman, G.M., Yahara, I., Wang, J.L.: Proc. Nat. Acad. Sci. <u>70</u>, 1442-1446 (1973)
37. Doherty, P.C., Zinkernagel, R.F.: Nature (Lond.) <u>256</u>, 50-52 (1975)
38. Edelman, G.M.: Science <u>192</u>, 218-226 (1976)

Immune Response Region Associated (Ia) Antigens. Some Structural Features[1]

M. Hess

Introduction

Ia antigens form an integral part of the lymphocyte cell membrane, and are direct gene products of the $\underline{I}$- or Immune Response Region of the Major Histocompatibility Complex (MHC) of the mouse. Twenty one serological Ia specificities or genetic (and structural?) determinants have already been identified by cross-immunization of appropriate inbred mouse strains (4, 25). Thus, a degree of polymorphism is already apparent which closely resembles the genetic polymorphism encountered in either $\underline{H-2K}$ or $\underline{H-2D}$ gene products, i.e. the classical transplantation antigens. $\underline{K}$, $\underline{I}$, and $\underline{D}$ denote three of the five genetic regions which form the H-2 complex on the 17th chromosome of the mouse (18).

Solubilization and Purification of Ia Antigens

The solubilization method determines to a certain extent the mode of purification. Hence, after solubilization with nonionic detergents, which releases the intact antigen from the cell membrane, the isolation of the antigenic substance is achieved by immune coprecipitation (3, 8, 9, 11, 13, 28); enzymic solubilization, however, lends itself to the conventional biochemical purification techniques of the serologically active degradation product (5, 13, 16, 27).

The detergent solubilization is accomplished on whole cells (or isolated cell membranes) after biosynthetic incorporation of radiolabeled amino acids or enzymic surface iodination.

The purification sequence of papain-solubilized Ia antigens (from mice which carry the Ia.11 antigens of the $\underline{H-2}^d$ haplotype, Ia.11^d) is shown in Figure 1. Antigenic material recovered from the active fraction in Figure 1c is homogeneous after poylacrylamide gel electrophoresis. The electrophoretic fraction, however, contains more than one gene product, namely products of at least two "subregions", $\underline{I-A}$ and $\underline{I-C}$. As a result of their very similar overall physicochemical properties, Ia substances are purified concomitantly. This behavior is reminiscent of H-2K and H-2D antigens (15). The final purification, therefore, requires the specific radiolabeling of this highly purified material and, after formation of a soluble antigen-antibody complex with monospecific Ia alloantisera, the isolation of this complex with an appropriate heteroantiserum.

---

[1] For a comprehensive discussion on the fundamental role assigned to I region products in alloimmune reactions see Transplantation Reviews, Vol. 30, Munksgaard, Kopenhagen, 1976.

249

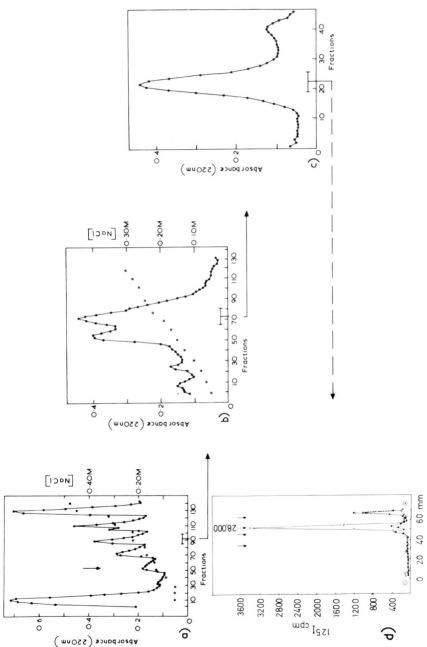

Fig. 1a-d. Purification of papain-solubilized Ia antigens. (a) Preparative Sephadex DEAE-A50 chromatography. (b) Analytical ion exchange chromatography. (c) Analytical gel filtration on Sephadex G 100. Fractions containing alloantigenic activity (measured by inhibition of complement dependent lysis) indicated by horizontal bars; ooo: NaCl gradient. (d) Lithium dode-cyl sulfate (LDS) electrophoresis in 10 % polyacrylamide gels of ^{125}I-labeled fraction from (c), after immune coprecipita-tion, and in absence of reducing agent. -ooo-: Ia.11^d; -●●●-: normal mouse serum control. Reference proteins are, from left, H chain (Ig), pepsin and L chain (Ig). As shown by similar figures obtained by electrophoresis of proteolytic fragment after reduction (Fig. 3 o—o), papain solubilized Ia antigens are present in monomeric form

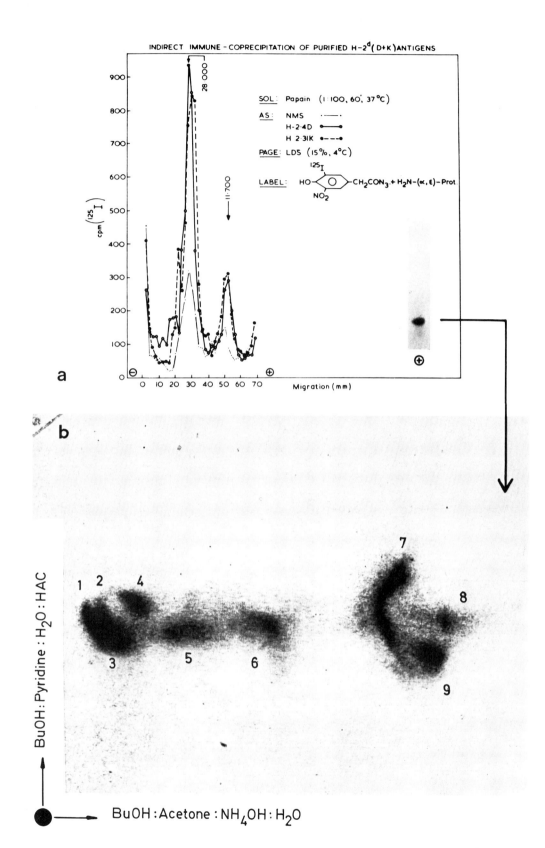

INDIRECT IMMUNE - COPRECIPITATION OF PURIFIED H-2^d(D+K)ANTIGENS

SOL: Papain (1:100, 60', 37°C)

AS: NMS ·—·—·
H-2·4D ●——●
H-2·31K ●---●

PAGE: LDS (15%, 4°C)

LABEL: HO—⬡—CH₂CON₃ + H₂N—(α,ε)—Prot.

cpm(^{125}I)

Migration (mm)

a

b

BuOH : Pyridine : H₂O : HAC

BuOH : Acetone : NH₄OH : H₂O

Radiolabeling by Chemical Modification

Methods which allow the specific introduction of radioactive label in
highly purified alloantigenic substances should assist the elucidation
of the primary structure of, for example, MHC gene products, and provide information complementary to the partial sequence data obtained
with the native molecule (7, 10, 26, 29).

At the present stage, the limiting factor with respect to the chemical
modification of Ia (or H-2) antigens is the consecutive loss of alloantigenic activity of these substances, since the indirect immune-pre-
cipitation, i.e. the isolation of the soluble alloantigen-antibody
complex with appropriate heteroantisera, is indispensable for the verification of the results obtained by more conventional methods of preparative biochemistry.

The specific labeling of various amino acid side chains with iodine
isotopes is meant as an alternative to the biosynthetic incorporation
of radiolabeled amino acid precursors. A few micrograms of highly
purified material are thus amenable to structural characterization by
peptide mapping and/or automated Edman degradation.

The rationale of this approach is exemplified in Figure 2 (see figure
legend), and shown by the peptide map obtained, in this case, after
immune coprecipitation (12). Similar results can be obtained with direct elution of the antigenic fragment from the gel, after electrophoresis and subsequent radiolabeling (unpublished).

The compounds used or synthesized in the author's laboratory are shown
in Table 1 and characterized with respect to the reactive side chain
and the retention of alloantigenic activity (14). (Needless to say, native proteins modified by acylation or amidination become refractory to
Edman degradation.)

Basic Structural Aspects of Ia Antigens

General Properties

Ia antigens are glycoprotein molecules composed of single polypeptide
chains with a molecular weight (M_r) of 35,000, or 26,000 after papain
solubilization. The papain degradation product is negatively charged

Fig. 2. Two-dimensional peptide map of acylated MHC gene products. (a) Electrophero-
gram of acetylated H-2 antigens (see Table 1, Reaction (1)). (b) Peptide map of
elastase digest showing lysine containing peptides present in H-2.4^d antigens. Specific coprecipitates fractionated on LDS-PAGE, radioactive H-2 component eluted
with 0.1 % LDS in 8M urea, spun, dialyzed, oxidized, freeze dried and subjected to
elastase hydrolysis for 4 h at 37° C in 50 mM NH_4HCO_3 containing 0.1M KCl. Total of
2-3000 cpm was then applied on a 5x5 cm thin layer sheet (Silica gel, Eastman Kodak),
and 2-dimensional chromatography performed with the following solvent systems: 1-
butanol/pyridine/water/acetic acid (204/143/143/25) in the 1st dimension (↑) and
acetone/1-butanol/ammonia/water (10/10/5/2) in the 2nd dimension (→). For autoradiography, the dry peptide maps were exposed to X-ray films (Kodak RP/R54) for ca. one
week. Work is now in progress with Ia antigens using compounds (2), (4), and (5)
of Table 1 (composite peptide maps)

Table 1. Chemical modification and radiolabeling of cell surface antigens

Compound	Labeled residue	Serologic activity
		Ia : H-2
- (1) ^-O—(diiodophenyl, I,I)—CH_2CON_3	Lysyl (and N-terminal) residue	n.d. +
- (2) ^-O—(diiodophenyl, I,I)—$C(^+NH_2Cl^-)(OCH_3)$	Lysyl (and N-terminal) residue	+ +
- (3) ^-O—(diiodophenyl, I,I)—$(CH_2)_2$-$\overset{O}{\overset{\|}{C}}$-O-N(succinimidyl)	Lysyl (and N-terminal) residue	+ +
- (4) ^-O—(diiodophenyl, I,I)—$(CH_2)_2$-NH_2	Gutaminyl (Transglutaminase)	+ +
- (5) ^-O—(diiodophenyl, I,I)—$(CH_2)_2$-NH_2	Aspartyl and Glutamyl (WCD promoted incorporation)	+ +
- (6) ^-O—(diiodophenyl, I,I)—Prot	Tyrosyl and Histidinyl	+ +

(1) 4-hydroxy-3,5-diiodophenyl acetyl azide. Compound synthesized as described by Brownstone et al. (2) and radiolabeled according to Mc Farlane (20) in a modification communicated by K. Himmelspach.

(2) 4-hydroxy-3,5 diiodo-1 methylbenzimidate hydrochloride. Reagent synthesized following procedure of Pinner (22), except that 4-cyanophenol was used as starting material instead of benzonitrile.

(3) 3(4-hydroxy-3,5-diiodophenyl) propionic acid N-hydroxysuccinimide ester. Radioiodination and acylation procedures used for hydroxysuccinimide ester were essentially those of Bolton and Hunter (1).

(4) Transglutaminase catalyzed reaction performed basically according to Dutton and Singer (6) using radiolabeled Tyramine as substrate.

(5) Tyramine also used for 1-ethyl-3-(3-diethylaminopropyl) carbodiimide promoted incorporation according to Hoare and Koshland (17).

(6) Tyrosyl and Histidinyl residues trace labeled according to Marchalonis et al. (21).

at pH 9.6 and displays an apparent isoelectric point of ca. 4.6. The complete loss of serological activity after incubation in 8M urea indicates that the antigenic determinants are of protein nature.

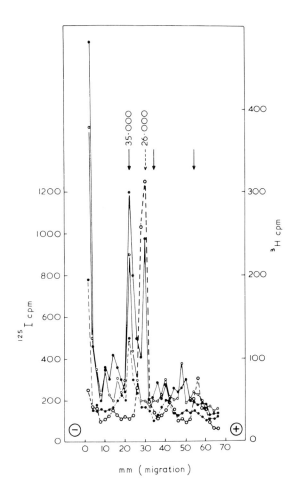

Fig. 3. Single polypeptide chain structure of Ia antigens. Radiolabeling. 5×10 viable cells from B10.D2 (H-2^d) mice were labeled with either [^{3}H]-lysine (●--●), [^{3}H]-histidine (●—●) or [^{3}H]-arginine (o—o) and isolated by immune coprecipitation after detergent-solubilization. The papain-solubilized fraction was trace labeled with ^{125}I (o--o). Lithium dodecyl sulfate (LDS) electrophoresis was performed according to Laemmli (19) in 14 % polyacrylamide gels after reduction of applied antigenic material with 2-mercaptoethanol. Migration is towards anode. Solid arrows: position of reference proteins (Pepsin 35,000 M_r, Ig L-chain 23,000 M_r and cytochrome c 11,700 M_r). Note absence of β2-m subunit (M_r 12,000) present in H-2 antigens (see Fig. 2a)

The Molecular Weight of Ia Antigens

The M_r of Ia.11^d antigens was determined on native, detergent-solubilized substances and after papain solubilization. In both instances, radiolabeled fractions were subjected to electrophoresis in polyacrylamide gels under dissociating conditions (Fig. 3).

1. Detergent-solubilized antigens have an M_r of 35,000 when electrophoresis is carried out under reducing conditions. In the absence of 2-mercaptoethanol, however, Ia antigens have a tendency to form dimers or higher aggregates (3).

2. Papain-solubilized Ia substances are monomeric fragments with M_r 26,000, as shown by the presence of a single molecular entity in Figure 1d (electrophoresis in the absence of 2-mercaptoethanol) and Figure 3, where the radiolabeled fragment was fractionated under reducing conditions. In analogy to H-2 antigens (10, 24) one can therefore assume that the polymerisation observed in detergent-solubilized molecules results from free sulfhydryl group(s) present in the 9,000 M_r Ia-fragment partially buried in the membrane (Fm) (7). The N-terminal sequence analysis on detergent- and papain-solubilized Ia antigens

will show whether the cleavage point(s) of activated papain occurs near
the amino- or carboxyl-terminal part of the molecule.

3. A final observation in Figure 3 is of particular interest: Despite
the use of enzyme inhibitors (Trasylol) during the detergent-solubilization procedure, the histidine labeled fraction shows, in addition
to the main Ia-peak with M_r 35,000, a second smaller peak (M_r 26,000).
This fragment is a proteolytic degradation product. The relative migration value of this fragment, however, is similar to the values obtained for papain-solubilized Ia (and H-2!) substances.

As has been shown for H-2 antigens (13, 14, 23) this is not mere coincidence, but could be the result of basic structural features common
to some MHC gene products. These structural properties account for the
preferential cleavage points observed after papain hydrolysis in native
Ia and H-2 antigens. (With regard to the evolutionary and genetic implications see Ref. 14).

Conclusion

Ia antigens represent the first gene product of the Immune Response
Region which could be isolated and characterized by immuno- and/or
biochemical methods. Although markedly different from H-2 antigens by
various physicochemical criteria (M_r, overall charge, pI, absence of
β2-m subunit) Ia, like H-2 antigens, show preferential cleavage points
after papain hydrolysis. Furthermore, the resulting proteolytic fragments as well as the F_m fragment in both Ia and H-2 antigens show a
striking size similarity. This suggests that a structural relationship
exists between some MHC gene products.

References

1. Bolton, A.E., Hunter, W.M.: Biochem. J. 133, 529-539 (1973)
2. Brownstone, A., Mitchison, N.A., Pitt-Rivers, R.: Immunology 10, 465-479 (1966)
3. Cullen, S.E., David, C.S., Shreffler, D.C., Nathenson, S.G.: Proc. Nat. Acad. Sci. 71, 648-652 (1974)
4. David, C.S.: In: Organization of MHC in Animals. Götze, D. (ed.). Berlin-Heidelberg-New York: Springer 1976, in press
5. Davies, D.A.L., Hess, M.: Nature (Lond.) 250, 228-230 (1974)
6. Dutton, A., Singer, S.J.: Proc. Nat. Acad. Sci. 72, 2568-2571 (1975)
7. Ewenstein, B., Freed, J.H., Mole, L.E., Nathenson, S.G.: Proc. Nat. Acad. Sci. 73, 915-918 (1976)
8. Finkelman, F.D., Svehach, E.M., Vitetta, E.S., Green, I., Paul, W. E.: J. Exp. Med. 141, 27-41 (1975)
9. Goding, J.W., Nossal, G.J.V., Shreffler, D.C., Marchalonis, J.J.: J. Immunogenetics 2, 9-25 (1975)
10. Henning, R., Milner, R.J., Reske, K., Cunningham, B.A., Edelman, G.M.: Proc. Nat. Acad. Sci. 73, 118-122 (1976)
11. Hess, M.: Folia Biol. (Prague) 21, 428-430 (1975)
12. Hess, M.: Folia Biol. (Prague) 21, 420-423 (1975)
13. Hess, M.: Europ. J. Immunol. 6, 188-193 (1976)
14. Hess, M.: Transpl. Rev. 30, 40-63 (1976)
15. Hess, M., Davies, D.A.L.: Europ. J. Biochem. 41, 1-13 (1974)
16. Hess, M., Davies, D.A.L.: Transpl. Proc. 7, 209-212 (1975)

17. Hoare, D.G., Olson, A., Koshland, D.E.: J. Am. Chem. Soc. $\underline{90}$, 1638-1643 (1968)
18. Klein, J.: Biology of the Mouse Histocompatibility-2 Complex. New York: Springer 1975
19. Laemmli, U.K.: Nature (Lond.) $\underline{227}$, 680-685 (1970)
20. McFarlane, A.S.: J. Clin. Investigation $\underline{42}$, 346-361 (1963)
21. Marchalonis, J.J., Cone, R.E., Santer, V.: Biochem. J. $\underline{124}$, 921-932 (1971)
22. Pinner, A.: Die Imidoaether und ihre Derivate. Berlin: Oppenheim 1892
23. Rask, L., Östberg, L., Lindblom, B., Fernstedt, U., Peterson, P.A.: Transpl. Rev. $\underline{21}$, 85-105 (1974)
24. Schwartz, B.D., Kato, K., Cullen, S.E., Nathenson, S.G.: Biochemistry $\underline{12}$, 2157-2164 (1973)
25. Shreffler, D.C., David, C.S.: Advan. Immunol. $\underline{20}$, 125-195 (1975)
26. Silver, J., Hood, L.: Proc. Nat. Acad. Sci. 73, 599-603 (1976)
27. Terhorst, C., Parham, P., Mann, D.L., Strominger, J.L.: Proc. Nat. Acad. Sci. $\underline{73}$, 910-914 (1976)
28. Vitetta, E.S., Klein, J., Uhr, J.W.: Immunogenetics $\underline{1}$, 82-90 (1974)
29. Vitetta, E.S., Capra, J.D., Klapper, D.G., Klein, J., Uhr, J.W.: Proc. Nat. Acad. Sci. $\underline{73}$, 905-909 (1976)

General Aspects of the Immune System

The Immune System: A Network of Lymphocyte Interactions

N. K. Jerne

So far, at this Colloquium, we have been hearing about molecules and
cells, about the tertiary conformations of antibody combining sites,
and about gene clusters, but we have not yet considered the immune
system as a total system. To study the immune system, one has to analyze its elements, of course, down to the molecular level; but finally
these elements will have to be put together, in order to describe the
development, the organization, and the functioning of the system as a
whole. At that level, we leave biochemistry and ascend into higher
regions of biology!

Clonal Selection

The clonal selection theory, which has enjoyed wide-spread acceptance
during the past ten years, represents such an attempt. This theory is
based on the postulated law that all antibody molecules that are synthesized by a given lymphocyte are identical and, moreover, that all
antibody molecules synthesized by cells of a proliferating clone of
lymphocytes are identical. What is meant by identical is that all these
molecules display the same variable regions in their polypeptide chains,
i.e., the same variable domains, without necessarily possessing the
same constant regions. Antibody molecules can be placed at the surface
of a lymphocyte to serve as receptors that can recognize a fitting antigen, or may be secreted by lymphocytes at a more mature stage of
differentiation. In other words, a clone of lymphocytes consists of
cells in different states of differentiation, which all produce antibody molecules that display the same combining sites and the same idiotypic determinants.

As we heard the other day, and as has of course often been suggested
in the literature, a reasonable estimate of the number of different
antibody molecules that one person can make may be of the order of 10^7.
There are various arguments, into which I will not go here, that indicate that the number is larger than a million. It can, at any moment,
not be larger than 10^{12} as this is about the number of B lymphocytes
that one person possesses. If we accept these figures, and disregard
the T cells for a moment, the population of lymphocytes would be made
up of 10^7 different clones, each consisting on the average of 10^5
cells. The clonal selection theory thus proposes that antigen entering
the body is confronted by 10^7 such clones, and that it selects those
clones that display receptors with combining sites that show reasonable
fit to the antigen. This recognition will then drive the selected
clones into further proliferation and to antibody secretion. If we
add nothing else to the theory, we could say that a clone is an element of the immune system consisting of about 10^5 cells in various
states of differentiation plus the antibody molecules and other factors that the cells of this clone may have secreted. As long as the
clones do not interact between one another, we could say that the
total immune system consists of 10^7 small immune systems, any of which

may react whenever a fitting antigen arrives with cell proliferation, antibody secretion, or with tolerization, depending on regulatory mechanisms operating within every clone.

Network Notions

The network notion goes one step further. It regards the immune system as a network of interacting elements, where one element consists of a clone of about 10^5 cells plus the molecules these cells have released. This notion came to me a few years ago from what was then known to me about idiotypes and anti-idiotypic antibodies. Dr Helmreich has already said in his introduction that the variable domains of antibody molecules display not only combining sites that can recognize an antigen, but also antigenic determinants that function as antigens to which other antibodies can be made: anti-antibodies and anti-anti-antibodies, and so forth. This makes it possible to view the entire system as a network of idiotypes and anti-idiotypic interactions. It struck me that the enormous diversity of idiotypes — I shall return to this point in a moment — makes a network notion obviously correct. Also, because it fitted in with my old preconceived idea that there is a deep-seated analogy between the immune system and the central nervous system, the brain. The way we speak, as immunologists, is already suggestive. An animal learns to make antibodies against certain environmental antigens, viruses, bacteria, etc., so there is a learning element. Then, the animal has a memory of what it has learned; next time the same antigen or agent arrives, the animal is much better prepared to deal with it. Finally, this learning cannot be transmitted to the offspring. A newborn has to start from scratch and learn again. The analogy between the immune system and the central nervous system — though it may be somewhat fortuitous — is this lack of genetic fixation of learning and memory. It is different from the hormonal network, for example. We could regard the interactions between hormone-producing cells as a network which is genetically fixed and not subject to learning. It would of course make immunology even more intriguing if it could serve as a stepping stone towards understanding the brain.

Idiotypes

Let us return to considering idiotypes. A classical antibody molecule is symmetrical and possesses two identical variable domains. A variable domain contains, as has been demonstrated in earlier lectures at this Colloquium, a combining site to which an antigen can fit. This combining site does not, however, occupy the entire surface of the variable domain. There is a lot more surface, and this displays antigenic determinants to which animals can and do make antibodies. The set of these determinants that occur on one antibody was named the idiotype of the antibody by Oudin. He has agreed with me that an individual idiotypic determinant be called an idiotope. It is like the distinction between antigen and antigenic determinant, or epitope. Sometimes the distinction is necessary, sometimes it does not matter. We could then ask: what is the repertoire of idiotopes in the immune system of one person? It was suggested earlier that the repertoire of different combining sites is perhaps of the order of 10^7. As both combining site and idiotopes occurring on a given variable domain are determined by the amino acid sequences of the variable regions of the polypeptide

chains that constitute this variable domain, a first approximation
would be to suggest that the repertoire of idiotypes or idiotopes is
also of the order of 10^7. Early experiments, in 1963 and the following
years, by Oudin and his colleagues and by Kelus and Gell, already
showed that in the rabbit the repertoire of idiotypes is enormously
large. Oudin immunized fifty rabbits with a given Salmonella vaccine,
so that every rabbit produced a large concentration of anti-Salmonella
antibodies. After isolating these antibodies from the serum of rabbit
number 1 and injecting them into a fresh rabbit, this rabbit produced
anti-idiotypic antibodies against the anti-Salmonella antibodies of
rabbit number 1. It could be shown that these anti-idiotypic antibodies
precipitated the anti-Salmonella antibodies of rabbit number 1, but
that they did not react with the serum of rabbit number 1 that was set
aside before immunization with Salmonella, nor with antibodies produced
by rabbit number 1 against other antigens. The anti-idiotypic antibodies did not precipitate anti-Salmonella antibodies produced by any
of the remaining 49 rabbits. It was thus demonstrated that every one
of fifty rabbits had produced antibodies with anti-Salmonella-combining
sites but with different idiotypes, showing that the repertoire of different idiotypes in the rabbit species must be very large. Many experiments of similar type suggest that it is reasonable to assume that the
repertoires of combining sites and of idiotypes are of the same order
of magnitude, say 10^7.

Let us now imagine that we could number all the different variable domains occurring within the immune system of one individual. Variable
domain number 1 has combining site 1 and idiotype 1; variable domain
number 2 has combining site 2 and idiotype 2, and so forth to 10^7. It
is inconceivable that the set of 10^7 combining sites can avoid recognizing the set of 10^7 idiotopes. We know that antigen recognition by
antibody is degenerate in the sense that there is no 1:1 relationship.
A given antigenic determinant fits with various degrees of precision
to many different combining sites. A given antibody fits with various
degrees of precision to many different antigenic determinants. Thus,
every member of our set of 10^7 combining sites will recognize several
members of the set of 10^7 idiotypes within the same immune system, and
every member of the set of 10^7 idiotypes will be recognized by several
members of the set of 10^7 combining sites. We are confronted with a
network. Every antibody molecule within an immune system is an anti-
idiotypic antibody against idiotypes of antibody molecules present in
the same immune system. Foreign antigens are only incidental. The antibody system contains its own antigens. The immune system is a self-
contained network even in the absence of foreign antigens. All this
might still be regarded as only a formal construction if a number of
beautiful experiments during the last two or three years had not demonstrated that it has functional importance.

Functional Network

For experimental purposes, anti-idiotypic antibodies are prepared by
injecting antibodies, made by one animal, into another animal. It has
been shown, however, that an animal which makes antibody against some
antigen will, with short delay, itself make anti-idiotypic antibodies
to this antibody. Clonal selection implies that all antibodies that
the immune system can make must already be present in the form of surface receptors on lymphocytes. Anti-idiotypic antibodies have therefore been shown to be already present. Furthermore, it has been shown —
by Eichmann and Eisen and Wigzell and their colleagues — that in some

way T cells both recognize idiotypes and display idiotypes on their surface. Unfortunately, we do not yet know the nature of the T cell receptor, nor whether the recognizing ability of T cells may be aided by passively absorbed molecules released by B cells. Antibody A made by one animal, when injected into another animal, has been shown to induce not only anti-idiotypic antibodies against antibody A, but also T cells, helper cells, that can assist in antibody formation against a hapten when antibody A is used as carrier. The injection of anti-idiotypic antibodies into a mouse has been shown to have dramatic effects. The dominant observation has been that the recipient mouse becomes suppressed with respect to its ability to produce antibodies of the corresponding idiotype. Eichmann has shown that this suppression is a T cell effect. The injection of anti-idiotypic antibodies results in a proliferation of T cells that are capable of inhibiting B cells from secreting antibodies of the corresponding idiotype. This suppression can be transferred to normal mice by transferring T cells from a suppressed mouse. Furthermore, he showed that the concentration of anti-idiotypic antibody needed for idiotype suppression, for the induction of T cells that can maintain this suppression, is small. One needs to inject only about 10^{10} molecules of anti-idiotypic antibody into a mouse to obtain complete suppression of that idiotype. As a mouse possesses more than 10^{16} molecules of gammaglobulin already in its blood, the addition of one anti-idiotypic molecule to every million other molecules is sufficient to suppress one of the potentialities of the immune system. If we admit that every gammaglobulin molecule is an anti-idiotypic molecule against some idiotype within the system, then the average concentration of every anti-idiotypic molecular species should be less than 10^{10} molecules per mouse — another argument for a repertoire exceeding one million.

The situation is more complex, however, because it has been shown that anti-idiotypic antibodies of a certain class do not suppress, but on the contrary enhance the expression of the corresponding idiotype.

Network Difficulties

If, on the basis of these considerations and experiments, we admit that the immune system can be described as a network of clones that are interrelated by idiotypic interactions, we are faced with the difficulties of defining this network more precisely and of devising a model that retains its main features and that does not disregard other selective and regulatory forces that impinge upon the system. First, there is the difficulty of formulating a manageable mathematical description of a network. Even simple networks present great difficulties. Imagine, for example, an aquarium in which there are large fish that feed on small fish that in turn feed on some abundant food. As the small fish are eaten, they decrease in number. This causes the large fish to die of lack of food until the small fish have had a chance to multiply. The result is some sort of oscillatory steady state, that can be mathematically described. If, however, we imagine ten different species of fish with predatory interactions, the situation becomes mathematically unmanageable. How then to approach the idiotypic interactions among 10^7 lymphocyte clones? On a chess-board, every square can be denoted by the row and the column to which it belongs. Similarly, we can think of a board of 10^7 rows, each assigned to a lymphocyte clone, and of 10^7 columns, also representing each of these clones. A square would then relate to a pair of clones, and we might imagine to have filled in this matrix by writing the degree of clonal interaction on every square. We might then ask what the consequences would be for this entire system, if some of the clones proliferate after stimulation by foreign antigens, and in what sense the steady states that might perhaps ensue would retain a memory of these occurrences. It does not seem possible, however, to approach the network in this all-encompassing way. The second difficulty results from cell differentiation within each clone. There may be 10^5 B cells within a given clone, all expressing a given combining site and a given idiotype on the antibody molecules they synthesize. Some of these B cells may carry receptors of the IgM class, others (in a subsequent state of differentiation) may carry both IgM and IgD receptors, as has been described in this Colloquium. Other cells again may secrete IgM antibodies, or may have switched to the secretion of IgG antibodies, and so forth. Cells in different states of differentiation will respond to different signals, and they may emit different signals of interaction with cells of other clones. We have already seen that different classes of anti-idiotypic antibodies can have inhibitory or excitatory effects. The interaction between two clones therefore depends not only on cell numbers and on the degree of fit between the idiotype of one clone and the combining site of the other clone, but also on the cellular composition of the clones.

The third difficulty is the place of the T cells in a network context. We do not know whether clones comprise T cells as well as B cells, or whether T cell clones should be viewed as a separate set. The resolution of this question must await the clarification of the nature of the T cell receptor and the mechanism by which T cells recognize histocompatibility-associated antigens. The existence of different classes of T cells (helper cells, suppressor cells, killer cells) and of active factors secreted by T cells further complicates the situation. This is the more serious since it has become clear that T cells are not idiotypically inert and that they have a major regulatory effect on all cells of the immune system. In the face of all these complexities it seems a prohibitive task, at present, to make viable assumptions about the essential interactions in the immune network, and to determine the importance of these interactions relative to other, non-idiotypic regulatory mechanisms.

Attempts have recently been made, however, which I think deserve the careful attention of experimental immunologists, because our aim must remain to achieve an integrated description of the immune system and not to be satisfied with the hope that a large enough collection of data and phenomena will fall into shape by itself. These early models were developed in the last couple of years and what they try to do is to stimulate what seem to be the major features and responses of the immune system. The models are to a large extent the work of non-immunologists who may be expected to be less inhibited by experimental data than immunologists. I find this no good reason, however, not to take their work seriously. Each of these models introduces simplifications which many of us would regard as inadmissible; but even unsatisfactory models may provide, for certain phenomena, interesting explanations which were not considered by the immunologist that designed the experiments. Models should also make predictions which could be tested.

I should now like to finish by making a few comments on three models proposed by Richter, Göttingen, by Hoffmann, Basel, and by Adam and Weiler, Konstanz. Each of these models has been laid down in sets of differential equations, but I shall avoid mathematics in these brief outlines.

The Richter Model

Richter starts out by considering heterogeneous populations which he
calls functional units and denotes by antibody 1, antibody 2, antibody 3, etc. (Ab-1, Ab-2, Ab-3). The functional unit that can respond
to a given antigen is Ab-1. It consists of all lymphocyte clones that
contain cells that recognize and respond to the antigen, T cells and
B cells, as well as the antibody molecules and factors produced. The
elements of Ab-1 have in common their recognition of the antigen.
These elements also display a set of $\underline{n}$ idiotypes which are antigens
to each of which there exists a functional unit Ab-2, anti-idiotypic
to the idiotypes of Ab-1. The total functional unit Ab-2 may therefore
contain $\underline{n}$ times as many elements as Ab-1. Ab-3 (anti-idiotypic to Ab-2)
may be $\underline{n}^2$ times larger than Ab-1. A large part of the entire immune
system may thus be encompassed by Ab-4 and Ab-5. Richter's model now
assumes a simple overall interaction between these functional units.
Thus, Ab-1 stimulates the proliferation of Ab-2, which stimulates Ab-3,
etc. Reversely, Ab-3 suppresses Ab-2 which suppresses Ab-1 which suppresses Ab-0, etc. Ab-0 is the functional unit to which Ab-1 is anti-
idiotypic (Ab-0 is the "internal image" of the antigen in my original
exposition). Furthermore, it is assumed that the concentration of elements needed for stimulation is larger than the concentration needed
for suppression. By introducing what he thinks are reasonable parameters in the differential equations describing this simple scheme,
Richter concludes that the basic mode of response of the immune system,
in this model, is low-zone tolerance. This comes about when antigen
stimulates Ab-1 which stimulates Ab-2 to an extent that is sufficient
for inhibition of Ab-1, but not sufficient for stimulation of Ab-3.
Finally, Ab-1 is totally suppressed, or disappears. A normal immune
response ensues if a higher dose of antigen is given. Ab-2 will now
rise to a sufficient level to stimulate Ab-3 to an extent that is not
yet sufficient however for stimulation of Ab-4. Ab-3 will suppress
Ab-2, leaving Ab-1 free to expand and persist. In his latest version
of this model, Richter suggests that the most active elements within
each functional unit are the T cells. Thus, as low-zone tolerance is
the suppressive domination by Ab-2, it should be transferrable with
T cells. Also, low-zone tolerance can be expected to be the main response mode in the developing immune system in early ontogeny, leading
to the establishment of suppressor Ab-2 populations which can maintain
protection against antibody formation to self-antigens. Richter also
suggests that accessory network interactions come into play when Ab-1
cells become coated by antigen. These cells will stimulate a functional
unit Ab'-2 whose elements recognize not the idiotypes of Ab-1 but the
"carrier"-determinants of the coating antigen. To obtain an antibody
response we now need Ab'-3 for anti-idiotypic suppression of Ab'-2.
In this scheme, this functional unit Ab'-3 would contain the helper
T cells, and these would not recognize the carrier but the Ab'-2 idiotypes. It would follow that helper cells act by suppressing suppressor
cells, and that they do not act at the B cell level. This is probably
wrong, but can be experimentally tested. There are, of course, experiments indicating that certain suppressor T cells do not act at the B
cell level, but by suppressing helper T cells. That would be the reverse situation. The enormous simplification achieved by Richter's
model is to disregard the complexities and interactions within a functional unit which may comprise thousands or millions of cells. By
shifting the elements of interpretation from cells to functional units,
the happenings within a functional unit become of marginal interest,
just as the happenings within a cell are of marginal interest to those
for whom the cell is the unit of interpretation. Richter remarks that,
in the absence of an apparently complex network of clonal interactions,

there would be the need for a more complex signalling and regulation within each clone.

The Hoffmann Model

This distinguishes Richter's approach from that of Hoffmann, whose model simplifies and reduces the network ramifications in order to concentrate more on details of cellular interactions which Richter disregards. Hoffmann does not distinguish between idiotopes and combining site. Both are properties depending on the configuration of the variable domains, or "sticky ends", of an antibody molecule. If in some way we could average the variable domains of Ab-1 recognizing an antigen, the resulting shape would be complementary to the antigen. If we could average the variable domains of Ab-2, recognizing the variable domains of Ab-1, then we would obtain a shape identical to the antigen. In this way, the functional unit Ab-3 becomes identical to Ab-1. Furthermore, it follows that there is no difference between the effect of an antibody molecule that recognizes the idiotype of a cell receptor, and the effect of an antibody molecule whose idiotype is recognized by the combining site of this cell receptor. Hoffmann's model, in this way, reduces the network by considering, in respect to a given foreign antigen, only two functional units. One (Ab-1) consists of "plus"-elements (T cells, B cells, factors, antibodies) all of which recognize the antigen, the other (Ab-2) consists of "minus"-elements displaying variable domains which recognize and are recognized by the variable domains of the "plus"-elements.

Within this symmetry of Ab-1 and Ab-2, or of plus and minus elements, Hoffmann looks for the most simple set of components of interaction which would give the system its main observed properties. The elements he considers are T-plus cells and B-plus cells that display receptors and can be stimulated by cross-linking antigen. T-plus cells and B-plus cells are also stimulated by T-minus cells and B-minus cells, and vice versa. T cells are assumed to be more easily triggered and to proliferate more rapidly than B cells. The stimulation of T-plus cells leads to the secretion of a monovalent antigen-specific and minus cell specific T cell factor which can block minus cell receptors, and vice versa. The stimulation of B-plus cells leads to the secretion of B-plus antibodies which can suppress minus cells, and vice versa.

From these assumptions, Hoffmann deduces three quasi-stable states: (1) a virgin state, with respect to a given antigen, in which the concentrations of plus and minus cells are below the threshold of mutual stimulation, (2) a suppressed state (tolerance to a given antigen) in which both plus and minus cells have proliferated, but have become mutually blocked by the respective T cell factors, and (3) an immune state in which the plus cells have proliferated rapidly enough for the B cell antibodies to have suppressed the minus cells. Hoffmann's model also involves accessory cells, and his attempt to integrate in the simplest manner the enormous diversity of components and phenomena revealed by current experimentation, is admirable. In the process, however, he seems to deemphasize the network features of the immune system in favor of a description of interactions within antigen-defined and quasi-autonomous dual sets of lymphocytes, uncoupled to the remainder of the immune system. Many would agree that this is a necessary task on the road towards full systemic integration.

The Adam and Weiler Model

Dr Weiler will be speaking after me about this model, so I shall only
briefly mention that the model attempts to describe the early ontogeny
of the immune system on the basis of idiotypic interactions between
emerging cell clones. It is clear that a series of selective forces
impinge on the immune system throughout ontogeny. I could mention,
first, the self-antigens, particularly the histocompatibility antigens;
second, maternal antibodies that have passed the placenta; third, the
idiotypic selective interactions within the immune system; and fourth,
the selective response to invading foreign antigens. The Adam-Weiler
model considers only idiotypic selection. The idea is that lymphocyte
proliferation, from its early starting conditions in the embryo, is
accompanied by the emergence of lymphocyte mutants with respect to
the genes encoding the variable domains of the antibodies that are
synthesized. These mutants can be thought of as drawn from a very large
pool of possible mutants. The survival of each mutant clone is assumed
to depend on its idiotypic interactions with other existing clones.
The process embodied in this model leads to a final dynamic steady
state of clone sizes and number of clones when the rate of appearance
of new mutant clones is balanced by the rate of clone disappearance.
The final repertoire available to one mouse will differ from that
which has arisen in another mouse, even within an inbred strain.

On the whole, the three models that I have briefly summarized try to
come to grips with different aspects of integration in immunology.
I think that these synthetic attempts should be encouraged to proceed
alongside with the analytical experimental work so as to ensure a balanced progress in our field of biology.

References

The reader is referred to the references listed in

Jerne N.K.: The Immune System: A Web of V-Domains, Harvey Lectures
    Ser. 70. New York: Academic Press 1976, pp. 93-110

Clonal Selection and Network Regulation

E. Weiler, G. Adam, W. Schuler, and I. J. Weiler

This article will be concerned with the process of diversification
during the ontogenic development of the immune system. We shall advance
the thesis that diversification is not only a problem of the phylogenetic and ontogenetic origin of multiple V-region genes, but also
a problem of the population dynamics of the set of unispecific lymphocyte clones which compose the repertoire of the immune system.

We shall pose the question, whether the type of idiotypic-antiidiotypic interactions as postulated by Jerne (1, 2, and this vol.) might
be instrumental in regulating the population dynamics of the immune
system during development and in driving it in the direction of diversity. This possibility is explored in a theoretical model (3, 4)
which will be reviewed briefly; then some experimental evidence from
several areas of research, related to the question discussed, will be
cited.

We shall limit the following discussion to the immune system of mice,
which has been analyzed in considerable detail. As early as the seventh
day of gestation, a small number of multipotent primordial immune cell
precursors can be observed as haemopoietic islands in the yolk sac meso-
derm. These will give rise, in the course of development, to somewhat
over 5×10^8 lymphocytes in the adult mouse. Diversity must evolve during this phase of growth of the lymphocyte population.

Proliferation of lymphocytes continues in the fetal liver. From around
day 11 onwards in intrauterine life the thymus becomes a major organ
of lymphocyte proliferation. By day 16 of gestation there are about
10^6 lymphocytes in the thymus, which enter into a phase of extremely
rapid proliferation (5, 6, 7) and begin to express surface markers typical of thymocytes (8). By four weeks of age, the thymocyte population
of the mouse has reached a size of about 4×10^8 cells.

B-cells, characterized by dense surface immunoglobulin, appear around
the 16th day of gestation in the liver and spleen. The total number
was found to be around 10^4 at that time. This population expands to
2×10^8 B cells in the adult mouse (9-11).

The B-cell population is a set of monospecific cell clones, each expressing (and potentially secreting) one V-domain (i.e. one L-chain,
one H-chain V sequence), and thus one antigen-binding site (the para-
tope, 1) and one idiotype. In the following discussion each clone so
defined will be called a species. The T-cell population also appears
to consist of monospecific clones; in fact, T cells have been shown
to share idiotypic surface markers with B cells (12, 13) and Wigzell,
this vol.). Ontogenetic diversification must therefore occur in both
cell compartments, and both may be interrelated during ontogeny in a
way as yet unknown. For the purpose of the minimal model that we shall
discuss, a formal distinction between B cells and T cells need not be
made at the present time.

The presentations and discussions on Thursday gave a lucid account of
the mechanisms that are most likely involved in producing the high

number of V-region genes needed for a diverse immune system: translocations of V genes to become contiguous with C-genes; somatic mutations
in the hypervariable regions; and permutations in the phenotypic pairing of germ-line H-chain and L-chain in V-region loci (see Milstein,
Tonegawa, and Eichmann in this vol.).

The ontogenetic evolution of diversity, however, poses questions not
only about the relative importance of each species within the system
as a whole. It may be useful to view the population of lymphocyte species in a way analogous to the population of taxonomic species in a
natural ecosystem. Textbooks on population biology (e.g. 14, 15) define the diversity of ecosystems by the Shannon-Weaver expression:

$$H_S = -\sum_{i=1}^{s} p_i \log_2 p_i \tag{1}$$

H_S is the measure of diversity of a system containing s different species. p_i is the relative abundance of the i th species expressed as
the number of individuals of species i versus the total number of individuals in the system. Inspection of the equation will show that diversity is maximal when each of the s species occupies an equal fraction 1/S of the whole population; it will be smaller when one or few
of the species hold the major share, while the others are represented
scantily.

Consider that we buy 20 bottles of wine, of five different kinds. If
the purchase includes four bottles of each kind, the random sampler
would consider the collection fairly diverse. If 16 bottles of one kind
are bought, but only one bottle of each of the remaining four, the set
will appear rather monotonous. In fact, H_S will be 2.33 in the former
case, and 1.34 in the latter. Analogous comparisons can be made between biological diversity of a rain forest and a wheat field, or between the high diversity (in terms of amino acid representation) of
hemoglobin and the low diversity of collagen (16).

The Shannon-Weaver expression is a product of information theory. H_S
is a measure of the uncertainty of predicting the species of the next
individual encountered. Uncertainty also stands for the degree of
choice. Uncertainty and choice seem to be intrinsic and necessary at-
tributes of the immune system; the uncertainty of encounter with antigens from without has to be matched by a commensurate choice between
clones present within.

Returning to the ontogeny of the immune system, we may ask what mechanism provides not only for the generation of numerous species of V
domains, but also of an equitable abundance of each. Two mechanisms
might be proposed:

1. Cell clones expressing the different V-domains contained in this
genetic repertoire of the animal strain appear in sequential order.
A theory of this kind has been proposed by Dreyer et al. (17): V-genes,
contained in the chromatin in linear order, are inserted one after another next to the C gene during DNA replication. As cells divide in
an asymmetric fashion, new V-domain phenotypes are thus created. This
proposal requires a fairly sophisticated chromosomal organelle. Difficulty arises when one demands that the permutative phenotypic pairing of H-chain and L-chain variable sequences also follow a programmed,
predetermined order which would be a prerequisite to the full realiza-
tion of the repertoire. The ordered ontogenetic evolution of diversity

would require that all V genes be contained in the genome (germ-line hypothesis). It would achieve a balanced relative abundance of species, and thus maximal diversity in the sense of the Shannon-Weaver definition.

2. Clones expressing different V-domains may appear at random during ontogeny. Point mutations, translocations, and permutative pairings of V_H and V_L gene transcripts may each or all contribute to random variability. If we lump together all these (or other) mechanisms as random events, we must ask how effective random variability might be in achieving species diversity as defined by the Shannon-Weaver expression.

In the population biology of phylogenetic evolution it has long been recognized that random variability per se ("mutation pressure") is extremely inefficient in achieving species diversity as defined by the Shannon-Weaver expression. Most clones will belong to the original non-variant (germ-line) species; few early variants will attain a relatively high abundance: but the great majority of variants will be represented each by very few members. The p_i values (Eq. 1) of species will be distributed with an extremely high variance (18).

The need for some selective mechanism-obvious in phylogenetic diversification — has also been felt with respect to the ontogenetic diversification of the immune system. Selection by antigens encountered early in life has been eloquently invoked by Cohn (19, 20), who also discusses in depth the conceptual difficulties inherent in an antigen selection model.

We feel, however, that selection of clones early in ontogeny by ubiquitous antigens, such as intestinal flora, foodstuffs, bedding etc. might tend to reduce, rather then increase, diversity. One could put it this way: the diversity of the set of ubiquitous antigens seems too low to select a set of clones of high enough diversity to be complementary to the whole antigenic universe, which includes rare pathogens, esoteric synthetic chemicals or specific proteins.

The argument so far has been on an intuitive level; but some experimental observations also appear to be at variance with the concept of antigenic selection as the driving force in diversification. (1) Germ-free mice, raised not antigen-free, but under a significantly lower antigenic load than their conventional counterparts, have as many precursor cells towards a given antigen (21, 23), or as heterogeneous an antibody population towards a given ligand, as do normal mice (24). The low exposure to antigens is reflected by a greatly reduced antibody concentration in the serum (25) but not by a decreased diversity of the immune system. (2) Animals do possess virgin immunocompetent B cells, presumably not previously stimulated by antigen, which can be distinguished qualitatively from stimulated B cells (memory cells) (26): the former have a high turnover, the latter are long-lived; the former have low surface Ig, the latter expose much Ig.

Prompted by such considerations, we wondered if a mechanism that regulated the diversity of lymphocytes could be operating within the immune system, rather than from without. Any postulated interactions within the system would have to have the same high degree of specificity as antibodies themselves. This requirement is fulfilled by the specificity of recognition between idiotype and anti-idiotype, exactly as Jerne has outlined in his Network Hypothesis (1, 2).

It appeared of interest to investigate the consequences of idiotypic-antiidiotypic interactions on the population dynamics of a growing and

diversifying immune system. We have formulated a theoretical model
which is based on the distinction between the <u>potential repertoire</u> P
of idiotypes of the animal strain, and the <u>available repertoire</u> M realized in the individual, as proposed by Jerne (27). The set P is considered to be much larger than the set M, and the species that constitute M appear out of P in <u>random</u> fashion during development. It is
further postulated that at a given time any two species in M(t) may
interact as idiotype and antiidiotype, with a finite probability; as
a first approximation, this interaction is thought to result in the
elimination of one or both interacting partners. We shall examine
later in how far these premises can be accepted as true. We shall not
here give a detailed description of the model, which is being published
(3,4), but only outline the main parameters and properties.

Since the element of mutual recognition is idiotype, this term may
stand for immunoglobulin species, or clone, in the following discussion.

We consider a growing immune system that starts out with a relatively
small number of lymphocytes N_O, and a correspondingly small number of
idiotypes. The lymphocyte number increases exponentially with a rate
constant r, which contains the rate constants of proliferation and of
random disappearance by indiscriminate death or emigration from the
system. As the system grows, somatic variants will arise with a probability proportional to the number of lymphocytes present. Concomitantly
there will be an increase in the probability that any two clones in
the system are idiotypically incompatible. For any given clone this
probability will depend on the average number υ, in the potential repertoire P, of antibodies antiidiotypic to the given one; it will also
depend on the number of species present in the immune system at a
given time: M(t). This probability will then be $k_e' \cdot \upsilon \cdot M(t)/P$ (k_e' =
rate constant of elimination).

Since all idiotypes present in the system are considered to be statistically equivalent with respect to their chance of being recognized
by anti-idiotypic clones, idiotypes generally will be eliminated with
a rate that is proportional to the square of the number of species $M_{(t)}$
present at that time; it will be $K'_e \cdot \upsilon \cdot M^2/P$.

Introducing the rate constant of somatic variation k_m, and n=N/M for
the average number of lymphocytes per clone, kinetic equations can be
written, which describe the time courses of N and M during ontogeny:

$$\frac{dM}{dt} = k_m N - k_e' \cdot \upsilon \cdot M^2/P \tag{2}$$

$$\frac{dN}{dt} = rN - k_e' \cdot \upsilon \cdot M^2 n/P \tag{3}$$

Three independent parameters have to be specified. The rate constant
of increase in cell number was taken to be r = 0.5 per day from data
on thymus cell proliferation (7). The steady-state number of lymphocytes in the adult was set as N = 2 x 10^8 (for a direct comparison
with published measurements on one thymic lobe), and the initial number N_O was set as 100 cells, each expressing a different idiotype (M_O =
100).

The solution of the two coupled differential equations led to the prediction of growth curves (a) of the total number N of lymphocytes in
the system; (b) of the number of different species (idiotypes); (c)
of the average number of lymphocytes per clone (which is simply N/M),
as shown in Figure 1.

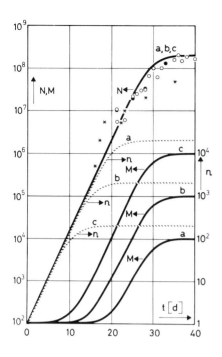

Fig. 1. Plot of the number of lymphocytes N, and of the number of idiotypes (clones) M during the ontogeny of the immune system as predicted by the theoretical model discussed in the text. (From Ref. 3). N and M are given on the left ordinate, and the average number of lymphocytes per clone (= N/M) on the right. For the theoretical curves the following parameters were chosen: $r = 0.5\ d^{-1}$; N (adult number) $= 2\cdot10^8$; and M (adult number) $= 10^4$, or 10^5, or 10^6, for the curves labeled a, b, or c, respectively. Counts of lymphocyte numbers in one thymic lobe from Ref. 5 (*) and Ref. 6 (o and ●) were entered for comparison, with t = 20 days: time of birth, as the reference point on the abscissa

The model has the following main features:

1. All idiotypes (clones) are considered statistically equal, independent of their time of appearance during ontogeny. Clone sizes are distributed around a mean value, all clones having an equal chance of gaining abundance in the population.

The predicted increase in the number M of antibody species correlates reasonably well with reports on the maturation of immunocompetence towards specific antigens during early postnatal life (28, 31) and especially with data on the limited heterogeneity of antibody populations elicited in newborn animals (24, 31).

2. The model predicts that the lymphocyte population is self-limiting, and attains a steady-state size at an adult level. The growth curves may be compared directly to observational data. Lymphocyte counts in one thymic lobe (5, 6) have been entered in Figure 1; the predicted growth curve appears to be in good agreement with these observations.

B cells, as defined by surface immunoglobulin, are fewer in number than thymic lymphocytes during ontogeny; but their number rises in proportion to T cells (9, 11). In a more recent version, the model was applied specifically to the B cell compartment; and it was found that it can be compared in a satisfactory manner to experimental data on the growth and turnover of the B cell population (4).

3. The model predicts that the immune system is in a steady state of flux throughout life: lymphocytes proliferate, new somatic variants arise out of the potential repertoire, and clones disappear by anti-idiotypic elimination at all times. This appears to be well borne out by experiments on the incorporation of tritiated thymidine: both thymic lymphocytes (7, 32) and B cells (26, 33) are in rapid turnover in developing animals as well as in adults.

272

4. Somatic variants have to occur with a frequency correlated to the size M of the available repertoire of V-domains (idiotypes) in the adult; when this is assumed to be 10^6 per mouse, the "mutation rate" would be of the order of 10^{-3} per cell and generation. This is larger by about an order of magnitude than has been observed for genes in other eukaryotic systems (14), or with a myeloma cell line in vitro (34). One should take into account, however, that there are three groups of genes that contribute to idiotypic specificity: V_H, V_κ, and V_λ, and that random variability depends not only one point mutations within genes, but also on the variable phenotypic pairing of different germ line loci (including their somatic variant progeny) coding for light chain and heavy chain variable region subgroups (see Milstein and Tonegawa in this vol.). Then the required "mutation" frequency may be in an acceptable range, without the need of special "mutator" mechanism.

The anti-idiotypic elimination process, as proposed in this model, could be suspected of causing a significant "waste" of somatic variants. It should be remembered, however, that the rate of elimination of clones depends on the square of the available repertoire M present at a given time (Eq. 2). The sacrifice of clones reaches a high level only as the system approaches the steady-state diversity of the adult. Indeed, Adam has recently constructed a basically similar model, in which anti-idiotypic interactions do not cause elimination of clones, but rather an arrest in proliferation; they are functionally conserved, not lost (4). Yet, required mutation rates are similar in both models, when the proliferation rate and the adult repertoire M are set for the same values.

The value of a theoretical model lies in the critical questions it provokes. We shall briefly consider these three: randomness, repertoire of antiidiotypic species within the system; biological consequence of idiotype recognition.

1. Is the available repertoire a random sample of the potential repertoire? Antibody populations reactive with a given ligand constitute a heterogeneous set within each animal; different animals express different sets, even when they have virtually the same genetic constitution, such as inbred mice. This has been shown to be the rule in numerous experiments, when functional characteristics (affinity, reaction kinetics), physical properties (isoelectric spectrum), or idiotypes were analyzed. This individuality of the available repertoire is displayed even by allotypic subsets in heterozygous animals (35), showing that it is an inherent property of the B cell population, not a result of inadvertent variations in the immunization process. The method of isoelectric focusing has permitted a measurement of the repeat frequency of clones, and thereby an estimate of relative sizes of the available and the potential repertoires of the mouse strain. The potential repertoire proved to be two orders of magnitude larger than the available reservoirs, when antibodies to 4-hydroxy-5-iodo-3-nitrophenylacetyl (NIP) (36), or antibodies cross-reacting with the dinitrophenyl-and trinitrophenyl groups (37) were analyzed.

The phenomenon termed "isogeneic barrier" (38) should be mentioned at this point: attempts to elicit a secondary response in a mouse that had received cells from an immunized donor mouse usually fail, even when donor and recipient belong to the same inbred strain. Using congeneic strains differing in H-chain allotype, and priming both donor and recipient, one can easily show that the antibodies produced in the recipient are of its own, not of the donor's type (unpublished experiments). This apparent ability of the immune system to distinguish between self and nonself, even in an isogeneic situation, is routinely overcome by irradiating the recipients; recipients that lack a thymus (strain Balb/nu) have also proved to be permissive towards congeneic memory cells (39). It is tempting to speculate that normal recipients recognize and eliminate (or suppress) the idiotypes of the donor. Attempts to show this directly have, however, not been successful so far.

While the individuality of available repertoires appears to be the rule, there are notable exceptions: antibody species (idiotypes) may in some cases recur in every animal of a given strain. These are the genetic markers of V-domains reviewed by Eichmann in this volume.

As evidence against randomness one may cite experiments showing that reactivity against members of an arbitrary set of antigens appear in sequential order during ontogeny (40): the fetus can produce first antibodies against antigen A, then against antigen B, and so on; and every fetus shows the same sequence. This, however, is not incompatible with random somatic diversification. It could, for example, mean that by chance antibodies against A are removed from the germ-line repertoire by fewer mutational steps than antibodies against B, and are thus more likely to appear at an earlier time. This point has been discussed thoroughly on an earlier occasion (see Refs. 27, 40). Further experiments on the variance, with time, of the ontogenetic appearance of specified idiotypes would be of value in this context.

2. What is the size of the antiidiotypic repertoire? This question is crucial for the model just reviewed, for the rate of elimination (or suppression) of clones depends on the average number of antibodies (out of the potential repertoire P) that recognize a given idiotype (Eq. 2). This question is amenable to experimental test; mice can be immunized against idiotypes of their own inbred strain (41); rabbits, in fact, against their own antibodies (42). In principle, then, it should be possible to enumerate the number of molecular specics produced by a given animal against a given idiotype, and to compare different animals with respect to the individuality of anti-idiotypic repertoires, in analogy to the analysis with DNP/TNP or NiP haptenic groups cited above (36, 37). W. Schuler in our laboratory is conducting an analysis of this kind, using the technique of isoelectric focusing to characterize antibodies produced by Balb/c mice against the idiotpye of Balb/c myeloma J558 (with antibody activity to the α-(1,3) dextran linkage); the results show that antiidiotypic antibodies appear as an oligoclonal set in each mouse, and that different animals of the Balb/c strain exhibit different sets (in preparation).

3. Which are the biological consequences of antiidiotypic recognition? This subject has been reviewed recently by Nisonoff and Bangasser (43). Idiotypes of inbred strains of mice can be suppressed readily by the passive administration of antiidiotypic antibodies produced in guinea pigs (44) or rabbits (45). In the case of the A5A idiotype discovered by Eichmann, guinea pig antiidiotypic molecules may have either suppressive or enhancing activity depending on the immunoglobulin class to which they belong (44). The interpretation of all these experiments, however, is made difficult by the fact that the antiidiotypic antibodies were foreign antigens for mice. That guinea pig or rabbit determinants do play a role in the effects observed is suggested by the failure of F(ab)$_2$ molecules to cause suppression (45), or by the lack of effect in vitro on cell populations derived from mice tolerant to rabbit immunoglobulins in experiments where antiidiotypic antibodies from rabbits induced an idiotypic response by mouse cells (46).

However, Eichmann could obtain, from mice suppressed by heterologous sera, T-cell populations which were able to transfer specific idiotypic

suppression serially to other mice, to an extent which made it virtually impossible that the originally suppressive antiidiotypic molecules
from guinea pigs could still play a role (47).

There exist observations which, at first sight, seem to be completely
incompatible with the model discussed in this article, as has been
noted before by Jerne (2). Idiotypic and antiidiotypic molecules have
been shown to coexist peacefully in the serum of the animal that produced both (reviewed in Ref. 48). We must note, however, that the idiotype concerned is reactive with phosphorylcholine, is represented by
several myelomas (e.g. TEPC 15), and is present in the serum of every
Balb/c mouse raised conventionally (25). It was stimulated, presumably,
by early contact with bacteria from the gastrointestinal tract; germ-
free mice do not express it in their serum (25). One is dealing, then,
not with naive (or virgin) cells in this case, but with memory cells
of long standing. These appear to be refractory towards antiidiotypic
suppression. Other experiments suggest that memory B cells and virgin
cells differ profoundly in their susceptibility to idiotypic suppression. When antiidiotypic serum, raised in A/J strain mice, was adminis-
trated passively to adult Balb/c mice, true suppression of the idiotype
could not be achieved; in newborn mice, suppression of long duration
proved possible (43, 48). Other qualitative differences between virgin
B-cells and long-lived memory B-cells have already been mentioned above
(cf. 26). It would be an interesting experiment to try idiotypic suppression of the anti-phosphorylcholine response (or the anti-dextran
response, see below) in adults raised germ-free.

This point is made also in recent experiments by I.J. Weiler and R.
Sprenger (in preparation). A family of cross-reacting idiotypes in
Balb/c (myelomas J558 and MOPc 104E) are connected with antibody activity against the α-(1,3)-glucosidic linkage of dextran; they are
present in every Balb/c mouse, again presumably because of early environmental induction. Antiidiotypic sera (from strain A/J mice) were
quite ineffective in suppression, when given to adults; they produced
moderate suppression, lasting several weeks, when injected into newborn animals. When, however, idiotype-negative (SJL/J) mothers were
immunized against the idiotype, and mated to idiotype-positive fathers
(Balb/c), the progeny proved to be exquisitely idiotype-suppressed for
the duration of the experiment, four months so far. This result is
analogous to experiments producing long-lasting allotype suppression
(reviewed in 49).

This last experiment may suggest a note of caution when we try to regard the developing immune system as closed and self-contained. Maternal antibodies are present throughout early ontogenetic evolution,
and might influence the development of the individual repertoire of
the fetus and newborn animal in a profound way (50). This is an area
in which many meaningful experiments may yet be done.

Acknowledgments. Work done in this laboratory was supported by the Deutsche Forschungsgemeinschaft, Sonderforschungsbereich 138.

References

1. Jerne, N.K.: Ann. Immunol. (Inst. Pasteur) 125C, 373 (1972a)
2. Jerne, N.K.: In: The Harvey Lectures, Ser. 70. New York: Academic
   Press 1975

3. Adam, G., Weiler, E.: In: The Generation of Antibody Diversity. Cunningham, A.J. (ed.). London: Academic Press 1976 (in press)
4. Adam, G.: In: Theoretical Immunology. Bell, G.I., Perelson, A.S., Pimbley, G.H. (eds.). New York: Marcel Decker 1976 (in press)
5. Ball, W.D.: Exp. Cell. Res. 31, 82-88 (1963)
6. Axelrad, A.A., van der Gaag, H.C.: J. Nat. Cancer Inst. 28, 1065-1093 (1962)
7. Metcalf, D.: CHS Symp. Cold Spring Harbour Symp. Quant. Biol. (1967) Vol. XXXII
8. Goldschneider, L.: In: Cell Surface in Development. Moscona, A.A. (ed.). New York: John Wiley and Sons 1974, pp. 165-185
9. Owen, J.J.T., Raff, M.C., Cooper, M.D.: Europ. J. Immunol. 5, 468-473 (1975)
10. Sidman, C.L., Unanue, E.R.: J. Immunol. 114, 1730 (1975)
11. Weissmann, I.L.: Transpl. Rev. 24, 159-176 (1975)
12. Binz, H., Kumura, A., Wigzell, H.: Scand. J. Immunol. 4, 413 (1975)
13. Eichmann, K., Rajewsky, K.: Europ. J. Immunol. 5, 661-667 (1975)
14. Wilson, E.O., Bossert, W.H.: Einführung in di Populationsbiologie. Berlin-Heidelberg-New York: Springer 1973, pp. 132-134
15. Pianka, E.R.: Evolutionary Ecology. New York: Harper and Row 1974
16. Yockey, H.P.: In: Locker A. (ed.): Biogenetics, Evolution, Homeostasis. Berline-Heidelberg-New York: Springer 1973
17. Dreyer, W.J., Gray, W.R., Hood, L.: Cold Spring Harbor Symp. Quant. Biol. 353-367 (1967) Vol. XXXII
18. Luria, S.E., Delbrück, M.: Genetics 28, 491 (1943)
19. Cohn, M.: Cellular Immunol. 1, 461-467 (1970)
20. Cohn, M., Blomberg, B., Geckeler, W., Raschke, W., Riblet, R., Weigert, M.: In: The Immune System Genes, Receptors, Signals. Sercarz, E.E., Williamson, A.R., Fox, C.F. (eds.). New York, London: Academic Press 1974, p. 89
21. Bosma, M.J., Makinodan, T., Walburg, Jr., H.E.: J. Immunol. 99, 420 (1967)
22. Press, J.L., Klinman, N.R.: Europ. J. Immunol. 4, 155
23. Sigal, N.H., Gearhart, P.J., Klinman, N.R.: J. Immunol. 114, 1354 (1975)
24. Goidl, E.A., Siskind, G.W.: J. Exp. Med. 140, 1285 (1974)
25. Liebermann, R., Potter, M., Mushinski, E., Humphrey, W., Rudikoff, S.: J. Exp. Med. 139, 983-1001 (1974)
26. Strober, S.: Transpl. Rev. 24, 84-112 (1975)
27. Jerne, N.K.: In: Ontogeny of Acquired Immunity. Porter, R., Knight, J. (eds.). Ciba Found. Symp. Amsterdam: Elsevier 1972b, p. 1-15
28. Arrenbrecht, S.: Europ. J. Immunol. 3, 506 (1973)
29. Spear, P.G., Edelman, G.M.: J. Exp. Med. 139, 249-263 (1974)
30. Rosenberg, Y.J., Cunningham, A.J.: Europ. J. Immunol. 5, 444-447 (1975)
31. Klinman, N.R., Press, J.L.: Transplant. Rev. 24, 41-83 (1975)
32. Metcalf, D., Wiadrowski, M.: Cancer Res. 26, 483-491 (1971)
33. Osmond, D.G., Nossal, G.J.V.: Cell Immunol. 13, 132-145 (1974)
34. Cotton, R.G.H., Secher, D.S., Milstein, C.: Europ. J. Immunol. 3, 135 (1973)
35. Weiler, I.J., Bosma, G., Dittrich, B., Weiler, E.: Europ. J. Immunol. 5, 399-406 (1975)
36. Kreth, H.W., Williamson, A.R.: Europ. J. Immunol. 3, 141 (1973)
37. Pink, J.R.L., Askonas, B.A.: Europ. J. Immunol. 4, 426 (1974)
38. Celada, F.: J. Exp. Med. 124, 1 (1966)
39. Kobow, U., Weiler, E.: Europ. J. Immunol. 5, 628-632 (1975)
40. Silverstein, A.M.: In: Ontogeny of Acquired Immunity. Porter, R., Knight, J. (eds.). Amsterdam-London-New York: Elsevier/North Holland 1972, pp. 16-25
41. Sakato, N., Eisen, H.N.: J. Exp. Med. 141, 1411 (1975)
42. Rodkey, L.s.: J. Exp. Med. 139, 712-720 (1974)

43. Nisonoff, A., Bangasser, S.A.: Transpl. Rev. $\underline{27}$, 24-56 (1975)
44. Eichmann, K.: Europ. J. Immunol. $\underline{4}$, 296 (1974)
45. Pawlak, L.L., Hart, D.A., Nisonoff, A.: J. Exp. Med. $\underline{137}$, 1442 (1973)
46. Trenkner, E., Riblet, R.: J. Exp. Med. $\underline{142}$, 1121-1131 (1975)
47. Eichmann, K.: Europ. J. Immunol. $\underline{5}$ (1975)
48. Köhler, H.: Transpl. Rev. $\underline{27}$, 24-56 (1975)
49. Herzenberg, L.A., Okumura, K., Metzler, C.M.: Transpl. Rev. $\underline{27}$, 57
50. Kindred, B., Roelants, G.E.: J. Immunol. $\underline{113}$, 445 (1974)

Considerations on Immunity and Disease

B. Pernis

If there were no other reason to justify our interest in the immune
system, and there are many, it should be sufficient to state that the
immune system is essential for our survival. This is clearly shown by
what happens when, in man, the immune system does not function in whole
or in part, as happens in a rather heterogeneous group of diseases
called immunodeficiencies.

These conditions can be subdivided into three main groups according
to the cellular basis of the immune deficiency. The most severe form
is believed to be due to lack of functional development of the stem
cells of the immune system. This occurs when both main branches of
immunocytes, the bone-marrow-derived (B) and the thymus-derived (T),
do not develop to an appreciable extent. One such condition is due to
a defect in one our autosomal gene (therefore autosomal recessive) and
has been termed the Swiss-type immune deficiency. Children affected by
this disease fail to thrive and are subject, soon after birth, to a
variety of infections, bacterial, fungal and viral, so that they usually do not survive more than a few months. A similar, although slightly
less severe, deficiency of both thymus-dependent and bone-marrow-de-
pendent immunity is a sex-linked recessive that has been defined as
"thymic alymphoplasia". Here too infections of different kinds drastic-
ally limit the survival time.

Bacterial, viral and fungal infections are also prevalent in individuals affected by immune deficiencies primarily of the thymus-dependent
system, and eventually lead to death. This is the case in the Di George
syndrome (aplasia of the thymus and parathyroids) and in the Nezelof
syndrome (defective thymus). In these diseases, whose genetic basic is
unknown, the T-dependent immune responses are grossly deficient; the
levels of serum immunoglobulins may be normal, but of course the production of specific antibodies after an antigenic challenge is relatively inefficient due to lack of cooperation between T and B cells.

The longest survival time amongst the patients with immune deficiencies
(up to adult age with antibiotic and immunoglobulin treatment) is in
those patients who have a selective defect of B cells, of which the
prototype is the sex-linked recessive agammaglobulinemia defined as
Bruton's. This group is certainly very heterogeneous and includes cases
in which B lymphocytes in the blood are absent or very rare, and others
in which they are normal in numbers but fail to mature to plasma cells
(see, for instance, Siegal et al., 1971). Selective deficiencies of B
immunocytes may also appear among adults and one may wonder about the
cause of these diseases, usually classified in the "variable" group of
immune deficiencies. It is interesting that recent evidence (Waldmann
et al., 1974) supports the possibility that at least some cases of B
deficiencies may be due to an abnormal function of suppressor T cells.
And it is interesting here to point out that the immune system may fail
to function properly, not because of an inborn defect of some of its
elements, but because of an alteration in the normal balance and integration between the different components. That is, it fails to function as a system.

Whichever the etiology and pathogenesis of a given B cell immunodeficiency may be, it is important to point out that the pathology is again due to infections, mainly of bacterial origin. In fact, the resistance to virus infections is normal in these patients, so they recover normally from measles, mumps and varicella, for instance. Also smallpox vaccination is followed by a normal primary take and not by generalized vaccinia. Here the experiments of nature, that is the immune deficiencies, teach us a very interesting lesson in the protective function of the thymus-dependent immune system. They tell us that this system is probably pre-eminent in the defense against infections of viral origin. This phenomenon may be connected with the fact that, in order to stop the progress of viral infection that would spread within the organism by direct intercellular contact, the immune system needs to actually kill the virus-infected cells. It is evident that the specific elimination of virus-infected cells is most efficiently performed by T lymphocytes. These cells perform this function with very interesting restrictions established by histocompatibility antigens (see Doherty et al., 1976) that closely tie this latter system to the defense against viral infections, and perhaps more generally with the immune recognition of non-self. Of course, humoral antibodies (the product of B lymphocytes) are also not without value in the protection against some viral infections, particularly when free virus particles travel from one infected cell to a relatively distant uninfected one, as may happen, for instance, in viral infections of the mucosal surfaces where soluble antibodies (of the IgA class) may be the most important element of the immune defense.

It is often stated in immunology lectures for students that immunoproliferative diseases represent the other side of the coin with respect to immune deficiencies inasmuch as they are accompanied by a hyperproduction of immunoglobulins or lymphoid cells. Of course this is not so, inasmuch as the immunoproliferative disorders (lymphosarcoma, lymphoid leukemia, Waldenström's macroglobulinemia, plasmacytoma and other related ones) are due to the uncontrolled proliferation and biosynthetic activity of only one clone of immunocytes; in most cases, in man, this is a clone of B immunocytes. Now it is clear that the essence of the function of the immune system is the possibility of responding with the stimulation of an appropriate set of clones to a given antigenic challenge. The uncontrolled activity of only one clone can never perform this function, and is in fact expected to produce the opposite effect, namely a defect of the immune system, either by feedback inhibition or simply because the proliferating clone crowds out the variety of normal ones in the central and peripheral lymphoid tissues. Therefore it is not surprising that the clinical picture, with regard to resistance to infections, also shows a profound deficiency in many immunoproliferative diseases.

The general lesson that we learn, both from the immune deficiencies and from the immunoproliferative disorders, is that the immune system is essential for protection against all sorts of infections. Of course, this has been known for a long time and has been the basis for the widespread use of prophylactic immunization against infections diseases. This practice has been extremely successful and is probably responsible for at least one part of the present extraordinary increase of the Earth's human population. It would be easy to claim this as a more than sufficient contribution of the science of immunology to human health, but perhaps this is not quite fair since empirical procedures often preceded the fundamental knowledge; everyone knows that Jenner and Pasteur established procedures of active immunization before the existence of antibodies was discovered.

It is obvious that without a functioning immune system, man cannot
lead an extra-uterine life for an appreciable period of time without
succumbing to infections, and there is good evidence that the same is
true for the other species, such as the usual laboratory animals,
that have an immune system. But how is the problem of defense against
infections solved by those species (i.e., the invertebrates) that do
not possess an immune system? They must, of course, possess some form
of recognition of not-self from self and it is customary to attribute
this function, in the above-mentioned species, to phagocytic cells.
This may well be true, but then we are left with the problem of what
was the evolutionary advantage of the complex immune system, since
apparently the problem of resistance to infectious agents could be
solved in other, simpler, ways. Of course this is the sort of question
to which no definite answer can be given, but perhaps some speculations
are allowed. I am inclined to consider the possibility that on the
whole the membranes of the different cells that make up the body of
the vertebrates carry a series of receptors (for complementary molecules involved in intercellular signals of all sorts) whose variety
is at least one order of magnitude larger than that represented on the
cell membranes in more primitive species. It is conceivable (see Ohno,
1976) that there has been a parallel evolution of infectious agents
(chiefly viruses) which can use this variety of cell receptors to infect the cells themselves. Therefore the number of different potential
infectious agents for the vertebrates is much higher than the number
of agents that can infect the cells of simpler organisms. It is thus
conceivable that up to a given level in phylogeny the problem of the
defense against infections might have been solved by the evolution of
appropriate biochemical reactions or by the synthesis of a relatively
limited set of pathogen-specific molecules (for instance enzymes),
but that when the number of different potential pathogens became too
great, then a general solution for their recognition and discrimination
was necessary. This solution might well have been provided by the emergence of the immune system, with its extraordinary variety of specific
molecules (the antibodies) that have a range of diversity, generated
by somatic mutation, larger than any that might be generated by the
evolutionary process amongst potential pathogens.

Several investigators have proposed that the function of the immune
system is not limited to the protection against external pathogens
(infectious agents), but that it also includes the detection and elimination of abnormal cells arising within the organism. The area of
immunology related to this problem (tumor immunology), as well as the
possibly related area of histocompatibility reactions, have undergone
an enormous expansion under the influence of both conceptual and practical motivations. It seems to me that at the present time it is difficult to predict what will be the ultimate contribution of immunology
to the solution of the practical problems in this area of medicine,
namely the prevention and control of tumors and organ transplantation.
It is fair to say that, at the present time, the practical results, for
instance in tumor immunology, are not at all proportionate to the amount
of effort put into this field. But we may hope that this ratio may
change in the near future; eventually further gains will depend on the
actual importance of the role of the immune system in detecting as antigens possible tumor-specific molecules that might be present on the
membranes of the neoplastic cells. A note of caution in accepting this
function of the immune system as certain is implicit in the recent observation that the incidence of spontaneous tumors is not significantly
greater in thymus-less nude mice than in the corresponding normal heterozygotes; this in spite of the fact that nude mice have a severely
impaired immune system (due to a deficiency of T cells) so that they
can accept skin grafts not only from other strains of mice but also
from other species.

In view of the complexity of the immune system, with its different branches, and of the operational problems involved in the distinction of self from non-self, it is not surprising that the function of the system may sometimes result in effects that are detrimental to the host. This is the field of immunopathology that is largely superimposable with that of the so-called "hypersensitivity" states. These hypersensitivity states have been classified by Gell and Coombs (1968) into four main types: type I includes the allergies of the immediate kind due to the release of active substances, upon contact with the antigen, by mast-cells that are passively sensitized by antibodies of the IgE class. It has been established (see Carson and Metzger, 1974) that the triggering of mast cells for degranulation and release of active material is the consequence of cross-linking by antigen of a certain number (probably not very many) of the passive IgE antibodies, and consequently of the cell receptors that anchor them onto the cell membrane. It is noteworthy that, since the passive IgE molecules bound by a single mast cell must be heterogeneous with regard to the specificity of the combining site, cross-linking by antigen (allergen) of a sufficient number of IgE molecules will only take place if a given mast cell has a sufficient aliquot of its passive IgE capable of reacting with the set of antigenic determinants that happen to be on the allergen molecule. From this simple consideration it is predictable that allergic symptoms, corresponding to type I hypersensitivity, will only become manifest, after contact with antigen, in those individuals in whom an appreciable proportion of IgE immunoglobulins specifically reacts with that given antigen.

I shall not comment on type II hypersensitivity, which includes pathology due to cytotoxic, complement-fixing antibodies that can react with the cells of the organism itself either directly or through the intermediate of a hapten. I also have nothing to contribute to the definition or mechanism of type III hypersensitivity, that is hypersensitivity mainly due to the pathogenic effect of immune complexes dealing with a large area of immunopathology ranging from serum sickness to lupus erythematodes and nephritis.

I should rather like to discuss some implications of recent progress in basic immunology for the understanding of type IV hypersensitivity. This type of allergy is due to thymus-dependent lymphocytes and is of the delayed type; a typical condition of type IV hypersensitivity is contact dermatitis that may be originated by repeated or prolonged contact of the skin with a variety of haptens. This condition is a frequent cause of industrial diseases, amongst which dermatitis due to contact with chromium that develops among mason workers is very common. For a long time dermatologists have investigated, without coming to a definite conclusion, the nature of the protein or proteins of the skin that might serve as carriers for the haptens involved in contact dermatitis. The recent work of Shearer et al. (1976) may offer a clue for the solution of that problem; these investigators have seen that it is possible to elicit in vitro the appearance of cytotoxic T lymphocytes directed against autologous cells to which haptens had been coupled. The specificity of the reaction was dependent on histocompatibility antigens and it appeared likely that the cytotoxic T cells were capable of recognizing a complex formed by the hapten and the self histocompatibility antigens. It seems likely to me that the same may happen in human pathology in the case of contact dermatitis, and more precisely, that in this condition the relevant carrier molecules for the sensitizing hapten will be found to be the histocompatibility antigens present on the membranes of the epidermal cells. This hypothesis is easy to test by looking for the possible presence, in the blood of the patients suffering from contact dermatitis, of cytotoxic T cells directed against <u>autologous</u> hapten-coupled lymphocytes. At a difference from the experiments performed in the mouse by Shearer et al., these lymphocytes should not require a period of sensitization in vitro, having already been sensitized in vivo.

In conclusion I should like to express my confidence that immunology will continue to contribute to medicine, both at the level of basic knowledge and of practical applications, and that perhaps a next burst of practical dividends, comparable to those provided by prophylactic immunization, will emerge when basic immunology has overcome some of its present stumbling blocks such as the molecular basis of T cell specificity and the mechanisms whereby genes belonging to the major histocompatibility complex control immune reactivity to specific antigens. Perhaps the advancing knowledge on the connections between the human histocompatibility antigens and different diseases (that is dealt with elsewhere in this volume) is the first glimpse of this new field of medical immunology.

References

Carson, D., Metzger, H.: J. Immunol. <u>113</u>, 1271-1277 (1974)
Doherty, P.C., Blanden, R.V., Zinkernagel, R.M.: Transplant. Rev. <u>29</u>, 89-125 (1976)
Gell, P.G.H., Coombs, R.A.A.: Clinical Aspects of Immunology, 2nd ed. Oxford: Blackwell 1968
Ohno, S.: Transplant. Rev. <u>33</u>, in press (1976)
Shearer, G.M., Rehn, T.G., Schmitt-Verhulst, A.M.: Transplant. Rev. <u>29</u>, 222-229 (1976)
Siegal, F.P., Pernis, B., Kunkel, H.G.: Europ. J. Immunol. <u>1</u>, 482-486 (1971)
Waldmann, T.A., Broder, S., Blease, R.M., Durm, M., Blackman, M., Strober, W.: Lancet <u>I</u>, 609-613 (1974)

Two Immunological Worlds: Antibody Molecules and Lymphoid Cell-Surfaces.
An Overview of the Mosbach Colloquium

H. N. Eisen

My assignment in this meeting is to summarize the proceedings of the past two days. Most of the talks were themselves summaries. To compress them further might well result in gibberish. Nevertheless, I intend to attempt an overview, focusing mainly on those issues that seem to present conflicts or unusual challenges especially for biochemists.

Let me begin with a general comment. As is obvious to most of us, this Colloquium has been held at an unusual time in the history of Immunology. What is unusual is not so much that new publications are appearing at an unprecedented rate, but rather that a major shift in direction is under way. The resulting challenges and opportunities are so remarkable that it seems not too much of an exaggeration to suggest that what we are witnessing is the emergency of a second world of immunology. The first world centers on antibody molecules; the second on cell interactions via their surface glycoproteins. I hope to refer repeatedly to the contrasts between these two worlds throughout this talk.

Two principal lessons emerged from the 1st session. Firstly, the structural basis for antibody specificity is virtually solved in a geometrical sense. It is now clear that seven short, discontinuous stretches of hypervariable amino acid sequences in variable domains of light (L) and heavy (H) immunoglobulin (Ig) chains make up the boundaries of antigen-combining sites and provide contact amino acids for bound ligand.

The chemical basis for specificity is, however, still obscure. What are the general rules that lead particular combinations of amino acids in hypervariable regions to specify the binding of particular ligands? The empirical approach to an answer — correlating particular residues in certain positions with the binding of particular ligands — is likely to require the accumulation of an enormous amount of information before the rules can be deciphered. I am not optimistic that enough people will feel impelled to carry out enough Ig sequences, even with Kabat's powers of persuasion, to provide the required amount of raw data. Whether theoretical approaches are more promising, I do not know, but I am skeptical that they will provide much insight in the short term. The solution is thus likely to be extraordinarily difficult. And one cannot overlook the added problem of degeneracy: i.e., a single hapten on a carrier protein (e.g. Dnp-ovalbumin) elicits a variety of anti-hapten Abs. The diversity is evident from heterogeneity of the Abs in respect to (1) affinity for the hapten, (2) amino acid composition, and (3) partial amino acid sequences.

Some have insisted that degeneracy is due to heterogeneity of the immunogen: each hapten group, such as Dnp, located on a different amino acid residue of a carrier constitutes a different immunogenic element and elicits a different Ab. But this explanation cannot be the sole answer, because degeneracy also exists for the anti-Dnp Abs made in individual rabbits against mono-Dnp-RNase, with 1 Dnp group attached to lysine 41 of the RNase. Moreover, Haber's group has recently found that individual rabbits also produce several Abs to the simple repetitive oligosaccharides of a pneumococcal polysaccharide homopolymer. The amino acid sequences of the L chains of these Abs differ considerably in their hypervariable amino acid sequences.

Degeneracy must mean that there are many clones that can respond to a single, homogeneous antigenic determinant, yielding antibodies with diverse patterns of amino acid sequences in their hypervariable regions. It is likely, therefore, to be a long time before our understanding of the chemical basis for specificity reaches a point where it will be possible to specify the ligands that can be bound by an Ab molecule directly from amino acid sequences in the molecule's V domains. All biology has a major stake in understanding the rules by which biological specificity is determined by primary polypeptide structure. Will the day ever come when we can write from first principles or empirical rules the sequences for a polypeptide chain that will bind a particular ligand? When that happens we will be designing new proteins rather than limiting ourselves to the isolation of what evolution has developed through hundreds of millions of years of trial and error.

The second take-home lesson of the first session concerns conformational changes of Ig C domains induced by ligand binding in V domains. It has long been suspected that such changes underlie a number of immunological effects; e.g., fixation of complement, binding of aggregated Igs to lymphocyte surface Fc receptors, but it has been extremely difficult to demonstrate conformational changes. Pecht et al. have now provided new kinetic and static stectroscopic evidence for changes in structure of the Fc domain following binding of large ligands in V domains. X-ray diffraction of whole protein Kol and its separate Fab and Fc fragments have led Huber et al. to suggest a "domino effect", whereby ligand-binding in the V domain could lead to increased contact of V with the CH_1 domain, which in turn could make altered contact with the CH_2 domain. The plausibility of the suggested transmission rests on strong evidence for flexibility of Ig chains at inter-domain junctions between V and CH_1 and at the hinge, between CH_1 and CH_2.

The impact of genetics on immunology dates from recognition of the germ-line/somatic mutation controversy over the origins of Ig diversity. The second session dealt with this issue. Is the great diversity of Ig molecules made by an individual due to a huge bank of genes for V domains carried in germ cells, or to a small number of germ-line V genes that are diversified in somatic cells? In reviewing the structure and genetics of antibodies, Porter reminded us that the a-allotypes of rabbit H chains figured conspicuously in early arguments. In contrast to simple allotypes, which differ from each other in approximately 1 or 2/100 amino acids, a-allotypes are "compound": they differ from each other at about 13 of the 85 N-terminal positions of rabbit H chains. It is easy to visualize their inheritance as a consequence of a single a-allotype germ-line gene that is diversified in somatic development. In contrast, many germ-line V genes, of a given type of a-allotype, would require many independent and parallel mutations to have occurred during evolution, and it is hard to see what pressures could be great enough to insure the selection of the particular patterns of residues that constitute particular a-allotypes. constitute particular a-allotypes.

These considerations may now have to be re-examined because of reports that genes coding for a-allotypes may not, afterall, be authentic alleles. Strossberg has found three a-allotypes in the serum of a single hyperimmunized rabbit, and about 50 % of normal rabbits have "latent" allotypes that, in view of the animals' pedigrees, ought not to be present. There is still considerable resistance to the iconoclastic implications of these findings. A second hyperimmunized rabbit with three a-allotypes, like Strossberg's has not turned up amongst the several hundreds examined in other laboratories. And the transient, trace-amounts of "latent" allotypes described in normal rabbits could reflect as yet undefined serological cross-reactions by anti-allotypic sera.

Supporters of the germ-line view can find comfort in data summarized by Eichmann. In nearly all mice of certain inbred strains a high proportion of the antibodies made against some antigens have in common the same idiotypes (V domain antigenic markers). These common (or shared or public) idiotypes have been found in antibodies to dextrans, to phosphorylcholine, to benzenearsonate, and to some other antigens. F1 hybrids made by crossing an Id+ with an Id- strain are always Id+. In back crosses of the F1 to the minus strain, the idiotype segregates with the Ig1 locus for C_H genes, indicating that the shared idiotypes in an inbred strain are associated with H chains.

Partial amino acid sequences provide additional evidence for the H chain localization of shared idiotypes. H chains of antibodies with shared idiotypes have the same N-terminal amino acid sequence, so far as has been determined, but their L chains are as dissimilar as L chains from unrelated molecules. Shared idiotypes are thus quite different from private idiotypes, which are uniquely associated with a singular L-H pair, representing (it is thought) a particular clone of B cells. The public or shared idiotypes probably represent particular framework sequences in the V domains of H chains. They probably have the same significance as V_H subgroups.

The fine specificities of Abs to some nitrophenyls provide additional assays for following the heritability in mice of V-region markers. Mäkela and his associates have shown that Abs made to NP in C57 black mice (and in some other strains) are heteroclitic: they bind the heterologous ligands NIP or NMP better than the homologous NP. These Abs also behave like public idiotypes: i.e., the capacity to produce Abs with heteroclitic binding sites in response to an immunization with NP, is specified by a dominant autosomal gene linked to genes for C_H allotypes of H chains.

Crosses between mice that possess and others that lack the capacity to form Abs with particular V domain markers has yielded a rather large number of recombinants between V_H and C_H genes. From the recombinant frequencies a linear map of V_H genes has been constructed. It has been reasonably assumed that what are mapped are structural V_H genes rather than regulatory genes. If so, it is likely that there are a large number of V_H genes. This follows because already four or five V_H genes have been found in a short time in just a few laboratories in examining the response to rather conventional antigens (e.g. benzenearsonate, dextran). Moreover, two different idiotypes (the A5A type and S117 type) have been identified in just the response to one streptococcal carbohydrate, and both genes can exist as linked pseudo-alleles in the same haplotype. The 5 V_H genes already identified may constitute a large proportion of the total pool of V_H genes in the mouse. It is more likely, however, that there are a great many V_H genes; those already found probably represent the tip of a large iceberg.

A different view of the frequency of genes for V domains of L chains is emerging from nucleic acid hybridization. Milstein et al. reported that cDNA made from the mRNA of a mouse k chain hybridizes to excess mouse DNA with the same kinetics (Cot curve) as does cDNA from globin mRNA. Oligonucleotide fingerprints indicate that the k chain probe is large enough to include the V-region. It thus appears that there are fewer than 5 V_k genes per mouse haploid genome. Moreover, it is likely that the L chain probe can cross-hybridize very effectively with genes of various L chains. For instance, under the conditions used almost complete cross-hybridization was seen with cDNAs corresponding to mouse, human, and rabbit k chains, and C-region amino acid sequences for these chains differ by approximately the same extent as the V-region sequences of diverse mouse k chains.

Similar conclusions were reached by Tonegawa with ^{125}I-labeled mRNA corresponding to a mouse λ chain. Mouse λ chains are especially interesting. Of the 25 sequenced, 12 have identical V-region sequences from which the others differ by what could be one to four base changes in hypervariable sections. The mRNA probes from different λ chains cross-hybridize: they block indistinguishably the hybridizing activity of a reference λ mRNA. With this probe Cot kinetics were also indistinguishable from that of globin mRNA. It appears, therefore, that there are fewer than five gene copies for mouse λ chains. On the other hand, from the number of different amino acid sequences, it can be estimated that there are likely to be on the order of at least 25 to 40 different λ chains in mice, assuming that myeloma proteins provide a reasonable sample of the mouse λ chain inventory.

Because bulk DNA seems to have fewer V λ genes than the number of different V λ sequences made in mice, these results point strongly to the existence of some mechanism for the diversification of germ-line genes. The search for this mechanism presents molecular geneticists with an unusual challenge and opportunity to uncover somatic gene-diversification processes that could have profound and general importance.

Another provocative observation was reported by Tonegawa. The mRNA for a mouse λ gene was tested as a complete probe (V plus C regions) and as a half-probe (lacking the V message) with DNA derived either from a mouse embryo or from the particular myeloma tumor that synthesized the mRNA probe. The probes hybridized to different DNA fragments from the embryo but to the same DNA fragment from the myeloma cells suggesting that V and C genes are separated in the genome of the embryo and contiguous or joined in the myeloma cell. The results seem to provide striking verification of the two gene-one chain hypothesis. They also raise a number of questions. If the myeloma cell DNA were tested with mRNA for another chain, say a k chain, would the V_k and C_k genes also be joined? Or would they be separated as in the embryo? What about other tissues of the adult mouse? Are V and C genes joined in liver, kidney, etc., or just in those cells that are committed to make particular immunoglobulin chains? Molecular geneticists and cell biologists are not likely to long overlook the inviting opportunities to hunt here for the mechanism by which genes are translocated from one part of the genome to another and to analyze differentiation at the DNA level.

According to the clonal selection hypothesis, the number of B-cell clones sets an upper limit to number of Igs that can be made. Working with small animals, Du Pasquier estimates that 1-g tadpoles have about 10^6 total lymphocytes or about 10^4 clones. Nevertheless, the diversity of Abs made in these animals in response to DNP seems to be just as great as those made in a rabbit or goat; and the repertoire of antigens to which they can respond also seems not to be restricted. Inapparent restrictions may eventually turn up in these small animals, or individual clones may have more potential multispecificity than has hitherto been suspected.

In the third session the conference focused on some of the key cellular aspects of immune responses. How, for example, do antigens activate B cells? The search for an answer has long centered on the fundamental observation that a hapten attached to a "carrier" macromolecule is immunogenic, whereas, the hapten alone is not. There are two main hypotheses. According to one, activation of a B cell requires aggregation of its antigen-binding receptors by multiple haptenic groups: helper T cells and macrophages are supposed to meet this requirement by binding and presenting antigen in such a way as to promote binding of multiple haptenic groups by individual B cells. A second theory suggests that activation of B cells requires two signals, one derived from the binding of a haptenic group and the other from T cells that, in turn, are activated by the carrier moiety of the antigen.

Moller suggests a third model for T-independent antigens, which are defined as antigens that can trigger B cells without the aid of T cells and macrophages. The T-independent antigens are generally high molecular weight polysaccharides. At appropriate concentrations they are also B-cell mitogens, stimulating a polyclonal B-cell response. At lower concentrations they trigger the response of just those clones that make antibodies to the mitogen itself or to haptenic groups attached to the mitogen, which can also act as carrier, e.g., Dnp on LPS. The presumption is that activation of B cells still needs two signals, one from a haptenic moiety and the other from a special carrier with mitogenic activity for B cells in general. Another unsolved fundamental problem concerns the mechanisms by which the same antigen can manage in different circumstances to exert opposing effects: (1) triggering cells to proliferate and differentiate into mature effector cells, e.g., antibody-secreting cells, or (2) deactivating the cells so as to produce tolerance (as of self-antigens). T-independent antigens and purified populations of B cells seem to offer a simplified system for analyzing details of triggering and tolerance-inducing mechanisms.

The antigen-binding receptors of B cells are clearly membrane-bound immunoglobulins. The corresponding receptors of T cells, however, have long been puzzling. Apparent differences in specificity of antibody-mediated and cell-mediated reactions (such as delayed-type hypersensitivity) suggested that the responsible cells (now known to be T cells) might have unusual antigen-binding receptors. Thus over 15 years ago in a symposium on Mechanism of Hypersensitivity (H.S. Lawrence, ed., 1959), it was remarked in the introduction to a session on delayed-type reactions that "the fascination which (these reactions hold for many immunologists) lies in the implication that underlying such reactions is an as yet undiscovered system of antibody-like substances which are cell-associated and which differ in significant manner from serum antibodies". I cite my old remark not as an indication of unusual prescience, but to emphasize the long duration of awareness of the possibility that Ag-binding receptors of cell-mediated immunity may not be conventional antibodies.

In its current form, this issue is expressed in different form: is the T-cell receptor an immunoglobulin or the product of Ir genes of the major histocompatibility locus? Work in the laboratories of Wigzell and of Eichmann and Rajewsky now strongly suggests that T cells, B cells, and Abs that react with the same antigen, have a common idiotype. What appear to be T-cell receptors have also been isolated in trace-amounts from serum. Except for their idiotype, they seem to have no known Ig markers. Their approximate molecular weight is 145,000; a light chain has not been detected. The receptor could be an H chain dimer. This would not be unprecedented: the H chains of conventional immunoglobulins (and also the L chains) can form dimers that have a

ligand-binding site. Whatever the detailed structure of the T-cell
receptor proves to be, the possession of an idiotype associated with
H chains means that products of V_H genes are used to form the receptors.
The total elucidation of the structure of T-cell receptors represents
a major problem whose solution will doubtless have great impact on
future directions of immunological research.

Tumor cells are generally less differentiated than their normal counterparts. But Potter's presentation reminded us that some lymphocytic
tumors can display highly differentiated activities. Everyone here is
familiar with plasma cell tumors, whose myeloma proteins have played
a powerful role in the elucidation of antibody structure. T-cell tumors
(lymphomas) can also show signs of differentiation: some are Ly1-2+,
like normal helper T cells, while others are Ly1+2+, like killer or
suppressor T cells. Identification of T lymphomas with recognizable
antigen-binding receptors would contribute greatly to the structural
analysis of T-cell receptors.

The final stages of the third session dealt with macrophages. As Jerne
has remarked aloud, and a majority of immunologists might silently
agree, macrophages have no function (or no clear function) except to
provide immunologists with a convenient scapegoat with which to try
to explain the inexplicable. But these cells refuse to go away, and
recent studies of T-cell/macrophage interactions with soluble antigens
emphasize the need for H-2 locus compatibility of macrophage and T
cell. This requirement suggests that macrophages play an important
physiological role in the earliest stages of the immune response, probably by binding and presenting antigen to T and/or B cells. How macrophages can bind a virtually limitless number of different foreign antigens, while not (apparently) binding self-antigens is a profound
mystery.

The fourth session dealt with genetic organization of the major histocompatibility complex (MHC) in man and mouse, and with structural studies on some of the main cell surface glycoproteins coded for by genes
of this complex. This locus and its products have long preoccupied
geneticists and immunologists interested in tissue transplantation.
But it has become apparent only in the past few years that besides the
cell surface antigens that determine responses to transplanted tissues
and organs, the linked genes making up this complex locus specify a
variety of important lymphocyte surface glycoproteins. These surface
proteins govern mixed lymphocyte reactions, graft vs host reactions,
immune responses to a great many antigens, susceptibility to many diseases, and also the level of a several complement components. It is
hardly surprising that with this battery of activities, much of immunology is now concerned with understanding MHC gene products and
their functions.

In all mammalian species so far studied, the genetic organization of
the MHC is marked by strong linkage disequilibrium, i.e., certain
alleles are linked much more frequently than would be expected from
their individual frequencies in the population at large. This suggests
that products of genes in this complex interact in important physiological functions. A hint of these functions is beginning to emerge
with recent findings reviewed by Albert and by Rüde in their respective
discussions of human and mouse systems.

Recognition of cell-surface products of the MHC seems to be required
in T-cell reactions with other cells. For instance, specific destruction of diverse target cells by killer T cells depends upon H-2D and
H-2K as well as on other antigens of the target cell surface. Similarly,

288

in secondary antibody responses, C. Pierce et al. have found that T
helper cells react only with antigen on macrophages that have the same
I region of the MHC as the antigen-pulsed macrophages that initiated
the primary immune response. Finally, there is considerable evidence
that T cell regulation of Ab-producing activity of B cells is limited
by MHC compatibilities between the interacting T and B cells. Whether
the T cell receptor is a doublet with one site for Ag and one for an
MHC product, or has a single site that detects an MHC-Ag complex is
still unclear. In either case, the key point is that reactions of T
cells with a variety of target cells require T cell recognition of
certain MHC products on the target cell surface.

What about the structure of MHC proteins? Strominger's description of
HLA-A and HLA-B shows how far recent structural work has gone with the
human antigens, and there was essentially close agreement with the
more preliminary findings on the homologous mouse H-2 glycoproteins
described by Hood et al., and by Henning et al. The surface glycoproteins that elicit the main T cell responses that lead to allograft
rejection (H-2K and H-2D in the mouse, HLA-A, B and C in man) have l
and h chains. The l chain (11,000 d) is β_2 microglobulin; by amino
acid sequence it has strong homology with Igs, especially with the CH_3
domain of γ chains. Limited analysis of N-terminal amino acid sequences
of H chains (ca 45,000) has so far revealed no definite homology with
Ig chains. The great similarity between products of different histocompatibility genes within the MHC (e.g. H-2K vs H-2D and HLA-A vs
HLA-B) supports the view that, like Igs, some MHC genes have also
evolved by gene duplication.

The final session opened with Jerne's general proposal that the immune
system is a network of idiotypes and anti-idiotypes. Much evidence now
supports this proposal. Individual animals seem to be able to make immune response to idiotypes of essentially any of their own Igs. Because
the anti-idiotypic Abs are themselves Igs, with their own idiotypes,
they can in principle elicit still other anti-idiotypes. Whether the
anti-idiotypes to self-idiotypes have a regulatory influence on Ig
synthesis is not yet established. Suggestive evidence in support of a
regulatory role has come from studies of the time-course of antibody
responses to certain antigens. Injection of Ag elicits Abs (call them
Ab≠1); as Ab≠1 diminishes with time other Abs (call then Ab≠2) appear and can react with Ab≠1. Limited studies of specificity suggest
that Ab≠2 is directed to the idiotype of≠1, but more specific
data is require. It is not completely excluded that Ab≠2 has "rheuma-
toid factor"-like activity against Ab≠1.

The anti-idiotypic responses against self-Igs include, besides antibody molecules, T helper cells and very likely other types of T cells.
Because one or more of these anti-idiotypic elements can enhance as
well as suppress the B cells that make the corresponding idiotype, the
development of a detailed network theory poses extraordinary difficulty. Some of the quantitative complexities were discussed by Weiler.
Finally, in a stimulating account of medical implications, Pernis outlined some recent advances in clinical immunology.

Concluding Remarks

Broadly speaking, the Symposium represents a microcosm of the two
worlds of contemporary immunology, one old, the other rapidly emerging.
The old world is that of B cell immunology. It began at the turn of

the century with serology, led to detailed analysis of antibody-ligand reactions and antibody specificity and finally, with the availability of homogeneous Igs, it has reached an understanding of the structure and structural diversity of antibody molecules. Not all problems have been solved, of course, and some of the more obvious fundamental ones that remain have been alluded to throughout the Symposium. Some of them are:

1. Amino acid sequences in hypervariable segments of Ig chains determine ligand-binding specificity, but we have only the most superficial understanding of how various combinations of residues within these segments produce a combining pocket of appropriate size, shape, charge, and hydrophobicity.

2. The alternate mechanisms by which antigens trigger B cells to differentiate into Ab-secreting cells or into a deactivated, tolerant state are still only vaguely perceived.

3. The molecular genetics for joining V and C genes.

4. Whether diversity of amino acid sequences in V domains originates in phylogeny or in ontogeny.

5. The differentiative steps undergone by a cell and its progeny as they move from the status of stem cells to that of committed, Ig-secreting B cells is not understood — during embryogenesis or in adult life.

6. The regulatory mechanisms that determine which C_H or C_L genes, or which allotypes, are expressed in a given mature B cell are also completely obscure.

To understand regulation in immune responses, it will eventually be necessary to deal with the controlled expression of appropriate genes. At present, however, the regulation studied by immunologists is at the level of whole cells: particularly at the level where T cells regulate the differentiation of B cells, of macrophages, and even other T cells. In reacting with other cells, recognition of MHC products by T lymphocytes seems to be an obligate step: T killers recognize products of the K or D genes of the MHC on diverse target cells, whereas T helpers seem to recognize I gene products on macrophages and on B cells. The obligate reactions with MHC products probably insure that T cells react only with surface of other cells, not with substances in solution.

As a broad generalization it appears that antibody molecules, the final product of B cell immunology, are designed to operate in solution, an aqueous 3-dimensional space, T cells, on the other hand, seem to react only with elements on cell surfaces, which approaches a hydrophobic 2-dimensional space. Because all cells express on their surface one kind of MHC product or another, the dual-recognition requirement of T cell receptors (i.e., for MHC products plus some other kind of surface Ag) seems a particularly apt device to insure the focusing of T cells on other cell surfaces.

This view, if valid, also suggests the great difficulties that probably lie ahead in the exploration of the molecular basis for T cell biology. For our present technology has been developed for reactions in aqueous solution, and an entirely different technology may be required to cope with the peculiarities of the relatively hydrophobic 2-dimensional space of cell surfaces. The development of new technology poses a tremendous challenge; if success is achieved the benefits will accrue to biology in general, not just immunology.

Bernal, the noted British crystallographer, once noted that it only
takes ingenuity to solve problems but that it takes real imagination
to discover their existence. If this is true of Immunology, then it
would seem to me that we could use more ingenuity and, for a time at
least, make do with a little less imagination. For this field is
blessed with an enormous number of fundamental problems; and they seem
to be emerging more rapidly than solutions are being found by immunologists. Before a steady state is reached Immunology may enjoy the
dubious distinction of encompassing a larger bank of unsolved problems
than any other science. Perhaps as scientists from other fields are
attracted by the challenges of Immunology, they may supply some of the
ingenuity that Bernal depreciated but which we now need desparately.

Subject Index

295

Colloquien der
Gesellschaft für Biologische Chemie
in Mosbach (Baden)

Biochemie des Sauerstoffes
19. Colloquium am 24. - 27. April 1968
Bearbeiter: B. Hess, H. Staudinger

Inhibitors. Tools in Cell Research
20. Colloquium am 14. - 16. April 1969
Editors: Th. Bücher, H. Sies

Mammalian Reproduction
21. Colloquium am 9. - 11. April 1970
Editors: H. Gibian, E.J. Plotz

The Dynamic Structure of Cell Membranes
22. Colloquium am 15. - 17. April 1971
Editors: D.F. Hölzl Wallach, H. Fischer

Regulation of Transcription and Translation
in Eukaryotes
24. Colloquium am 26. - 28. April 1973
Editors: E.K.-F. Bautz, P. Karlson, H. Kersten

Biochemistry of Sensory Functions
25. Colloquium am 25. - 27. April 1974
Editor: L. Jaenicke

Molecular Basis of Motility
26. Colloquium am 10. - 12. April 1975
Editors: L.M.G. Heilmeyer Jr., J.C. Rüegg, T. Wieland

Springer-Verlag   Berlin   Heidelberg   New York

Immunogenetics

Now in its third year of publication

Editor-in-Chief
George D. Snell, Bar Harbor, Maine

Managing Editor
Jan Klein, Dallas

Associate Editors
B. Benacerraf, Boston
H.O. McDevitt, Stanford
N.A. Mitchison, London
J.W. Uhr, Dallas

Information resulting from the rapid evolution of immunogenetics has been compiled into an up-to-date, concise, and highly scientific journal — Immunogenetics. Within the compilation of original full-length articles, brief communications, and reviews on research are such topics of vital importance to the field as immuno-genetics of cell interaction, immunogenetics of tissue differentiation and development, phylogeny of allo-antigens and immune response, genetic control of immune response and disease susceptibility, and genetics and biochemistry of alloantigens.

Sample copies available upon request.

Springer-Verlag   Berlin   Heidelberg   New York